CAD/CAM:
Computer-Aided Design and Manufacturing

CAD/CAM:
Computer-Aided
Design
and Manufacturing

MIKELL P. GROOVER

EMORY W. ZIMMERS, JR.

Department of Industrial Engineering
Lehigh University

PRENTICE-HALL, INC., Englewood Cliffs, New Jersey 07632

Library of Congress Cataloging in Publication Data

Groover, Mikell P.,
    CAD/CAM: computer-aided design and manufacturing.

    Bibliography: p.
    1. CAD/CAM systems. I. Zimmers, Emory W. II. Title.
III. Title: C.A.D./C.A.M.
TS155.6.G76   1984   670′.28′54   83-11132
ISBN 0-13-110130-7

Editorial/production supervision: *Mary Carnis*
Interior design: *Barbara Cassel*
Cover design: *Edsal Enterprises*
Manufacturing buyer: *Gordon Osbourne*

Printed in the United States of America

10   9   8   7

ISBN   0-13-110130-7

Prentice-Hall International, Inc., *London*
Prentice-Hall of Australia Pty. Limited, *Sydney*
Editora Prentice-Hall do Brasil, Ltda., *Rio de Janeiro*
Prentice-Hall Canada Inc., *Toronto*
Prentice-Hall of India Private Limited, *New Delhi*
Prentice-Hall of Japan, Inc., *Tokyo*
Prentice-Hall of Southeast Asia Pte. Ltd., *Singapore*
Whitehall Books Limited, *Wellington, New Zealand*

To

The CAD/CAM Program

at Lehigh University

Contents

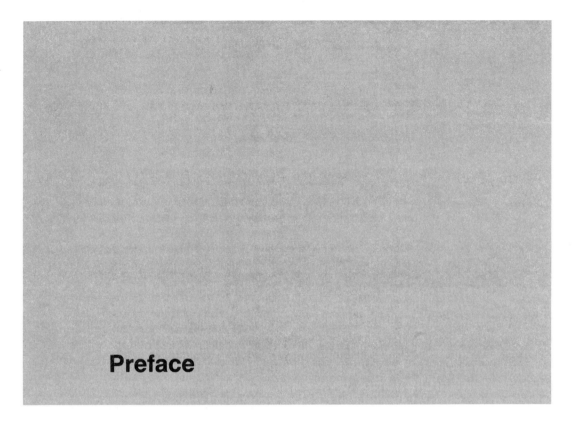

Preface

This book is intended to provide a comprehensive survey of the technical topics related to CAD/CAM (Computer-Aided Design/Computer-Aided Manufacturing). These topics include interactive computer graphics and CAD, numerical control, computer process control, robotics, group technology, computer-integrated production management, and flexible manufacturing systems. Many of these topics are covered in greater detail in other publications, and we have attempted to include a listing of important references at the end of each chapter. The unique feature of this book is that it brings all of these specialized topics together in one volume and attempts to demonstrate how they are all related. Our viewpoint is that these various topics represent a continuum of activity that must take place in a manufacturing firm rather than a collection of separate functions. With CAD/CAM, it should be possible to integrate and automate virtually every aspect of the design and production operations of the firm, thereby increasing the efficiency and the productivity of these operations.

The emphasis in the book is on computerized systems used in the discrete product manufacturing industries. It is designed for engineers, computer specialists, and managers who wish to learn about the technology, applications, and scope of CAD/CAM. The book should be suitable as a technical reference for the practicing professional who must make engineering and financial decisions about CAD/CAM projects. The technology of computer-aided design and manufacturing has undergone significant evolution during the last 10 to 15 years. This book should be useful in exposing engineers, managers, and computer scientists to the latest technology.

The book has also been designed as a textbook for college courses and industry continuing education courses in CAD/CAM. There is a good chance that we will use the book for these purposes at Lehigh University. More and more educators are recognizing the importance of this technology in their curricula. We anticipate that there will be a significant growth in the number of CAM and CAD/CAM courses offered by engineering schools during the present decade. It is our hope that this book will be an important contribution towards satisfying the need for a text in this growing area.

When I first started negotiations with Prentice-Hall in 1980 regarding a book on CAD/CAM, my thought was to develop a companion to my previous book entitled *Automation, Production Systems, and Computer-Aided Manufacturing*. The first book emphasizes automation and manufacturing, and it was my plan that the second book would emphasize computer-aided design and its integration with manufacturing using computer systems. It has turned out that the two books overlap each other to some extent. Readers of my first book might recognize in this second book a number of topics that are familiar. This overlap is not inappropriate since the two areas of automation and CAD/CAM do indeed include many common topics. My hope is that the two books will constitute a substantial and complementary set of volumes in the general field of computer-aided design, computer-integrated production systems, and factory automation.

In this CAD/CAM book, it seems appropriate that I acknowledge the contribution of Emory Zimmers, my coauthor, colleague, and contemporary in the Industrial Engineering Department at Lehigh. At the time I began to develop the outline for the book, it was apparent to me that the subject of CAD/CAM encompassed a very large and varied set of topics and technologies. If the full scope of CAD/CAM were to be covered in the desired detail, it seemed to me that the subject represented more material than could be adequately documented by a single author. Meanwhile, Emory had also been entertaining thoughts about writing a book on CAD/CAM, perhaps facing up to some of the same kinds of difficulties I was encountering. In the interest of producing a thorough, high-quality treatment in the field of CAD/CAM, Emory and I decided to collaborate on the project. His knowledge of the field would fill in the gaps of my knowledge, and vice versa. Specifically, Emory has made significant contributions in Chapters 2, 3, 4, 5, 6, 17, 21, and 22. The task of actually preparing the final manuscript for the book fell to me.

Acknowledgments

In a project as large as this book, the authors must usually rely on the assistance of other persons. In our case, contributions were made by many individuals and many companies, and we would like to hereby acknowledge their help. For their valuable assistance in either providing technical input and/or reviewing portions of the manuscript, we are indebted to the following individuals: John W. Adams, our colleague and statistician/mathematician at Lehigh; Frank Bibas, our former student now working in the robotics industry; Jim Buskirk, CAD/CAM specialist at Air Products and Chemicals, Inc.; Arthur Gould, our colleague and former department chairman at Lehigh; Jack Hughes, my good friend and sometimes coauthor from Air Products and Chemicals; Herb Ketcham, formerly of our Manufacturing Processes Laboratory at Lehigh; Marvin Kreithen of the Bridgeport Controls Division of Textron; Mark Lang of Lehigh's Computer-Aided Design Laboratory and faculty member in Mechanical Engineering and Mechanics; Lance Leventhal, P-H author on computers; Ron Lovetri of the General Electric Company; Roger Nagel, Director of Lehigh's Institute for Robotics; Glenn Offord, one of my master's candidates at Lehigh and simultaneously an engineer for Western Electric Company; Louis Plebani, Jr., our colleague in the Industrial Engineering Department; Paul Quantz, who runs his own CAD/CAM consulting company; Ron Sherertz of the McDonnell-Douglas Corporation; Tom Shank of the General Electric Company; Theodore Terry, one of our colleagues in the mechanical engineering department; Mitchell Weiss of United States Robots; Bob Wolfe of the IBM Corporation; and Nello Zuech of Object Recognition Systems, Inc.

Much of the research and preparation of source materials for the book was accomplished by some of our students at Lehigh. Major contributions were made by Thomas Costello, Robert Gervis, Donna Harle, Robert Kimball, Carol Richardi (now Mrs. Glenn Riggin), and Jonathan Ripsom.

We are also indebted to a large number of companies in the CAD/CAM or related industries, which supplied us with a wealth of technical information, photographs (many of which appear as figures in the book), and other resource materials. These companies include Applicon, Bendix Corp. (Automation and Measurement Division), Bridgeport Machines Division of Textron, Inc., Cincinnati Milacron, Computervision Corp., Digital Equipment Corp., General Electric Company, Gerber Scientific Instrument Company, Heath Corp., IBM Corp., Kearney & Trecker, McDonnell-Douglas Automation Corp., MDSI, MTM Association, Numeridex, Inc., Object Recognition Systems, Inc., Organization for Industrial Research, Inc., Prab Conveyors, Inc., Scans Associates, Inc., Threshold Technology, Unimation, Inc., and Warner & Swasey, Inc.

A number of figures from my previous book, *Automation, Production Systems, and Computer-Aided Manufacturing,* were also used in the present book, and it is required that I acknowledge that these figures were reproduced from the previous book with the permission of Prentice-Hall. These figures are (figure numbers refer to the previous book): 7.3, 7.4, 7.5, 7.6, 7.7, 7.8, 7.10, 8.1, 8.4, 8.5, 8.6, 8.7, 8.8, 8.9, 8.10, 8.11, 8.12, 8.13, 8.14, 8.15, 9.1, 9.2, 9.3, 9.4, 9.6, 11.1, 11.2, 11.3, 11.4, 11.5, 11.6, 11.7, 11.8, 11.12, 11.14, 11.20, 12.2, 14.2, 15.1, 15.2, 15.3, 15.9, 16.3, 16.4, 17.1, 17.2, 17.4, 17.7, 17.8, 17.9, 18.1, 18.2, 18.3, 18.4, 18.5, 18.6, 18.7, 18.8, 18.9, 18.13, 18.14, and 18.15. In addition, Table 16.2 from the automation book was reproduced with the permission of Prentice-Hall for the present book.

Finally, I would like to express my appreciation to Mrs. Marcia Mierzwa and her company, Information Processing Systems, for an excellent typing job on the manuscript for the book.

Dedication

The CAD/CAM Program at Lehigh University had its beginnings in the Industrial Engineering Department with some very modest project work on a single interactive computer graphics terminal in the early 1970's. I remember one of those projects dealt with the graphics modelling of a cutting tool in a machining operation. In 1975, the Department formed the Computer-Aided Manufacturing (CAM) Laboratory, under the direction of Emory Zimmers. At that time, the growing importance of computers in manufacturing for numerical control, process monitoring, MRP, and other applications was clearly understood by those of us in industrial engineering whose specialty was related to manufacturing, but Emory had been largely responsible for the development of this area within our department. By around 1979, the Department of Mechanical Engineering and Mechanics had recognized the importance of computer graphics for design, and there were efforts to integrate computer graphics into the activities of that department. These efforts were led by their department chairman, Doug Abbott, and a new faculty member at that time, John Ochs. Ochs had done his doctoral work in acoustics at The Pennsylvania State University, and had used computer graphics as a tool in that research. Abbott, Zimmers, Ochs, and George Kane, Chairman of Industrial Engineering, formed a small group which included Mike Bolton of Lehigh's Development Office to begin to develop the CAD/CAM area at the University. The decision to include Bolton in the group turned out to be a decision of significant merit. One of the principal activities of the group was to visit companies to solicit their participation in the University's CAD/CAM Program. Bolton was very effective at making the right contacts and organizing these visits to industry. To merely say that many days were spent away from the University promoting the new program is an understatement of the amount of time and effort that was contributed by this group. The result of their efforts was a multimillion dollar development program that has provided some of the finest university CAD/CAM laboratory facilities in the country. In addition, and perhaps far more important, the CAD/CAM Program has also produced an awareness, an involvement, and an excitement about computer-aided design and manufacturing at Lehigh University that has motivated and benefited both our faculty and our students. At the time of this writing, the CAD/CAM activity has expanded well beyond industrial and mechanical engineering to include many other departments of the University. It is largely a result of the Lehigh CAD/CAM Program that this book was made possible, and it seems quite appropriate that our book be dedicated to the program.

Mikell P. Groover

CAD/CAM:
Computer-Aided Design and Manufacturing

chapter 1

Introduction

1.1 CAD/CAM DEFINED

CAD/CAM is a term which means computer-aided design and computer-aided manufacturing. It is the technology concerned with the use of digital computers to perform certain functions in design and production. This technology is moving in the direction of greater integration of design and manufacturing, two activities which have traditionally been treated as distinct and separate functions in a production firm. Ultimately, CAD/CAM will provide the technology base for the computer-integrated factory of the future.

Computer-aided design (CAD) can be defined as the use of computer systems to assist in the creation, modification, analysis, or optimization of a design. The computer systems consist of the hardware and software to perform the specialized design functions required by the particular user firm. The CAD hardware typically includes the computer, one or more graphics display terminals, keyboards, and other peripheral equipment. The CAD software consists of the computer programs to implement computer graphics on the system plus application programs to facilitate the engineering functions of the user company. Examples of these application programs include stress–strain analysis of components, dynamic response of mechanisms, heat-transfer calculations, and numerical control part programming. The collection of application programs will vary from one user firm to the next because their product lines, manufacturing processes, and customer markets are different. These factors give rise to differences in CAD system requirements.

Computer-aided manufacturing (CAM) can be defined as the use of computer systems to plan, manage, and control the operations of a manufacturing plant through either direct or indirect computer interface with the plant's production resources. As indicated by the definition, the applications of computer-aided manufacturing fall into two broad categories:

1. *Computer monitoring and control.* These are the direct applications in which the computer is connected directly to the manufacturing process for the purpose of monitoring or controlling the process.
2. *Manufacturing support applications.* These are the indirect applications in which the computer is used in support of the production operations in the plant, but there is no direct interface between the computer and the manufacturing process.

The distinction between the two categories is fundamental to an understanding of computer-aided manufacturing. It seems appropriate to elaborate on our brief definitions of the two types.

Computer monitoring and control can be separated into monitoring applications and control applications. Computer process monitoring involves a direct computer interface with the manufacturing process for the purpose of observing the process and associated equipment and collecting data from the process. The computer is not used to control the operation directly. The control of the process remains in the hands of human operators, who may be guided by the information compiled by the computer.

Computer process control goes one step further than monitoring by not only observing the process but also controlling it based on the observations. The distinction between monitoring and control is displayed in Figure 1.1. With computer monitoring the flow of data between the process and the computer is in one direction only, from the process to the computer. In control, the computer interface allows for a two-way flow of data. Signals are transmitted from the process to the computer, just as in the case of computer monitoring. In addition, the computer

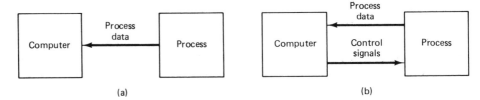

FIGURE 1.1   Computer monitoring versus computer control: (a) computer monitoring; (b) computer control.

issues command signals directly to the manufacturing process based on control algorithms contained in its software.

In addition to the applications involving a direct computer-process interface for the purpose of process monitoring and control, computer-aided manufacturing also includes indirect applications in which the computer serves a support role in the manufacturing operations of the plant. In these applications, the computer is not linked directly to the manufacturing process. Instead, the computer is used "off-line" to provide plans, schedules, forecasts, instructions, and information by which the firm's production resources can be managed more effectively. The form of the relationship between the computer and the process is represented symbolically in Figure 1.2. Dashed lines are used to indicate that the communication and control link is an off-line connection, with human beings often required to consumate the interface. Some examples of CAM for manufacturing support that are discussed in subsequent chapters of this book include:

Numerical control part programming by computers. Control programs are prepared for automated machine tools.

Computer-automated process planning. The computer prepares a listing of the operation sequence required to process a particular product or component.

Computer-generated work standards. The computer determines the time standard for a particular production operation.

Production scheduling. The computer determines an appropriate schedule for meeting production requirements.

Material requirements planning. The computer is used to determine when to order raw materials and purchased components and how many should be ordered to achieve the production schedule.

Shop floor control. In this CAM application, data are collected from the factory to determine progress of the various production shop orders.

In all of these examples, human beings are presently required in the application either to provide input to the computer programs or to interpret the computer output and implement the required action.

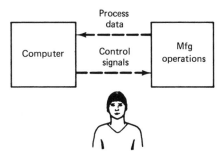

FIGURE 1.2 CAM for manufacturing support.

1.2 THE PRODUCT CYCLE AND CAD/CAM

For the reader to appreciate the scope of CAD/CAM in the operations of a manufacturing firm, it is appropriate to examine the various activities and functions that must be accomplished in the design and manufacture of a product. We will refer to these activities and functions as the product cycle.

A diagram showing the various steps in the product cycle is presented in Figure 1.3. The cycle is driven by customers and markets which demand the product. It is realistic to think of these as a large collection of diverse industrial and consumer markets rather than one monolithic market. Depending on the particular customer group, there will be differences in the way the product cycle is activated. In some cases, the design functions are performed by the customer and the product is manufactured by a different firm. In other cases, design and manufacturing is accomplished by the same firm. Whatever the case, the product cycle begins with a concept, an idea for a product. This concept is cultivated, refined, analyzed, improved, and translated into a plan for the product through the design engineering process. The plan is documented by drafting a set of engineering drawings showing how the product is made and providing a set of specifications indicating how the product should perform.

Except for engineering changes which typically follow the product throughout its life cycle, this completes the design activities in Figure 1.3. The next activities involve the manufacture of the product. A process plan is formulated which specifies the sequence of production operations required to make the product. New equipment and tools must sometimes be acquired to produce the new

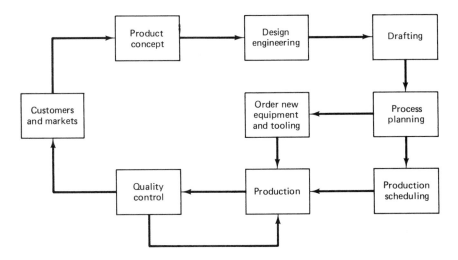

FIGURE 1.3 Product cycle (design and manufacturing).

product. Scheduling provides a plan that commits the company to the manufacture of certain quantities of the product by certain dates. Once all of these plans are formulated, the product goes into production, followed by quality testing, and delivery to the customer.

The impact of CAD/CAM is manifest in all of the different activities in the product cycle, as indicated in Figure 1.4. Computer-aided design and automated drafting are utilized in the conceptualization, design, and documentation of the product. Computers are used in process planning and scheduling to perform these functions more efficiently. Computers are used in production to monitor and control the manufacturing operations. In quality control, computers are used to perform inspections and performance tests on the product and its components.

As illustrated in Figure 1.4, CAD/CAM is overlaid on virtually all of the activities and functions of the product cycle. In the design and production operations of a modern manufacturing firm, the computer has become a pervasive, useful, and indispensable tool. It is strategically important and competitively imperative that manufacturing firms and the people who are employed by them understand CAD/CAM.

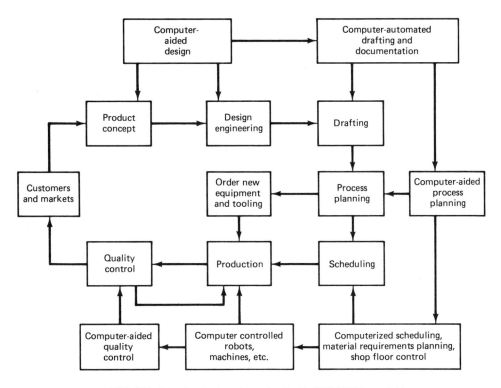

FIGURE 1.4 Product cycle revised with CAD/CAM overlaid.

1.3 AUTOMATION AND CAD/CAM

In a previous book [3], automation was defined as the technology concerned with the application of complex mechanical, electronic, and computer-based systems in the operation and control of production. It is the purpose of this section to establish the relationship between CAD/CAM and automation.

As indicated in Section 1.2, there are differences in the way the product cycle is implemented for different firms involved in production. Production activity can be divided into four main categories:

1. Continuous-flow processes
2. Mass production of discrete products
3. Batch production
4. Job shop production

The definitions of the four types are given in Table 1.1. The relationships among the four types in terms of product variety and production quantities can be conceptualized as shown in Figure 1.5. There is some overlapping of the categories as the figure indicates. Table 1.2 provides a list of some of the notable achievements in automation technology for each of the four production types.

One fact that stands out from Table 1.2 is the importance of computer technology in automation. Most of the automated production systems implemented today make use of computers. This connection between the digital computer and manufacturing automation may seem perfectly logical to the reader. However, this logical connection has not always existed. For one thing, automation technology

TABLE 1.1 Four Types of Production

	Category	Description
1.	Continuous-flow processes	Continuous dedicated production of large amounts of bulk product. Examples include continuous chemical plants and oil refineries.
2.	Mass production of discrete products	Dedicated production of large quantities of one product (with perhaps limited model variations). Examples include automobiles, appliances, and engine blocks.
3.	Batch production	Production of medium lot sizes of the same product or component. The lots may be produced once or repeated periodically. Examples include books, clothing, and certain industrial machinery.
4.	Job shop production	Production of low quantities, often one of a kind, of specialized products. The products are often customized and technologically complex. Examples include prototypes, aircraft, machine tools, and other equipment.

Source: (Adapted with permission from *Industrial Engineering* Magazine, November, 1981. Copyright © Institute of Industrial Engineers, 25 Technology Park/Atlanta, Norcross, GA 30092.)

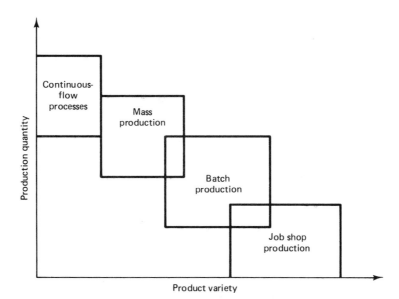

FIGURE 1.5 Four production types related to quantity and product variation. (Reprinted with permission from *Industrial Engineering* Magazine, November, 1981. Copyright ©Institute of Industrial Engineers, 25 Technology Park, Atlanta Norcross, GA 30092.)

TABLE 1.2 Automation Achievements for the Four Types of Production

	Category	Automation achievements
1.	Continuous-flow processes	Flow process from beginning to end
		Sensor technology available to measure important process variables
		Use of sophisticated control and optimization strategies
		Fully computer-automated plants
2.	Mass production of discrete products	Automated transfer machines
		Dial indexing machines
		Partially and fully automated assembly lines
		Industrial robots for spot welding, parts handling, machine loading, spray painting, etc.
		Automated materials handling systems
		Computer production monitoring
3.	Batch production	Numerical control (NC), direct numerical control (DNC), computer numerical control (CNC)
		Adaptive control machining
		Robots for arc welding, parts handling, etc.
		Computer-integrated manufacturing systems
4.	Job shop production	Numerical control, computer numerical control

Source: (Adapted with permission from *Industrial Engineering* Magazine, November, 1981. Copyright © Institute of Industrial Engineers, 25 Technology Park/Atlanta, Norcross, GA 30092.)

historically preceded modern computer technology.[1] There are examples of mechanized and partially automated flow lines which date back to the early days of the automobile industry. Another factor is that the early applications of the computer were mostly extensions of applications performed by electronic accounting machines. Computer applications in manufacturing during the 1950s and early 1960s were almost nonexistent. However, as the cost of computers has decreased and their capabilities have increased, the economic feasibility of using computers in manufacturing and design has developed.

Another observation from Table 1.2 is that the economics of high production quantities (categories 1 and 2 in the table) have tended to stimulate some of the most productive achievments in automation. The technology of production automation has traditionally focused on the manufacturing operations and the equipment that performs them. One might argue that the scope of production automation is limited to the operations and associated machine tools. By contrast, CAD/CAM, in addition to its particular emphasis on the use of computer technology, is also distinguished by the fact that it includes not only the manufacturing operations but also the design and planning functions that precede manufacturing.

To emphasize the differences in scope between automation and CAD/CAM, let us consider the mathematical model of the product life cycle developed by Groover and Hughes [4]. This is a model of the amounts of time expended in designing, planning, and producing a typical product. Let T_1 be the time required to produce 1 unit of product. This would be the sum of all the individual process times for each component in the product plus the time to assemble, inspect, and package a single product.

Let T_2 be the time associated with planning and setting up for each batch of production. T_2 would include the ordering of raw materials by the purchasing department, time required in production planning to schedule the batch, setup times for each operation, and so forth. If the batch size is very large (at or near the mass production level), the batch time can be spread among many units of production. If the batch size is very small (at or near the job shop level), the batch time would become relatively important.

Finally, let T_3 be the time required for designing the product and for all the other activities that are accomplished once for each different product. These include process planning, cost estimating and pricing, building of special tools and fixtures, and various other functions which must be done to get the product ready for production.

The preceding definitions thus provide three distinct levels during the product's life cycle when time is spent on the product. Two additional parameters are needed to complete the model. Let B equal the number of batches produced throughout the product's life cycle. And let Q be the number of units produced in each batch. We assume, for simplicity, that the batch size, Q, will always be the

---

[1] We are dating the start of computer technology at 1951 (the UNIVAC I) and ignoring the early contributions of Charles Babbage (1792−1871) and others.

same for each batch. Accordingly, the total number of units produced during the life cycle is BQ.

The aggregate time spent on the product throughout its life cycle can be defined as

$$TT_{LC} = BQT_1 + BT_2 + T_3 \qquad (1.1)$$

where TT_{LC} is the total time during the product life cycle. This total time can be allocated evenly among the total number of units produced, BQ, to determine the average time spent on each unit of product during its life cycle. Calling this average time T_{LC}, we have

$$T_{LC} = T_1 + \frac{T_2}{Q} + \frac{T_3}{BQ} \qquad (1.2)$$

In mass production and batch production the T_2 and T_3 terms can be spread out over a large number of units. Their relative values, therefore, become less important as the production quantities increase. The T_1 term becomes the most important term. In job shop manufacturing, the T_2 and T_3 terms can become very significant because the quantities are so low. The engineering and planning expense of a complex piece of equipment, when 1 unit is made, can become the dominant cost of the product.

The relationship between CAD/CAM and automation with regard to the product life cycle model is this. The goal of both CAD/CAM and automation is to reduce the various time elements in the product life cycle. It is by achieving this goal that we are able to increase productivity and improve our standard of living. The difference is that automation technology is concerned principally with reducing the T_1 and T_2 terms, with emphasis on the unit production time (T_1). CAD/CAM technology is concerned with all three terms but is perhaps focused on the T_3 and T_2 terms in the life cycle model. The emphasis in CAD/CAM includes the design and planning functions of the product life cycle. A list of some of the important achievements in CAD/CAM are presented in Table 1.3 for comparison with the automation achievements in Table 1.2. What is perhaps missing in Table 1.3 is the important contribution that CAD/CAM has made toward integrating the functions of design and manufacturing. CAD/CAM promotes the viewpoint that design and manufacturing represent a continuum of activities in the product life cycle rather than the two separate and sometimes opposing functions that they have traditionally been.

1.4 ORGANIZATION OF THIS BOOK

Table 1.3 reads almost like the table of contents for this book. The text is organized into eight major parts plus this introductory chapter. Part I, which follows, gives an introduction to computers and contains two chapters (Chapters 2 and 3). Chapter 2 is concerned with general computer technology: how the computer

TABLE 1.3 Achievements in CAD/CAM

Interactive computer graphics systems
Color computer graphics
Animated computer graphics
Design engineering analysis (stress–strain, etc.) software
Parts classification and coding (group technology)
Design retrieval systems
Automated drafting
Design and manufacturing data base
Computer-assisted process planning
Computerized machinability data systems
Computer-generated work standards
Computer-assisted NC part programming
Interactive graphics NC part programming
Finely tuned production scheduling
Material requirements planning (MRP)
Shop floor control
Computer numerical control (CNC)
Direct numerical control (DNC)
Computer-controlled robots
Microprocessor control applications
Computer-aided inspection (CAI)
Computer-aided quality testing (CAT)

---

works and how it communicates with the outside world. Chapter 3 concentrates on minicomputers, microcomputers, and programmable controllers, which have made the development of computer-aided design and computer-aided manufacturing economically viable. To understand CAD/CAM, a basic understanding of computers is helpful.

Part II deals with computer-aided design (Chapters 4, 5, and 6). Chapter 4 covers the fundamentals of CAD, its purposes, and its possibilities. Chapter 5 describes the hardware components in a CAD system and how they work. Chapter 6 presents an explanation of the graphics software in a computer-aided design system. The mathematical algorithms for manipulating images on the graphics screen are described in this chapter as well as some of the important application programs in design analysis and manufacturing planning.

Part III (Chapters 7, 8, and 9) treats the subject of numerical control (NC) with emphasis on the use of the computer in NC. Chapter 7 provides a brief introduction to numerical control technology and applications. Chapter 8 considers how the NC machine tools are programmed. In this chapter we describe the powerful APT programming language (APT stands for Automatically Programmed Tools). The development of APT in the late 1950s is considered to be one of the pioneering efforts in the evolution of CAD/CAM. We also examine how NC programming is accomplished on an interactive graphics system. Chapter 9 describes the various uses of computer systems to control the NC machine tools.

Industrial robotics is the subject of Part IV (Chapters 10 and 11). Chapter 10 defines the technology of robots and how they are programmed. Chapter 11 describes some of the interesting industrial applications performed by these fascinating machines.

Part V deals with group technology (GT) and process planning (Chapters 12 and 13, respectively). We attempt to define the role of group technology in both design and manufacturing. Process planning is related to group technology because many of today's computer-aided process planning schemes rely on GT principles.

Part VI (Chapters 14, 15, and 16) describes the operation of computer-integrated production management systems. Basically, these are systems which incorporate the computer into production planning and control. Chapter 14 provides an overview of these systems. Chapter 15 describes MRP. Traditionally, MRP stands for material requirements planning. Recently, it is being used in a broader sense to represent manufacturing resource planning (known as MRP II). Both meanings are explained in Chapter 15. Chapter 16 discusses shop floor control and computer process monitoring. Part VI is not meant to be a treatise on production planning and control. Instead, it is intended to define how computer-aided manufacturing applies to production management.

Part VII is concerned with computer process control (Chapters 17 through 20). Chapter 17 describes how the computer is interfaced with the manufacturing process. Chapter 18 provides a descriptive overview of process control with emphasis on the use of computer systems. Chapter 19 describes how computers are utilized for quality inspection and testing. Finally, Chapter 20 gives a good example of computer process control in the discrete-products industries. Specifically, it deals with computer-integrated manufacturing systems, which include the highly automated flexible manufacturing systems (FMS).

Finally, Part VIII contains two chapters. The first (Chapter 21) presents a general procedure for implementation of a CAD/CAM system. Chapter 22 is a summary chapter which explores some of the probable future developments in this exciting technical field.

REFERENCES

[1] BYLINSKI, G., "A New Industrial Revolution Is on the Way," *Fortune*, October 5, 1981, pp. 106–114.

[2] Computervision Corp., *The CAD/CAM Handbook*, Bedford, Mass., 1980. (A collection of articles.)

[3] GROOVER, M. P., *Automation, Production Systems, and Computer-Aided Manufacturing*, Prentice-Hall, Inc., Englewood Cliffs, N.J., 1980.

[4] GROOVER, M. P., and HUGHES, J. E., JR., "A Strategy for Job Shop Automation," *Industrial Engineering*, November, 1981, pp. 66–76.

[5] GROOVER, M. P., AND ZIMMERS, E. W., JR., "Automated Factories in the Year 2000," *Industrial Engineering*, November, 1980, pp. 34–43.

[6] INGLESBY, T., "CAD/CAM: Should We or Shouldn't We?" *Assembly Engineering*, March, 1982, pp. 48–50.

[7] KINNUCAN, P., "Computer-Aided Manufacturing Aims for Integration," *High Technology*, May/June, 1982, pp. 49–56.

[8] KROUSE, J. K., "Automated Factories: the Ultimate Union of CAD and CAM," *Machine Design*, November 26, 1981, pp. 54–60.

[9] LERRO, J. P., JR., "CAD/CAM System: Start of the Productivity Revolution," *Design News*, November 16, 1981, pp. 46–65.

[10] ZIMMERS, E. W., JR., AND PLEBANI, L. J., "Using a Turnkey Interactive Graphics System in Computer-Aided Manufacturing," *Industrial Engineering*, November, 1981, pp. 98–104.

PART I

Computers, The Foundation of CAD/CAM

chapter 2

Computer Technology

2.1 INTRODUCTION

The central and essential ingredient of CAD/CAM is the digital computer. Its inherent speed and storage capacity have made it possible to achieve the advances in image processing, real-time process control, and a multitude of other important functions that are simply too complex and time consuming to perform manually. To understand CAD/CAM, it is important to be familiar with the concepts and technology of the digital computer. In this first part of the book, we focus on computers as a foundation for computer-aided design and manufacturing.

The modern digital computer is an electronic machine that can perform mathematical and logical calculations and data processing functions in accordance with a predetermined program of instructions. The computer itself is referred to as hardware, whereas the various programs are referred to as software.

There are three basic hardware components of a general-purpose digital computer:

1. Central processing unit (CPU)
2. Memory
3. Input/output (I/O) section

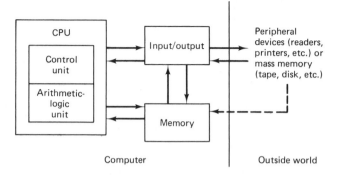

FIGURE 2.1   Basic hardware structure of a digital computer.

The relationship of these three components is illustrated in Figure 2.1. The central processing unit is often considered to consist of two subsections: a control unit and an arithmetic–logic unit (ALU). The control unit coordinates the operations of all the other components. It controls the input and output of information between the computer and the outside world through the I/O section, synchronizes the transfer of signals between the various sections of the computer, and commands the other sections in the performance of their functions. The arithmetic–logic unit carries out the arithmetic and logic manipulations of data. It adds, subtracts, multiplies, divides, and compares numbers according to programmed instructions. The memory of the computer is the storage unit. The data stored in this section are arranged in the form of words which can be conveniently transferred to the ALU or I/O section for processing. Finally, the input/output provides the means for the computer to communicate with the external world. This communication is accomplished through peripheral equipment such as readers, printers, and process interface devices. The computer may also be connected to external storage units (e.g., tapes, disks, etc.) through the I/O section of the computer. Figure 2.2 shows a modern large computer with associated peripheral equipment including storage units, card reader, and printer.

The software consists of the programs and instructions stored in memory and in external storage units. It is the software that assigns to the computer the various functions which the user desires the system to accomplish. The usefulness of the computer lies in its ability to execute the instructions quickly and accurately. Because the contents of the computer's memory can be easily changed, and therefore different programs can be placed into memory, the digital computer can be used for a wide variety of applications.

Regardless of the application, the computer executes the program through its ability to manipulate data and numbers in their most elementary form. The data and numbers are represented in the computer by electrical signals which can take one of two alternative states. This form of representation is called the binary system. The more familiar decimal number system and a whole host of software

FIGURE 2.2 A modern large mainframe computer and associated peripheral hardware. The computer is the IBM 3033 processor. (Courtesy of IBM Corporation.)

languages can utilize the binary system to permit communication between computers and human beings.

2.2 CENTRAL PROCESSING UNIT (CPU)

The central processing unit (CPU) regulates the operation of all system components and performs the arithmetic and logical operations on the data. To accomplish these functions, the CPU consists of two operating units:

1. Control unit
2. Arithmetic–logic unit (ALU)

The control unit coordinates the various operations specified by the program instructions. These operations include receiving data which enter the computer and deciding how and when the data should be processed. The control unit directs the operation of the arithmetic–logic unit. It sends data to the ALU and tells the ALU what functions to perform on the data and where to store the results. The capability of the control unit to accomplish these operations is provided by a set of instructions called an executive program which is stored in memory.

The arithmetic and logic unit performs operations such as addition, subtractions, and comparisons. These operations are carried out on data in binary form. The logic section can also be used to alter the sequence in which instructions are executed when certain conditions are indicated and to perform other functions, such as editing and masking data for arithmetic operations.

Both the control unit and the arithmetic–logic unit perform their functions by utilizing registers. Computer registers are small memory devices that can receive, hold, and transfer data. Each register consists of binary cells to hold bits of data. The number of bits in the register establishes the word length the computer is capable of handling. The number of bits per word can be as few as 4 (early microcomputers) or as many as 64 (large scientific computers).

The arrangement of these registers constitutes several functional areas of the CPU. A representative configuration is given in Figure 2.3. To accomplish a given sequence of programmed instructions, the functions of these register units would be as follows:

Program counter.   The program counter holds the location or address of the next instruction. An instruction word contains two parts: an operator and an operand or a reference to an operand. The operator defines the type of arithmetic or logic operation to be carried out (additions, comparisons, etc.). The operand usually specifies the data on which the operation is to be performed. The CPU sequences the instructions to be performed by fetching words from memory according to the contents of the program counter. After each word is obtained, the program counter is incremented to go on to the next instruction word.

Memory address register.   The location of data contained in the computer's memory units must be identified for an instruction, and this is the function of the memory address register. This unit is used to hold the address of data

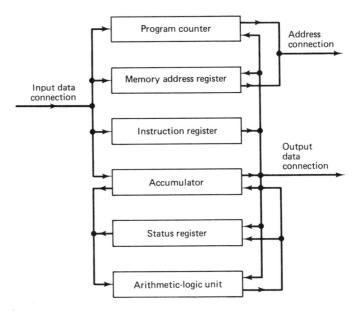

FIGURE 2.3 Typical arrangement of registers in the computer's CPU. (From Lance A. Levanthal, *Introduction to Microprocessors: Software, Hardware, Programming,* ©1978, p. 128. Reprinted by permission of Prentice-Hall, Inc., Englewood Cliffs, N.J.)

held in memory. A computer may have more than a single memory address register.

Instruction register.   The instruction register is used to hold the instruction for decoding. Decoding refers to the interpretation of the coded instruction word so that the desired operation is carried out by the CPU.

Accumulator.   An accumulator is a temporary storage register used during an arithmetic or logic operation. For example, in adding two numbers, the accumulator would be used to store the first number while the second number was fetched. The second number would then be added to the first. The sum, still contained in the accumulator, would then be operated on or transferred to temporary storage, according to the next instruction in the program.

Status register.   Status registers are used to indicate the internal condition of the CPU. A status register is a 1-bit register (often called a flag). Flags are used to identify such conditions as logical decision outcomes, overflows (where the result of an arithmetic operation exceeds the word capacity), and interrupt conditions (used in process control).

Arithmetic–logic unit (ALU). The ALU provides the circuitry required to perform the various calculations and manipulations of data. A typical configuration of the arithmetic–logic unit is illustrated in Figure 2.4. The unit has two inputs for data, inputs for defining the function to be performed, data outputs, and status outputs used to set the status registers or flags (described above).

The arithmetic–logic unit can be a simple adder, or its circuitry can be more complex for performing other calculations, such as multiplication and division. ALUs with simpler circuits are capable of being programmed to perform these more complicated operations, but more computing time is required. The more complex arithmetic–logic units are faster, but these units are more costly. Referring to Figure 2.4, the two inputs, A and B, enter the ALU and the logical or mathematical operation is performed as defined by the function input. Among the

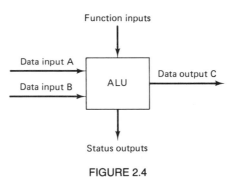

FIGURE 2.4

possible functions are addition, subtraction, increment by 1, decrement by 1, and multiplication. The ALU places the result of the operation on A and B in the output, C, for transfer to the accumulator.

2.3   TYPES OF MEMORY

The memory section consists of binary storage units which are organized into bytes (there are typically 8 bits per byte). A byte is a convenient size for the computer to handle. Computer words can typically be 4, 8, 12, 16, 32, or 64 bits long. Each word has an address in the memory. The CPU calls words from memory by referring to the word address. The time required to find the correct address and fetch the contents of that memory location is called the access time. Access time is an important factor in determining the speed of the computer. Access times range from 10^{-7}s (100 ns) to several microseconds.

The memory section stores all the instructions and data of a program. Thus the CPU must transfer these instructions and data (in the form of words) to and from the memory throughout the execution of the program.

Today, the technology and conceptual framework of computer storage represents a rapidly changing field. The type of memory is a very important consideration in the design of the entire computer system. We will adopt a very conventional organization of this topic by dividing computer memories into two basic categories:

1.   Main memory (primary storage)
2.   Auxiliary memory (secondary storage)

In present-day computer systems, there often exist a greater number of hierarchical levels of computer memory. The reader interested in this topic might want to refer to either of two books in our list at the end of the chapter [7, 14].

Main memory (primary storage)

The main memory or primary storage are terms which designate storage areas that are physically a part of the computer and connected directly to the CPU. It includes the working registers and memory devices closely configured to the CPU. Primary storage can be divided into three main categories:

1.   Main data storage, such as magnetic core or solid-state memory. This storage is characterized by its close proximity to the CPU, fast access rate, relatively low storage capacity, and very high cost compared to other forms of memory.
2.   Control storage, which commonly contains the microprograms that assist the CPU circuitry in performing its functions.

TABLE 2.1   Types of Computer Storage Technology for Main Data Storage

1. *Magnetic core storage*, where each data bit is represented by the magnetic state of a small ferromagnetic "doughnut." This type of storage is nonvolatile, which means that it retains its data when power is interrupted.
2. *Semiconductor storage*, which consists of memory cells made up of transistor circuits. This type of storage is volatile, which means the data in each cell is retained only as long as power is maintained.
3. *Semiconductor monolithic storage*, or large-scale integrated (LSI) memory circuitry, which contains the equivalent of thousands of microminiaturized transistor memory cells. LSI memory uses less power than conventional transistor memory. However, it is volatile, and it has a slower rate of data transfer than semiconductor storage.

   3.   Local storage, the high-speed working registers used in the arithmetic and logical operations.

Some of the principal types of storage technology commonly used in computer systems to accomplish the main data storage (category 1) are listed in Table 2.1.

Auxiliary memory (secondary storage)

Programs and data files are not generally kept in primary storage but are stored on large-capacity auxiliary devices and loaded into main memory as required. Main storage is very expensive, and has a rather limited capacity (the capacity is limited somewhat by its high cost). Also, some operations require more data than can be held in main storage at one time. As an example, an airline reservation system may use a file containing information about each scheduled flight (available seating, passenger information, arrival and departure times, ticket prices, etc.). The set of characteristics for each flight is called a record, and the set of all the flight records is called a file. It would be inefficient from a processing standpoint to tie up main memory with a file as large as this, especially when only one flight record is needed at any given time. To eliminate this problem, the entire file is stored on an auxiliary device, and individual records are accessed as needed by the reservation program. These auxiliary devices constitute the secondary storage and are physically external to the computer, which means that the programs and data files are not directly available to the CPU. There are two basic types of secondary storage:

   1.   *Sequential access storage.* A sequential access storage unit is distinguished by the fact that to read one particular record in the file, all records preceding it must also be read.
   2.   *Direct access storage.* With this storage method, individual records can be located and read immediately without reading any other records.

Because of its method of operation, the sequential storage method has a substantially lower access rate than that of direct access storage. On the other hand, the cost per bit of data stored is higher for the direct access method, and its technology is more complicated. These factors tend to define the applications of the two storage types. Sequential access storage is suitable for applications that do not require a high level of file activity. Direct access storage is best suited to files where a high level of activity is involved. The airline reservation system described earlier would be an appropriate application for direct access storage.

Table 2.2 presents a list of some of the hardware devices and storage technology used in computer systems. Most of this technology refers to secondary storage.

2.4   INPUT/OUTPUT

The purpose of the input/output section of the computer is to provide communication with the variety of peripheral devices used with the computer system. As the name implies, there are two inverse functions involved. First, programs and data are read into the computer. The I/O section must interpret the incoming signals and hold them temporarily until they are placed in main memory or into the CPU. Second, the results of the calculations and data processing operations must be transmitted to the appropriate peripheral equipment.

In the conventional applications of the computer (e.g., scientific and engineering calculations, business data processing, manufacturing support functions), the computer must communicate with people via the peripheral devices. In Table 2.3 we present a list of some of the traditional and modern peripheral devices used to communicate with the computer for these conventional applications.

In Chapters 4, 5, and 6 our attention will be directed to the topic of computer-aided design. The operation of a CAD system involves a great deal of interactive communication between the human designer and the computer system. The specialized input/output section and associated peripheral equipment required in computer-aided design will be discussed in those chapters (Chapter 5 in particular).

2.5   DATA REPRESENTATION

Computer systems depend on the capability to represent and manipulate symbols. When we communicate with the computer and its peripheral equipment, the data and programming instructions must be reduced to a set of symbols which can be interpreted by the system. The symbols used by the computer are based on electrical signals that can take one of two states. The smallest unit of data is the bit. It has two possible values, on or off (1 or 0). The bits can be arranged into groups and, depending on the sequential bit values, the group can be used to represent a more sophisticated symbol such as a number or alphabetic character. By arranging these numbers and letters into groups, information and programming instructions can be communicated to and from the computer.

The purpose of the peripheral devices described in Table 2.3 is to accomplish

TABLE 2.2   Hardware Devices Used for Computer Storage Technology

Magnetic Tape Storage

Magnetic tape storage is a prime example of sequential access storage technology. Data are stored on magnetically coated Mylar tape, similar to the magnetic tape used in audio systems. Recording and reading data are nondestructive, which means that the tape can be erased and reused. Since data are stored sequentially, access time is relatively slow. However, the low cost per bit and high capacity of magnetic tape make it ideal for system backup. In backup, the contents of other memory units are copied onto tape. In the event of a catastrophe which erases these other memory units, there would be a recent copy of the system on tape which can be reloaded.

Magnetic Drum Storage

The magnetic drum is a random access storage device with high capacity and high access rates. The magnetic drum consists of a magnetically coated cylinder. During operation, the drum is rotated at a constant speed and data are recorded in the form of magnetized spots. The drum can be read repeatedly without causing data loss. Read/write heads are used to read data to and from the drum as it rotates. The drum surface is divided into tracks, each with its own read/write head.

Magnetic Disk Storage

Magnetic disk storage is also a direct access storage device. The storage medium is a magnetically coated disk. There are several types and sizes of disks, each best suited to a particular set of applications.

> The flexible (floppy) diskette comes in several standard sizes (5.25 in., 8 in., etc.) and is packaged in a square plastic envelope to protect the magnetic surfaces. Reading and writing are accomplished through access holes in the envelope. Floppy disks are available with either one or both sides used for storing data.

> The hard disk is a thin metal disk which is coated on both sides with magnetic ferric oxide. Data are recorded in the form of magnetized spots on tracks on the disk surface. Several disks are combined into a single disk pack and these are separated by a fixed distance and joined by a vertical shaft. The disk pack is rotated at several thousand revolutions per minute by a disk drive unit. Data are transferred by moving a set of read/write heads (one per recording surface) to the appropriate track. Only one of the heads is used to transfer data at a time, although all heads are moved together. The particular track is read until the appropriate data are found. This means that the access time is dependent on the rotational speed of the disks and the capacity of the head to read from the disk surface. This rate is usually several thousand bytes per second.

Bubble Memory

Bubble memory consists of microscopic magnetic bubbles on a thin crystalline magnetic film. Bubble memory is impressive for its high storage density and random address capabilities. However, this technology is still in its infancy and is not yet price competitive with other memory technologies.

Laser Beam Storage

Optical data storage using the laser beam is becoming a feasible alternative to conventional magnetic memory technology. Data are stored as microscopic bits on a metallic surface. Data are read by directing a laser beam across the surface, and measuring the amount of light reflected. This reflected light can be converted into electrical pulses. The storage medium is metal film strips, arranged into tracks.

TABLE 2.2 *(cont.)*

Videodisk Memory Systems

Videodisk systems are currently being used in television video-recorder/player units. However, several manufacturers are developing computer storage systems based on this technology. The videodisk resembles a long-playing phonograph record with a glossy black or silvery surface. The disk has grooves and is capable of storing both analog and digital information on one disk. Data are read by several means, including mechanical stylus or laser beams. Videodisks have extremely high capacity because of the very small bit storage size of 1.5 μm^2. One videodisk unit is reported to have a capacity of 10 billion bits.

Electron Beam-Addressable Memory Systems

Electron beam-addressable memory (EBAM) relies on information storage in metal-oxide semiconductor (MOS) LSI memory chips placed inside and near the face of a cathode ray tube. An electron beam is used to read and write to each chip. EBAM storage has advantages of fast access time, high data-transfer rates, and long-term data integrity. Typical storage time without applied power consists of several months for EBAM memories.

TABLE 2.3 Common Peripheral Devices Used for Computer Input/Output

---

Card Readers

The punched card has been used more during its long existence than any other data-recording medium. There are two types of punched cards: the 80-column Hollerith card and the 96-column IBM card. A card reader transfers data from the punched card to the computer system. There are two types of card readers currently in use:

1. *The brush reader.* In this device the punched cards are moved past a set of electrically conductive brushes. Holes in the cards allow circuits to be closed, producing electrical pulses which correspond to binary data. Some readers use two sets of brushes, in which the second set is used as a check on the first set.
2. *The photoelectric reader.* This device utilizes a set of photocells and a light source which shines through the holes in the card to produce electrical pulses at the photocell junctions. This type of reader has certain advantages as a result of having fewer moving and contacting parts.

Card Punches

A card punch records the output from the computer onto punched cards. Cards are transferred from a hopper, punched, and then reread to assure correct punching. Card punching speeds typically range from 100 to 300 cards per minute. Card readers and punches are often combined into a single unit.

Magnetic Tape Units

Magnetic tape units were introduced because of the demand for faster I/O devices. They can be used for program and data storage, and they can be interfaced to the computer as both input and output units. The medium used is magnetic tape. This medium was described in Table 2.2. During operation, the tape is moved past a read/write head, usually at a constant speed (25 to 200 in./s).

TABLE 2.3 (*cont.*)

Punched Tape Readers

A punched tape reader reads data from punched holes on a strip of paper tape (other materials are also used) having five to eight channels. As the tape is moved through the reading head, the presence or absence of holes is sensed. Paper tape data entry is usually slower than magnetic tape. A common application of punched tape readers is in numerical control programming (Chapter 8).

Paper Tape Punches

Data from a computer system can be outputed onto punched paper tape. Data from main storage are converted into the appropriate code and punched on the tape as it is fed through the punching unit. Paper tape readers and punches are often combined into a single unit.

Keyboard Input Devices

Many input devices employ a typewriterlike keyboard which can be used by a typist with little additional training. Some of these devices input data and programs directly to the computer. Others produce data on a special medium for subsequent input to the computer system.

The Keypunch

The keypunch is an electromechanical keyboard device which converts operator keystrokes into machine-readable holes on cards. The cards are then submitted through a card reader to the computer.

Key-to-Tape Unit

The key-to-tape unit is an electronic typewriter device that converts operator keystrokes into machine-readable codes on magnetic tape. Two types of units are available. The first produces computer-compatible tape (1/2 in., seven or nine tracks), but is now considered to be obsolete. The other unit produces magnetic cartridges or tape cassettes.

Alphanumeric Displays

An alphanumeric display consists of a typewriterlike keyboard and a display screen, usually a cathode ray tube (CRT), which can be used to display data. Stand-alone CRT terminals include the following components: screen, keyboard, communications interface, buffer memory, and sometimes a local microprocessor used for editing. The CRT terminal can be connected directly to a computer for *on-line* operation, or it can be used with independent devices for *off-line* operation. Transmission speeds are usually selectable, from 110 to 9600 baud. A baud is a unit representing the number of discrete signal changes per second. For a binary system it is equal to the number of bits per second. Communication-line quality limits the speed of data transmission. Programming and data input on a CRT terminal are faster than for other keyboard entry devices because of enhancements in keyboard design, screen formatting, prompting capabilities, and local editing.

Teleprinters

A teleprinter consists of an electromechanical or electronic typewriter keyboard and a hard-copy printing device. It can function both as a remote data entry terminal and as an on-line output terminal. During input, data are usually transmitted character by character as keys are depressed, although some units have buffer memory available to permit batched continuous transmission. Transmission speed depends on the device design. Older electromechanical units print data at a rate between 110 and 300 baud. The newer electronic units operate at speeds of up to 9600 baud.

Magnetic Ink Character Recognition (MICR)

MICR readers are electronic devices that operate by interpreting the sensed waveforms of the individual magnetic ink characters. This technology was developed for the banking industry to facilitate the mass handling of checks and deposit slips. These devices are capable of reading up to 1600 documents per minute. The human operator is not a factor in the speed of these devices.

TABLE 2.3 *(cont.)*

Optical Character Recognition (OCR)

In the operation of the OCR reader, a mechanical drum is used to rotate documents past an optical scanning station. A light source and lens system can distinguish the patterns of the character. These patterns are converted into electrical pulses which are interpreted as each individual character. An OCR can be programmed to read a variety of character sets, and even handwriting. Entire pages can be scanned and read very quickly. Data transfer occurs at a maximum rate of 3600 baud.

Optical Mark Readers

Optical mark readers sense the physical position of marks on a document. The position is correlated to a previously defined character. The marks can be read from 80-column cards or full-page documents. Data transmission speed depends on the speed of the feeding device. A rate of 1500 forms per minute is not unusual. Many standardized tests, such as the Scholastic Aptitude Test (SAT), use optical mark techniques to grade responses to questions.

Optical Bar-Code Readers (OBR)

An optical bar-code reader senses the configuration of shaded bars of different widths and correlates them to previously defined characters. An OBR can read characters at 50 to 400 per second, and is limited mainly by the operator's ability to manipulate the items to be scanned.

Line Printers

In a typical data processing shop, the volume of output is sufficient to justify a high-speed line printer. These units print entire lines at one cycle (80 to 132 characters per line) at rates that may exceed 1000 lines per minute. These units are expensive and the cost is generally proportional to the printing rate. One of the recent innovations in high-speed printers combines laser and xerography technologies to achieve print speeds of about 10,000 lines per minute.

---

the conversion process between the higher-level characters (numbers and letters) and the basic units of data used by the computer (the bit).

The binary number system

The bit, or binary digit, is the basic unit of data which can be interpreted by the digital computer. It is easily adaptable to the binary number system because there are only two possible conditions which can take on the values of 0 or 1. The meaning of successive digits in the binary system is based on the number 2 raised to successive powers. The first digit is 2^0, the second is 2^1, the third is 2^2, and so forth. The two numbers, 0 or 1, in successive bit positions indicate either the absence or presence of the value. Table 2.4 shows how the binary number system can be used to represent numbers in the more familiar decimal system.

The conversion from binary to decimal systems makes use of the following type of computation. We will illustrate the conversion for the decimal number 5:

$$1 \times 2^0 + 0 \times 2^1 + 1 \times 2^2 + 0 \times 2^3$$
$$= 1 \times 1 + 0 \times 2 + 1 \times 4 + 0 \times 8 = 5$$

TABLE 2.4   Binary and Decimal
Number System Equivalence

Binary	Decimal
0000	0
0001	1
0010	2
0011	3
0100	4
0101	5
0110	6
0111	7
1000	8
1001	9

A minimum of four digits are required in the binary system to represent any single-digit number in the decimal system, as indicated in Table 2.4. By using more than four binary digits, higher-valued decimal numbers or other high-level data can be represented.

An alternative way to represent decimal numbers larger than nine involves separate coding of each digit, using four binary digits for each decimal digit. This coding system is known as binary-coded decimal (BCD). The binary-coded decimal system is explained in Table 2.5 together with two other common coding schemes.

TABLE 2.5   Common Binary Coding Schemes

Binary-Coded Decimal (BCD)

The binary-coded decimal system uses a total of 7 bits. The first 6 represent the data itself (alphabetic, numeric, or special character). The last bit position is used as a parity check. The first 4 bits are called numeric bits, and are assigned the values 1, 2, 4, and 8. By various combinations of these bits, the decimal numbers from zero to nine can be represented. The fifth and sixth bits are called the zone bits. The zone bits are both zero when numeric characters are represented. Combinations of the zone and numeric bits can be used to code the alphabetic and special characters.

EBCDIC

The maximum number of unique characters a computer code can represent is 2 raised to the power equal to the number of bits. Thus BCD allows for $2^6 = 64$ distinct characters. EBCDIC (Extended Binary-Coded Decimal Interchange Code) uses an 8-bit code plus a parity bit so that it can define 256 distinct characters. These include upper- and lowercase alphabetics, the numerals, many special characters, and control characters for I/O devices.

ASCII

ASCII (American Standard Code for Information Interchange) was developed for telecommunications to simplify machine-to-machine and system-to-system communications. It is a 7-bit code, which provides 128 bit patterns for character representation.

2.6   COMPUTER PROGRAMMING LANGUAGES

The preceding section demonstrated how the binary number system could be used to represent any decimal number, alphabetic letter, or other common symbol. Data and instructions are communicated to the computer in the form of binary words. In executing a program, the computer interprets the configuration of bits as an instruction to perform electronic operations such as add, subtract, load into memory, and so forth. The sequence of these binary-coded instructions define the set of calculations and data manipulations by which the computer executes the program.

The binary-coded instructions that computers can understand are called machine language. Unfortunately, binary-coded instructions and data are very difficult for human programmers to read or write. Also, different machines use different machine languages. To facilitate the task of computer programming, higher-level languages are available which can be learned with relative ease by human beings. In all there are three levels of computer programming languages:

1.   Machine language
2.   Assembly language
3.   Procedure-oriented (high-level) languages

Machine and assembly languages

The language used by the computer is called machine language. It is written in binary, with each instruction containing an operation code and an operand. The operand might be a memory address, a device address, or data. In machine language programming, storage locations are designated for the program and data, and these are used throughout the program to refer to specific data or program steps. In addition, the programmer must be familiar with the specific computer system since machine language instructions are different for each computer. Programming in machine language is tedious, complicated, and time consuming. To alleviate the difficulties in writing programs in binary, symbolic languages have been developed which substitute an English-like mnemonics for each binary instruction. Mnemonics are easier to remember than binary, so they help speed up the programming process. A language consisting of mnemonic instructions is called an assembly language. Figures 2.5 and 2.6 illustrate the difference between machine language and assembly language.

Assembly languages are considered to be low-level languages. The programmer must be very knowledgeable about the computer and equipment being programmed. Low-level languages are the most efficient in terms of fast execution on the computer, but there are obvious difficulties for the programmer in writing large programs for various applications using different computers. We consider programming with assembly language in Chapter 3 in our discussion of microcomputers and microprocessors.

Address	Contents
0	00100001
1	01000000
2	00000000
3	00010001
4	01100000
5	00000000
6	00000110
7	00001010
8	01111110
9	00010010
10	00100011
11	00010011
12	00000101
13	11000010
14	00001000
15	00000000
16	01110110

FIGURE 2.5

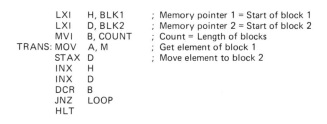

```
        LXI    H, BLK1      ; Memory pointer 1 = Start of block 1
        LXI    D, BLK2      ; Memory pointer 2 = Start of block 2
        MVI    B, COUNT     ; Count = Length of blocks
TRANS: MOV     A, M         ; Get element of block 1
        STAX   D            ; Move element to block 2
        INX    H
        INX    D
        DCR    B
        JNZ    LOOP
        HLT
```

FIGURE 2.6. Portion of an assembly language program. (From Lance A. Leventhal, *Introduction to Microprocessors: Software, Hardware, Programming*, ©1978, p. 129. Reprinted by permission of Prentice-Hall, Inc., Englewood Cliffs, N.J.)

Assembly language programs must be converted into machine language before the computer can execute them. The conversion is carried out by a program called an assembler. The assembler takes the assembly language program, performs the necessary conversions, and produces two new programs: the machine language version and an assembly listing. The assembly listing shows the mnemonic instructions and their associated machine language equivalents, and any errors the original assembly language program may have contained.

High-level language

Assembly languages are machine oriented. High-level languages, by contrast, are procedure oriented. They are to a large extent independent of the computer on which they are used. This means that a program written on one computer can be run on a different computer without significant modifications to the program.

High-level languages consist of English-like statements and traditional mathematical symbols. Each high-level statement is equivalent to many instruc-

```
       DO 100    I = 1, 10
100    BLK1    (I)  =  BLK2 (I)
```

FIGURE 2.7 Portion of FORTRAN program. (From Lance A. Levanthal, *Introduction to Microprocessors: Software, Hardware, Programming*, ©1978, p. 133. Reprinted by permission of Prentice-Hall, Inc., Englewood Cliffs, N.J.)

tions in machine language. To illustrate, Figures 2.5 and 2.6 present two lists of instructions, written in machine language and assembly language, respectively. Both sets of programming instructions accomplish the same task, which is to transfer the contents of one memory location into another memory location. This can be accomplished in FORTRAN (a high-level language) with two lines of instruction as shown in Figure 2.7.

The advantage of high-level languages is that it is not necessary for the programmer to be familiar with machine language. The program is written as an English-like algorithm to solve a problem. Like assembler languages, high-level languages must also be converted into machine code. This is accomplished by a special program called a compiler. The compiler takes the high-level program, and converts it into a lower-level code, such as the machine language. If there are any statement errors in the program (e.g., misspelled words), error messages are printed in a special program listing by the compiler.

There are many different high-level languages. Table 2.6 describes several of the common high-level languages used for business and engineering applications. In Chapter 8 we describe the APT language, which is a high-level language used to program automatic machine tools.

TABLE 2.6 Some Common High-Level Computer Programming Languages for Business and Engineering Applications

---

FORTRAN

FORTRAN stands for FORmula TRANslation. It is one of the oldest and most popular of the high-level languages. FORTRAN was developed in the mid-1950s for scientists, engineers, and mathematicians, but it has also been used for some business processing applications.

FORTRAN consists principally of mathematical notation. Data items and variables are given names, and these names are used as symbols in the program to be operated on by mathematical and logical operators. Other instructions are also allowed in FORTRAN, for example to form lists and tables, or to call various input/output functions, such as printing or accepting data from a file. FORTRAN is quite efficient for mathematical computations but is not very efficient for applications such as file processing or document production.

COBOL

COBOL (COmmon Business-Oriented Language) was developed around 1959. It has become a major computer language for business data processing applications. COBOL can be used for numeric and alphabetic data processing and in data-file applications.

TABLE 2.6   *(cont.)*

COBOL is similar to FORTRAN in the sense that it uses English-like statements. Indeed, COBOL statements are called sentences. A series of sentences is termed a paragraph. COBOL's vocabulary tends to make it self-documenting. It is very easy to read a program and understand what operations are taking place. However, the language presumes an understanding of information processing principles. Therefore, it is not intended for the inexperienced programmer. While COBOL is suitable for a file environment, it is not capable of the complex mathematical calculations that can be accomplished with FORTRAN.

BASIC

BASIC (Beginner's All-Purpose Symbolic Instruction Code) was developed in the 1960s at Dartmouth College to be an easy-to-learn programming language. Learning the "basics" of this language requires only a few hours.

BASIC was developed as an interactive language, where the user appears to be interacting directly with the computer rather than through punched cards or other off-line I/O format. Each line of code is interpreted as it is typed, so that the programmer is informed of many types of program errors immediately upon their entry. Programs are usually entered and executed by means of CRT terminals in a time-sharing network. BASIC is used heavily in the field of microcomputers, and is being used increasingly as an interactive teaching tool in schools.

APL

APL (A Programming Language) was designed for interactive problem solving. A significant feature is that it permits users to define complex algorithms efficiently. The primary data structure in APL is in the form of arrays and an extensive set of array operators is provided.

RPG

RPG (Report Program Generator) is a language designed for writing programs that produce printed reports as output. It is widely used in business environments, where it updates data files, performs analysis of data, and generates documents and reports.

RPG is relatively easy to learn and programs can often be written in a short time, making it suitable for the needs of upper-level managers, who often desire to have immediate access to information in a given format. It is most appropriate for simple, straightforward problems which do not require complicated logic or use of data files. It is a common language on minicomputers because of its relatively modest memory requirements.

PL/1

PL/1 is a general-purpose programming language designed to be a compromise between FORTRAN and business languages, each of which has obvious strong and weak points. It uses standard arithmetic operators, similar to FORTRAN. PL/1 is a flexible language which meets the needs of a wide variety of programmers. Default conditions assumed by the compiler when explicit alternative statements are omitted allow many functions to be performed in fewer steps than in FORTRAN or other languages.

PASCAL

Pascal is a high-level language developed in the early 1970s and named for the famous French mathematician Blaise Pascal. The objectives in developing the language were to facilitate the teaching of computer programming as a systematic discipline of knowledge and to accomplish programming implementations which are reliable and efficient on modern computers.

2.7   OPERATING THE COMPUTER SYSTEM

In the evolution of computer systems, one of the most important considerations has been the ease of operation of the system. Many technological improvements have been made which now make computer systems much easier to operate and much more efficient. For example, in early systems, the CPU and I/O operations could not take place independently. This meant that the CPU was forced to be idle while a slow input/output device transferred data to and from main memory. In modern computer systems, input/output and data processing operations can occur simultaneously to make the operation of the computer system more efficient. In this section we examine some of these techniques which facilitate the operation of the computer system by the user.

I/O control systems and operating systems

An input/output control system (IOCS) is a series of related programs which, when loaded into main storage, interpret I/O data transfer commands (such as READ and WRITE) and control the transfer of data to and from main storage. The IOCS also performs functions such as reading the labels on magnetic tape reels which are ready to be loaded, and it possesses features such as error-checking and recovery procedures for data-transfer operations.

An IOCS is designed to help improve computer operating efficiency. As the use of computers has increased, attention has also been focused on other areas that require control. One of these areas deals with the time wasted in the CPU between jobs. In early systems, the CPU was forced to remain idle while the operator manually staged the next job by mounting the correct tape or reading in the deck of cards. In present systems, processing of one job and input of another can occur concurrently by means of input/output channels and some sort of buffering system.

Another problem was the interaction of jobs requiring little CPU time but much I/O time, with those jobs that needed little I/O time but that monopolized the CPU. A solution to this problem involves placing both programs in memory simultaneously and executing parts of each for short intervals. To accomplish this, there must be a set of control programs to prevent the user programs from writing over each other in main memory, resulting in the destruction of both programs. This set of programs is called the operating system. Managing the various computer resources (CPU, memory, and I/O) and controlling peripheral devices are accomplished by the operating system. Unlike user programs, which support applications external to the computer system, an operating system supports the internal functions of the entire computing system. Its objective is to maximize the performance and efficiency of the system and to increase the ease with which the system can be used.

Virtual storage

A problem arises when a program requires more main storage than the computer possesses. For example, if there is 70K (1K equals 1024 bits) of main storage, and the user program requires 75K of storage, where does the computer acquire the additional 5K to execute the program?

With virtual storage, programs are not limited to memory locations in primary storage (main memory). Although each program instruction must be in main storage at the time it is executed, not all instructions must be there all the time. Maintaining only a portion of the program in real storage at any given time, and bringing in other portions as they are needed, increases the effective main memory available. This means that more user programs can be executing at one time; hence system throughput is increased.

Virtual storage is accomplished by one of two methods: segmentation and paging. In segmentation, the program is divided into variable-size blocks which represent logical program units such as subroutines and data groups. By monitoring each block's location in a file called a segment table, the operating system loads only those segments needed by the part of the program currently being executed, and exchanges program segments as required.

Paging methods divide the usable main memory into fixed-size blocks called page frames. Programs and data are separated into units of the same size called pages. Page size depends on system hardware, and ranges from 1K to 4K bytes. The operating system monitors page frame usage, and as a program completes the current page, the next page is exchanged so that execution continues almost uninterrupted.

In both methods, virtual storage is achieved by the use of disk storage devices. The transfer of program and data units to and from main storage is handled by a program called the disk operating system (DOS). Segmentation and paging systems are sometimes combined into a single system to yield the benefits of both methods. More programs can be executed at a given time, for higher CPU and I/O device utilization.

Time sharing

Time sharing is a function on some operating systems that permits more than one user to have simultaneous access to a computer system. To each user, it seems as if the computer is giving its undivided attention. In reality, the computer is sharing its resources among many separate users. When CPU time is requested by a user terminal, a portion of main memory is assigned to that terminal. During operation, the CPU switches from one terminal to another, executing small portions of each program. When a particular program has completed execution, a new program is entered into the newly available memory space. The concept of many programs executing almost simultaneously with one CPU is called multiprogramming.

Time-sharing systems are of three major types:

1. General-purpose time-sharing systems that permit programmers to write and execute their own programs in various languages.
2. Execution-only systems that only allow users to execute programs, but not to create, alter, or delete programs.
3. Single-application systems in which all programs are related to a certain application area. In these systems, the user supplies the appropriate inputs and the system responds accordingly.

Distributed processing

Consistent with modern management philosophies for decentralization in the control of businesses, advances in computer technology have made possible a procedure called distributed computer processing. Although there is no universally accepted definition of the term, distributed processing generally refers to the use of "intelligent" terminals which can perform local editing and data manipulation and which can transmit partially processed data to a central computer facility for further work. This type of configuration also occurs when a communications network exists to facilitate the transfer of data from terminals to a central computer. Distributed processing allows geographically separate groups of users to accomplish much of their processing locally, but providing each group with access to a large central computer for jobs requiring a large mainframe and access to corporate data bases.

REFERENCES

[1] BELZER, J., HOLZMANN, A. G., AND KENT, A., *Encyclopedia of Computer Science and Technology*, Vol. 11, Marcel Dekker, Inc., New York, 1978.

[2] BOHL, M., *Information Processing*, 2nd ed., Science Research Associates, Inc., Chicago, 1976.

[3] GOODSTEIN, D., "Output Alternatives," *Datamation*, February, 1980, pp. 122−130.

[4] GROOVER, M. P., *Automation, Production Systems, and Computer-Aided Manufacturing*, Prentice-Hall, Inc., Englewood Cliffs, N.J., 1980, Chapter 11.

[5] KLINE, R. M., *Digital Computer Design*, Prentice-Hall, Inc., Englewood Cliffs, N.J., 1977.

[6] LEVENTHAL, L. A., *Introduction to Microprocessors: Software, Hardware, Programming*, Prentice-Hall, Inc., Englewood Cliffs, N.J., 1978.

[7] MACEVEN, G. H., *Introduction to Computer Systems Using the PDP-11 and Pascal*, McGraw-Hill Book Company, New York, 1980.

[8]   MATICK, R. E., *Computer Storage Systems and Technology*, John Wiley & Sons, Inc., New York, 1977.

[9]   NASHELESKY, L., *Introduction to Digital Computer Technology*, 2nd ed., John Wiley & Sons, Inc., New York, 1977.

[10] "Peripheral Equipment for Computers," *Electrical Engineering*, Morgan-Grampian Ltd., London, November, 1978, pp. 107–116.

[11] RECICAR, S., *Selection of Data Entry Equipment*, Special Publication 500-55, National Bureau of Standards, U.S. Dept. of Commerce, Washington, D.C., 1970.

[12] ROLPH, S., "The Disk Revolution," *Datamation*, February, 1980, pp. 147–150.

[13] STEIN, D., "Videodisks—The Revolution in Information Storage," *Output Magazine*, May, 1981, pp. 22–26.

[14] TANENBAUM, A. S., *Structured Computer Organization*, Prentice-Hall, Inc., Englewood Cliffs, N.J., 1976.

PROBLEMS

2.1 Determine the decimal equivalent for each of the following binary numbers.
- (a)   001011
- (b)   000101
- (c)   010111
- (d)   111111

2.2 Determine the binary equivalent for each of the following decimal numbers.
- (a)   7
- (b)   11
- (c)   25
- (d)   17
- (e)   80

2.3 The decimal number system is based on the number 10. The binary number system is based on the number 2. Develop the framework (to four digits) for an octal number system (a number system based on the number 8). Use Table 2.4 as a framework for your model, but an abbreviated form would be adequate. Express the following decimal numbers in their octal equivalent.
- (a)   8
- (b)   73
- (c)   116
- (d)   1625
- (e)   8000

2.4 Do Problem 2.3 except for a hexadecimal number system (a number system based on the number 16).

chapter 3

Minicomputers, Microcomputers, and Programmable Controllers

3.1 INTRODUCTION

During the mid-1960s, small computers began to appear on the market. To distinguish them from the larger mainframes of that period, they were called minicomputers. The appearance of these minicomputers reflected an ever-continuing trend of miniaturization in computer technology that has now reached the point where a CPU, together with a small main memory and input/output circuitry, can be contained on a single LSI (large-scale integrated) circuit chip several millimeters square. Such a chip, when combined with input and output devices, is called a microcomputer.

In this chapter we describe the minicomputer and microcomputer, which are frequently used in CAD/CAM applications. We also discuss the programmable controller, which is a related device often used in process control applications. Before considering the details of these three devices, we provide definitions and an overview of their applications.

Computers are generally considered to fall into three size categories, all of which are based on roughly the same architecture as that described in Chapter 2.

The three size categories are:

1. The large mainframe computer
2. The minicomputer
3. The microcomputer

The large mainframe computer is distinguished by its cost, capacity, and function. The price of a new corporate general-purpose computer can run into millions of dollars. The main memory capacity is several orders of magnitude larger than the minicomputer, and the speed with which computations can be made is several times the speed of a minicomputer or microcomputer. The functions which are accomplished on a large mainframe computer can typically be classified into two categories, as described in Table 3.1. An example of a large mainframe computer was shown in Figure 2.2.

Minicomputers are smaller versions of the large mainframe computers. The trend toward miniaturization in computer technology provides two alternative approaches in the design of a computer. The first is to package greater computational power into the same physical size with each new computer generation. The second approach is to package the same computational power into a smaller size. The minicomputer manufacturers elected the second approach in developing their product lines. Traditionally, the cost of a minicomputer has ranged from less than a thousand dollars up to about $50,000. Minicomputers can be utilized for the same two general functions as a large mainframe computer (described in Table 3.1). However, the size of the jobs to be processed must be smaller to be within the capacity of the minicomputer. Smaller minicomputers often overlap with microcomputers in terms of the functions which they perform. These microcomputer/minicomputer functions are different from those performed by large mainframes. The characterisitics of typical minicomputer/microcomputer applications are described in Table 3.2.

A microcomputer uses a microprocessor as the basic central processing unit. The microprocessor consists of integrated circuits contained on LSI chips. The LSI

TABLE 3.1   Two Typical Applications Performed on Large Mainframe Computers and Large Minicomputers

1. Complex Engineering and Scientific Problems
    Examples include iterative calculation procedures often required in heat-transfer analysis, fluid dynamics analysis, or structural design analysis. These calculations are typical of computer-aided design applications.
2. Large-Scale Data Processing
    Examples include corporate accounting and payroll operations, production scheduling, compiling production costs, and maintenance of large information files. Some of these examples are found in indirect computer-aided manufacturing applications.

TABLE 3.2   Typical Characteristics of Microcomputer/Minicomputer Applications

1. The computer is a system component. The overall system, which might be a piece of test equipment, a machine tool, or a banking terminal, uses the small computer much as it might a switch, power supply, or display. The computer may not even be visible from the outside.

2. The computer performs a specific task for a single system. It is not shared by different users as a large computer is. Instead, the small computer is part of a particular unit, such as a medical instrument, typesetter, or factory machine.

3. The computer has a fixed program that is rarely changed. Unlike a large computer, which may solve a variety of business and engineering problems, many small computers perform a single set of tasks, such as monitoring a security system, producing graphic displays, or bending sheets of metal. Programs are often stored in a permanent medium or read-only memory.

4. The computer often performs real-time tasks in which it must obtain the answers at a particular time to satisfy system needs. Such applications include machine tools that must turn the cutter at the right time to obtain the correct pattern, or in missile guidance where the computer must apply thrust at the proper time to achieve the desired trajectory.

5. The computer performs control tasks rather than arithmetic or data processing. Its primary function might be managing a warehouse, controlling a transit system, or monitoring the condition of a patient.

Source: Compiled from Ref. [6].

chips can be manufactured in large quantities very inexpensively. The microprocessor is capable of performing virtually all the functions of the conventional CPU (e.g., arithmetic—logic operations, or fetching data from memory). Accordingly, the microprocessor can be connected to a memory unit and the appropriate input/output device(s) to form a microcomputer. Typical characteristics of microcomputer applications are described in Table 3.2.

The programmable controller (PC) was introduced in the late 1960s as a substitute for electromechanical relay logic systems. At that time relay control systems were typically used to regulate the operation of production equipment. The problems with electromechanical relays are their physical size and programming inflexibility. The programmable controller could perform the same functions as a relay logic system with greater flexibility and lower space requirements. The PC can be defined as a sequential logic device which generates output signals according to logic operations performed on input signals. The sequence of instructions that determines the inputs, outputs, and logical operations constitutes the program. The initial success of the programmable controller was due to its apparent similarity to relay control systems. Wiring charts called "ladder diagrams" have been used for many years to set up relay panels, and shop technical personnel are familiar with these diagrams for wiring and maintenance of the panels. Most PCs are programmed using the same type of ladder diagrams. Accordingly, shop people could easily adapt to the new control devices. In recent years, LSI technology in programmable controllers has reduced the differences between PCs and microcomputers/minicomputers. PCs can be interfaced with computers in production operations. The programmable controller is used to direct certain aspects of the operations while computers are used for other monitoring, control, and data processing functions.

The following three sections elaborate on the preceding discussion of minicomputers, microcomputers, and programmable controllers.

3.2 MINICOMPUTERS

A minicomputer is typically distinguished as a low-cost, general-purpose word-addressable computer with 4K to 32K of 8- to 16-bit words of main storage. Thirty-two-bit minicomputers are becoming more and more common. Two commercially available minicomputers are illustrated in Figures 3.1 and 3.2. Factors such as speed, ease of programming, and other operating characteristics are determined by the particular architecture of the minicomputer. The architecture of a minicomputer follows the general computer structure described in Section 2.1. The fundamental subunits are the memory, arithmetic and logic processor control unit, and input/output (I/O) unit.

The memory stores the instructions and data of both user and control programs. The arithmetic–logic processor receives data from memory and performs

FIGURE 3.1   A midrange-power minicomputer, the DEC PDP-11/24. Courtesy of Digital Equipment Corporation.)

FIGURE 3.2   The more powerful VAX-11/730 model (background), a 32-bit minicomputer capable of supporting interactive graphics workstations. (Courtesy of Digital Equipment Corporation.)

operations on it as directed by the program instructions. The control unit regulates the flow of data in the system, fetches program instructions from memory, and decodes the instructions in instruction registers. Together, the arithmetic–logic processor and the control unit (along with registers and the interconnecting buses) are known as the CPU of the system. The input/output unit provides the interface to peripheral devices, transferring data to and from the external world.

Minicomputers are sometimes designed with memory as a device external to the system (like a peripheral), and communication is established over an I/O bus. The resulting single-bus architecture is more efficient because the CPU is independent of memory timing. This allows for high-speed memory to be added to the system at a later date without extensive CPU modification. The single-bus structure also permits memory devices of different speeds to be used within the same system.

Minicomputer instructions

The basic minicomputer instruction consists of two parts: an operation code and an operand. The operation code specifies the function to be performed (arithmetic operation, data transfer, etc). This function is executed when its operation code is properly decoded by the CPU control logic. The operation is performed on data that are contained in memory. The operand typically consists of an address and an addressing mode. The address specifies the memory location of the data. The addressing mode refers to the method by which the data are accessed. The data might be accessed directly by going to the address specified in the instruction, or it could be accessed by any of several other schemes.

The instruction is carried out during the minicomputer instruction cycle. This consists of a series of events which can be grouped into two phases: fetch and execute. During the fetch phase, the control logic fetches the address of the instruction from the program counter. This address is decoded by the decoder circuitry. The instruction is brought from this address in memory and loaded into the instruction register, where it is decoded.

During the execution phase of the cycle, the particular operation (e.g., addition, subtraction, comparison, or other) defined by the contents of the instruction register is carried out. Each instruction has a unique sequence of steps associated with its execution.

Most of the processing that occurs in the minicomputer during program execution takes place in registers. Several of the various types of registers were described in Section 2.2. The number and type of registers in the computer architecture determine its performance. Generally, as the number of registers in a minicomputer is increased, the required program size and the execution time decrease. Thus an indication of computing power in a minicomputer is its number of registers.

A minicomputer is provided with the capability to perform a variety of operations and functions, such as data transfers, arithmetic operations, and logic functions. These operations and functions are known as the instruction set. Most minicomputer instruction sets consist of less than 200 instructions. However, only a portion of these are likely to be used in a typical program.

The instructions can be classified as memory or nonmemory. Memory instructions are used to transfer data to and from memory locations. Nonmemory instructions carry out the other functions of the computer. The nonmemory instructions can be divided into two categories:

Arithmetic and logic functions

Input/output instructions

The arithmetic instructions include functions such as addition, subtraction, multiplication, and comparisons. Logical functions include Boolean operations such as AND, OR, and EXCLUSIVE OR.

Input functions are concerned with the transfer of data between main storage and peripheral devices. The factors that determine the I/O capability of a minicomputer are its I/O transfer speed, the number and types of I/O software schemes available, and the priority control structure, which permits external devices to interrupt the execution of the CPU.

Input/output functions are either controlled or interrupt driven. Controlled I/O functions are usually accomplished by means of the system programming. In this case, data are transferred to and from peripheral devices through an I/O bus.

The I/O bus is equipped with circuitry that can decode the address of a specific peripheral in order to effect the transfer of data to it. In some computers, the device address is given directly, and the circuitry transfers the data to that device during a clocked time period associated with that specific device. This method of transfer is called time multiplexing. I/O transfer rates vary. The rate is limited by the fact that the CPU must constantly poll the peripheral channels to determine which devices are requesting service.

Interrupt-driven input/output occurs when the peripheral device is capable of stopping the execution of the current programs in the CPU and transferring program control to special service subroutines. These subroutines, when called, must save the contents of the working registers, the status register, and the program counter, so that when control is returned to the main program, no changes will have taken place and execution of the previous program can continue.

Microprogramming

In the conventional minicomputer, the architecture consists of memory, arithmetic–logic processor, a control unit, and an I/O section. A microprogrammed minicomputer differs from the conventional machine in the sense that the control unit, instead of being hard-wired logic, is a stored-program device. The control unit consists of two functional parts, the control decode and the control store. The control decode operates all elements of the computer system, including main storage and high-speed control store. The control store contains microprograms called firmware. The firmware corresponds to conventional computer sequences (software).

Microprogramming increases the speed with which the minicomputer can operate. For certain operations, a microprogrammed minicomputer can perform at considerably higher speeds than a conventional minicomputer. When a large microinstruction set is available, the machine can be optimized to perform a broad variety of high-speed tasks.

3.3 MICROCOMPUTERS

Microcomputers are generally smaller than minicomputers. As indicated in the introduction, the CPU, I/O circuitry, and a small amount of memory are all implemented by means of integrated circuits contained on a very small silicon base, called an LSI chip. Despite the small size of its circuitry, the microcomputer is capable of performing all the regular functions of a traditional computer. Because of its small size, the cost of a microcomputer is potentially much lower than the cost of a minicomputer. The cost of the microprocessor (the LSI circuit on a chip used as the CPU) can be as low as a few dollars when produced in large quantities. Because of its low cost, as well as limitations on its capabilities, microcomputers are typically used for functions different from those of a traditional computer. The

FIGURE 3.3   A microcomputer, the DEC MICRO/PDP-11. (Courtesy of Digital Equipment Corporation.)

characteristics of these microcomputer functions are described in Table 3.2. Figure 3.3 illustrates a commercially available microcomputer.

Microcomputer instructions

The microcomputer instruction cycle operates in much the same manner as the minicomputer instruction cycle. It consists of the fetch and execute phases, which are each composed of smaller steps.

The instruction set for a microcomputer consists of the instructions for operations which can be accomplished by the CPU. Different microcomputers possess varying capabilities and, accordingly, there are differences in their instruction sets. The types of instructions generally provided include:

1. Data transfer and input/output instructions
2. Arithmetic operations
3. Logical operations
4. Branching instructions

The following paragraphs describe these four categories of microcomputer instructions.

TABLE 3.3   Data Transfer and Input/Output Instructions for the Intel 8085A

IN data	This would input data from port number "data" to the accumulator.
OUT data	This would output data from the accumulator to port number "data."
STA addr	This would transfer data to the accumulator from a memory or output register, defined by "addr."
LDA addr	This would transfer data to the accumulator from a memory or output register, defined by "addr."
MOV r_2, r_1	This would transfer data from one register (r_1) to another register (r_2). The transfer would take place between internal registers and other internal or external registers.
MVI r, data	This would load data into register r.

DATA TRANSFER AND INPUT/OUTPUT INSTRUCTIONS. This type of instruction is concerned with the transfer of data between a microprocessor register and another component, such as memory, an I/O device, or another register. Each instruction specifies a source register and a destination register. If the destination register is part of the CPU, this is usually implied in the instruction. This type of register is called an internal register. External registers, on the other hand, must be explicitly named in the instruction. Table 3.3 presents a listing of instructions in this category for the Intel 8085A microprocessor.

ARITHMETIC OPERATIONS. Simple arithmetic operations (such as addition and subtraction), as well as more complex operations (multiplication, division, etc.), can be executed by means of the appropriate instructions in this category. Instructions for binary addition are always provided in the microprocessor's instruction set. Subtraction is usually provided also. Multiplication and division can be accomplished either directly in hardware by means of special hardware circuits or by software in the form of subroutines. These more complex operations are not typically included within the instruction sets for 8-bit microprocessors. Microprocessors with 16-bit and greater capacity would usually include multiplication and division instructions. Table 3.4 presents some typical basic instructions for executing arithmetic operations.

TABLE 3.4   Microprocessor Instructions for Arithmetic Operations

ADD r	This would add the contents of register r to the contents of the accumulator and store the result.
SUB r	This would subtract the contents of register r from the contents of the accumulator and store the result.
INR r	This would increase the contents of register r by one.
DCR r	This would decrease the contents of register r by one.
SUI data	This would subtract data from the contents of the accumulator and store the result in the accumulator.

TABLE 3.5   Microprocessor Instructions for Logical Operations (Intel 8085A Microprocessor)

Mnemonic	Description
ANI data	AND accumulator with operand data
ANA r	AND accumulator with register r
ANA M	AND accumulator with contents of memory at address specified in HL register pair
ORI data	OR accumulator with operand data
ORA r	OR accumulator with register r
ORA M	OR accumulator with contents of memory at address specified in HL register pair
XRI data	EXCLUSIVE-OR accumulator with operand data
XRA r	EXCLUSIVE-OR with register r
XRA M	EXCLUSIVE-OR with memory location
CMA	Complement contents of accumulator

LOGICAL OPERATIONS. Most microprocessors include instructions for carrying out the logical operations AND, OR, NOT, and EXCLUSIVE-OR. As an illustration, Intel Corporation's 8085A microprocessor includes the instructions presented in Table 3.5 to accomplish logical manipulations of data.

BRANCHING OPERATIONS. A branch instruction causes program control to be transferred from the current value of the program counter to another, nonconsecutive instruction. In the Intel 8085A, branching is implemented through the ''jump'' instruction. Branching operations make possible the use of program loops, in which a series of program instructions can be executed repeatedly.

Branching can be made conditional on a set of circumstances, or it can be made unconditional. In the unconditional jump instruction, the address specified in the instruction is loaded into the program counter, and control of execution is passed there. Conditional branches can be performed such that if the given condition is true, the operation is carried out. Otherwise, it is ignored, and program execution proceeds sequentially. Several branching instructions are given in Table 3.6.

TABLE 3.6   Branching Instructions (Intel 8085A)

Mnemonic	Description
JMP addr	Unconditional branch to statement ''addr''
JZ addr	Branch to statement number ''addr'' if zero flag set
JNZ addr	Branch if zero flag not set
JPO(E) addr	Branch if parity odd (even)
JP(M) addr	Branch if accumulator positive (negative)

EXAMPLE 3.1

We will use two examples to illustrate the use of these instructions. The first problem is to write a program in assembly language which will simulate a NOR gate. (A NOR gate is a hardware device with two inputs and one output which operates as follows: If both inputs are zeros, the output is one; otherwise, the output is zero.) In other words, we want to use software to duplicate the operation of a hardware device. The program, based on instructions from Tables 3.3, 3.4, and 3.5, is shown below with explanations at the right.

IN dataa   Input the first data (dataa)

MOV B, A   Move first data to register B

IN datab   Input the second data (datab)

ORA B   Use ''or'' on first and second data (dataa and datab)

CMA   Complement the accumulator

EXAMPLE 3.2

The second example shows how a microprocessor might be programmed to operate in an industrial situation. Suppose that an incoming conveyor has three limit switches vertically positioned so that it will sense three different sizes of boxes. Based on the input signal received from the limit switches, the microprocessor program should output a corresponding signal to a sorter so that each size of box is diverted to a separate conveyor. The input should be the first 3 bits of address data, with all other bits grounded to zero. We will assume that an IN command will accomplish the input of the data from the limit switches. The input data corresponding to three limit switches are:

0000 0001   One limit switch activated (corresponding to smallest box)

0000 0011   Two limit switches activated (corresponding to middle-size box)

0000 0111   Three limit switches activated (corresponding to largest box)

The program, using statements from Tables 3.3 through 3.6, would be as follows:

IN data   Input data to accumulator

SUI 0000 0001   Subtract first bit

JZ ONE   Test for one limit switch

SUI 0000 0010   Subtract second bit

JZ TWO   Test for two limit switches

MVIA, 0000 0100   Signal to divert largest box to its conveyor stored in accumulator

JMP END   Unconditional jump to statement END

ONE: MVIA, 0000 0001   Signal to divert smallest box to its conveyor stored in accumulator

JMP END   Unconditional jump to statement END

TWO: MVIA, 0000 0010   Signal to divert medium-size box to its conveyor stored in accumulator

END: OUT data Output the signal stored in the accumulator to the system

3.4 PROGRAMMABLE CONTROLLERS

The National Electrical Manufacturers Association defines a programmable controller (PC) in the following terms as NEMA Standard ICS3-1978, Part ICS3-304(5):

> A digitally operating electronic apparatus which uses a programmable memory for the internal storage of instructions for implementing specific functions such as logic, sequencing, timing, counting, and arithmetic to control, through digital or analog input/output modules, various types of machines or processes. A digital computer which is used to perform the functions of a programmable controller is considered to be within this scope. Excluded are drum and similar mechanical type sequencing controllers.

Programmable controllers were first introduced around 1969 and are available today in a wide variety of styles from a variety of manufacturers. One model is pictured in Figure 3.4. PCs have been used in a wide variety of industrial applications, including transfer machines, flow line conveyor systems, injection molding, grinding, welding, cement processing, food processing, energy management, and control of production testing equipment.

FIGURE 3.4 A series of programmable controller models with CRT terminal in foreground for programming. (Courtesy of General Electric Company.)

Components of a PC

The basic components of any programmable controller are the following:

Input/output interfaces
Processor
Memory
Programming device
Power supply

These components will be contained in suitable housings and cabinets to permit them to withstand the shop environment. The configuration of the PC system is illustrated in Figure 3.5.

The programmable controller is designed to be connected to industrial equipment. This connection is accomplished by means of the input/output interface. The input interface is designed to receive process and machine signals and convert them into an acceptable form for the PC. The output interface converts PC control signals into a form which can be used by the process equipment. The input interfaces are separate from the output interfaces, and both types are designed to be modular for flexibility. One I/O module might be designed for up to 16 circuits. The external signals which must be interfaced with the PC through the I/O modules might include [5]:

AC voltage, various levels
DC voltage
BCD (binary-coded decimal) inputs and outputs
Pulse data
Low-level analog signals such as thermocouple millivolt signals

The internal operation of a programmable controller would typically be based on low-voltage dc (e.g., 5 V dc).

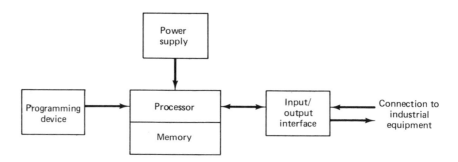

FIGURE 3.5 Typical programmable controller configuration.

The processor is the central component of the PC and is sometimes referred to as the central processing unit (CPU). It executes the various logic functions, performs operations on inputs, and determines the appropriate outputs. As microprocessor technology has developed, these devices have been incorporated in the PC processor design to increase its mathematical and decision-making capabilities.

The PC memory is used to store the program which specifies the logic of the input/output processing. Memory for a programmable controller is specified the same way as for a digital computer (e.g., 1K equals 1024 bits of storage). Memory capacities of commercial PCs range from less than 1K up to more than 48K.

The program is loaded into the PC memory by means of a programming device. Either of two types of programming devices can be used for this purpose. The first is the CRT terminal as pictured in Figure 3.4. The CRT permits the programmer to use either a relay ladder diagram or other programming language to input the control logic into memory. We will consider programming of the PC in the following subsection. The second type of programming device is a small, manual keyboard device. With this device, the control logic and other data are entered by means of special function buttons and thumbwheels. Manual programming devices are less expensive and more portable, but the CRT is more convenient for programming.

The power supply drives the PC and serves as a source of power for the output signals. It is also used to help protect the PC against noise in the electrical power lines.

Programming the PC

One of the attractive features of a PC is considered to be its ease of programming. Shop personnel who are familiar with relay ladder diagrams are not required to learn a totally new language in order to use the programmable controller. Various PC manufacturers offer different language formats but there are three basic types of PC programming languages:

Relay ladder diagrams
Boolean-based languages
Mnemonic languages similar to computer assembly languages

The three types are illustrated in Figure 3.6.

Relay ladder diagrams are currently the most popular type, owing to the fact that electricians, control engineers, and maintenance personnel are familiar with them. As shown in Figure 3.6, the ladder diagrams consist of symbols representing normally open and closed contacts and other components to control electrical equipment.

Boolean-based languages make use of logic statements to establish relationships among PC inputs and outputs. These Boolean statements can include logical

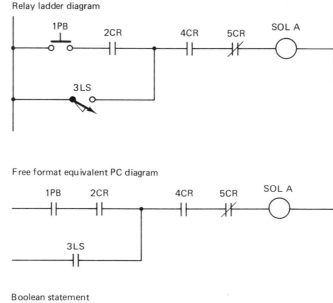

Relay ladder diagram

Free format equivalent PC diagram

Boolean statement

$$((1PB \cdot 2CR) + 3LS) \cdot 4CR \cdot \overline{5CR} = SOL\ A$$

Code or mnemonic language

LOAD	1PB
AND	2CR
OR	3LS
AND	4CR
CAND	5CR
STORE	SOL A

FIGURE 3.6   Three basic types of PC programming languages. The most popular is the relay ladder diagram illustrated in the top two sections of the figure. (Reprinted from Jannotta [5].)

"AND" (represented by a dot ·), "OR" (+), and equal (=). These symbols are shown in the Boolean statement of Figure 3.6.

The third type of PC language is quite similar to computer assembly language. The language would include statements such as LOAD, AND, OR, and STORE, as illustrated in the figure. The address attached to the statement would be used to identify a particular input/output signal. Programming the PC by means of one of these mnemonic languages often requires the assistance of a computer programmer since plant personnel are usually unfamiliar with them. However, use of these computer-type languages is expected to increase in the future as knowledge about computers becomes more widespread.

Programmable controller functions

Some of the basic functions performed by a programmable controller include the following:

1. *Control relay functions.* These functions involve the generation of an output signal from one or more inputs according to a particular logic rule contained in the PC memory. The example shown in Figure 3.6 is illustrative of the kind of logic rules used in control relay functions.
2. *Timing functions.* Most programmable controllers allow the programming of timing functions. This might be used to generate an output signal a specified delay time after an input signal has been received. Another example would be to maintain an output signal for a certain length of time and then shut off.
3. *Counting functions.* Counting functions are similar to timing functions. The counter adds up the number of input contact closures and generates a programmed output when the sum reaches a certain count. The counter would then be reset to repeat the cycle.
4. *Arithmetic functions.* As the use of programmable controllers in industry has grown, PCs have been equipped with more features and capabilities. In addition to the basic relay logic functions, timers, and counters, some PCs offer mathematical functions such as addition, subtraction, and in certain cases, multiplication and division. Hence it is possible to use the controller for more sophisticated processes where complex calculations are required.
5. *Analog control functions.* Analog control devices are used to accomplish proportional, integral, and derivative control functions. (An explanation of these control functions is provided in Section 13.5 of Ref. [2].) The capability to perform these control functions is available on some PCs, which permits it to be utilized to regulate analog devices directly.

The first three functions are common features on programmable controllers. Arithmetic and analog control functions are available on only the more powerful PCs.

Advantages of the programmable controller

There are many advantages of the programmable controller compared to the use of conventional relay controls. Among these are the following:

Programming the PC is generally much easier than wiring the conventional relay control panel.

The PC can be reprogrammed. Conventional controls must be rewired to alter the control logic, and they are often scrapped instead because of the time and expense involved in rewiring.

Programmable controllers often require less floor space than do conventional relay controls.

Maintenance of the PC is easier, and reliability is better.

The programmable controller can be interfaced with plant computer systems more easily than relays can.

In recent years, the use of microprocessors has increased the capabilities of PCs and has tended to reduce the differences between programmable controllers and minicomputers/microcomputers. Let us examine some of the differences.

Computers versus programmable controllers

The difference between programmable controllers and computers is more a difference in application than a difference in technology. The technologies have, in fact, become quite similar. A programmable controller can be thought of as a specialized computer. The intended applications of the PC have caused this specialization.

One difference between the PC and a computer is that the programmable controller is designed to be interfaced with industrial processes. The inputs and outputs of the PC can be wired directly to production equipment in the plant through the I/O modules. With computers, special arrangements have to be made to link up with the process, as we shall discuss in Chapter 17.

Another difference, related to the first, is that programmable controllers are intended to be placed in an industrial plant environment. The typical features of this environment include vibration, electrical noise, humidity, and a wide range of temperatures. The PC is designed to function in this type of environment. Its operating specifications typically call for a temperature range from 0 to 60°C (32 to 140°F) and a range of relative humidity from 0 to 95% [5].

Finally, a third important difference between a PC and a computer is in the programming. The PC uses a programming language (the relay ladder diagram) which is familiar to shop personnel. Similarly, the maintenance of the PC can also be accomplished by plant electricians since the system is modular and diagnostics are therefore made easier.

REFERENCES

[1] DELTANO, D., "Programming Your PC," *Instruments and Control Systems,* July, 1980, pp. 37−40.

[2] GROOVER, M. P., *Automation, Production Systems, and Computer-Aided Manufacturing*, Prentice-Hall, Inc., Englewood Cliffs, N.J., 1980, Chapter 11.

[3] HICKEY, J., "PCs: Reasons Why They're Used," *Instruments and Control Systems,* April, 1980, pp. 59−61.

[4] HICKEY, J., "PC Application Ideas," *Instruments and Control System,* September, 1980, pp. 67–70, and October 1980, pp. 55–59.

[5] JANNOTTA, K. L., "What Is a PC?" *Instruments and Control Systems,* February, 1980, pp. 21–25.

[6] LEVENTHAL, L. A., *Introduction to Microprocessors: Software, Hardware, Programming,* Prentice-Hall, Inc., Englewood Cliffs, N.J., 1978.

[7] SHORT, K. L., *Microprocessors and Programmed Logic,* Prentice-Hall, Inc., Englewood Cliffs, N.J., 1981.

PROBLEMS

The following programming problems are to be answered using only the instructions found in Tables 3.3 through 3.6. It is appropriate to assume that no values overflow the registers during the computations and that no negative numbers are used.

3.1. Write an assembly language program that will simulate a NAND gate. (A NAND gate is a device with two inputs and one output which operates as follows: If both inputs are one, the output is zero; otherwise, the output is one.)

3.2. Write a program to perform a multiplication operation using the contents of registers B and C as inputs. Disregard computational speed.

3.3. Write a program to perform a division operation using the contents of the accumulator to divide by the contents of register B. Assume that there will be no remainder and disregard computational speed. Also, assume that all other registers are zero.

3.4. Write a program that will perform the power operation. Disregard the issue of computational speed. (*Hint*: Use the program developed by problem 3.2 as a starting point.)

3.5. A heat sensor continually measures the temperature of the space shuttle. If the temperature is between 50 and 64°F inclusive, a heater is to be turned on. If the temperature is between 65 and 75°F inclusive, all units are turned off. If the temperature is between 76 and 100°F inclusive, an air conditioner is turned on. If it is greater than 100°F or less than 50°F, a status check is made of the shuttle's systems. If something is malfunctioning, an alarm is turned on. If nothing is found wrong during the status check, the temperature is checked again. If the temperature is still out of range, the alarm is sounded even though all systems seem to be functioning properly. Write a program that will accomplish the series of logical operations described. (*Hint*: Refer back to Example 3.2, a similar problem.)

3.6. Is the situation described in Example 3.2 more appropriate for a microprocessor or a programmable controller? List the relative advantages and disadvantages to support your answer.

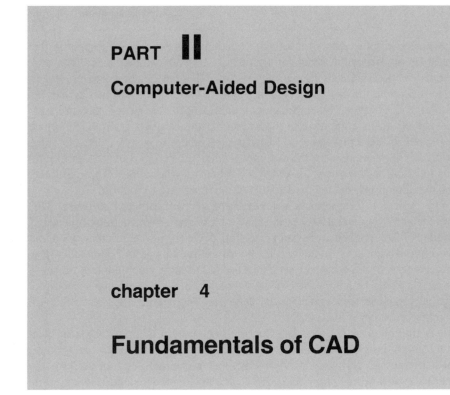

PART II

Computer-Aided Design

chapter 4

Fundamentals of CAD

4.1 INTRODUCTION

The computer has grown to become essential in the operations of business, government, the military, engineering, and research. It has also demonstrated itself, especially in recent years, to be a very powerful tool in design and manufacturing. In this and the following two chapters, we consider the application of computer technology to the design of a product. This chapter provides an overview of computer-aided design. Chapter 5 is concerned with the hardware components used in a CAD system. Chapter 6 describes some of the graphics software used for computer-aided design.

The CAD system defined

As defined in Chapter 1, computer-aided design involves any type of design activity which makes use of the computer to develop, analyze, or modify an engineering design. Modern CAD systems (also often called CAD/CAM systems) are based on interactive computer graphics (ICG). Interactive computer graphics denotes a user-oriented system in which the computer is employed to create, transform, and display data in the form of pictures or symbols. The user in the computer graphics design system is the designer, who communicates data and commands to the computer through any of several input devices. The computer communicates with the user via a cathode ray tube (CRT). The designer creates an image on the CRT screen by entering commands to call the desired software subroutines stored in the computer. In most systems, the image is constructed out of basic geometric elements—points, lines, circles, and so on. It can be modified according to the commands of the designer—enlarged, reduced in size, moved to another location on the screen, rotated, and other transformations. Through these various manipulations, the required details of the image are formulated.

The typical ICG system is a combination of hardware and software. The hardware includes a central processing unit, one or more workstations (including the graphics display terminals), and peripheral devices such as printers, plotters, and drafting equipment. Some of this hardware is shown in Figure 4.1. The software consists of the computer programs needed to implement graphics processing on the system. The software would also typically include additional specialized application programs to accomplish the particular engineering functions required by the user company.

It is important to note the fact that the ICG system is one component of a computer-aided design system. As illustrated in Figure 4.1, the other major component is the human designer. Interactive computer graphics is a tool used by the designer to solve a design problem. In effect, the ICG system magnifies the powers

FIGURE 4.1 Some of the important components in a computer-aided design system. (Courtesy of Computervision Corp.)

of the designer. This has been referred to as the synergistic effect. The designer performs the portion of the design process that is most suitable to human intellectual skills (conceptualization, independent thinking); the computer performs the task best suited to its capabilities (speed of calculations, visual display, storage of large amounts of data), and the resulting system exceeds the sum of its components.

There are several fundamental reasons for implementing a computer-aided design system:

1. *To increase the productivity of the designer.* This is accomplished by helping the designer to visualize the product and its component subassemblies and parts; and by reducing the time required in synthesizing, analyzing, and documenting the design. This productivity improvement translates not only into lower design cost but also into shorter project completion times.

2. *To improve the quality of design.* A CAD system permits a more thorough engineering analysis and a larger number of design alternatives can be investigated. Design errors are also reduced through the greater accuracy provided by the system. These factors lead to a better design.

3. *To improve communications.* Use of a CAD system provides better engineering drawings, more standardization in the drawings, better documentation of the design, fewer drawing errors, and greater legibility.

4. *To create a data base for manufacturing.* In the process of creating the documentation for the product design (geometries and dimensions of the product and its components, material specifications for components, bill of materials, etc.), much of the required data base to manufacture the product is also created.

Historical perspective

The evolution of computer-aided design has been largely related to developments in computer graphics. Of course, CAD encompasses much more than computer graphics, as we shall discuss in the remainder of this chapter. However, ICG forms the essential technological foundation for computer-aided design. An excellent history of the development of computer graphics is presented in an article by Chasen [3]. We discuss some of the important highlights of his article in this section.

One of the significant initial projects in the area of computer graphics was the development of the APT language at the Massachusetts Institute of Technology in the middle and late 1950s. APT is an acronym for Automatically Programmed Tools, and this project was concerned with developing a convenient way to define geometry elements for numerical control part programming using the computer. We discuss numerical control programming, and in particular the APT language, in Chapter 8. Although the development of APT was an important milestone in the field of computer graphics, the early use of APT was not accomplished interactively.

Another concept which took form during the late 1950s was the ''light pen.'' The idea for this device came about during research on the processing of radar data for a defense project called SAGE (Semi-Automatic Ground Environment system). The purpose of the project was to develop a system to analyze radar data and to present possible bomber targets on a CRT display. To save time in displaying the interceptor aircraft against the bombers, the notion of using a light pen to identify a particular sector of the CRT screen was developed.

During the early1960s, Ivan Sutherland worked on a project at MIT called "Sketchpad" and presented a paper on some of his results at the Fall Joint Computer Conference in 1963. The Sketchpad project is significant because it represents one of the first demonstrations of the creation and manipulation of images in real time on a CRT screen. To many observers, it marks the beginning of interactive computer graphics.

A number of large industrial concerns, including General Motors, IBM, Lockheed-Georgia, Itek Corp., and McDonnell (now McDonnell-Douglas), all became active in projects in computer graphics during the 1960s. Several of these projects eventually emerged in the form of commercial products (e.g., Unigraphics by McDonnell-Douglas and CADAM by Lockheed). In the late 1960s several CAD/CAM system vendors were also formed, including Calma in 1968 and Applicon and Computervision in 1969. These firms sell ''turnkey'' systems, which include all or most of the hardware and software components needed by the user. Other vendor firms have specialized in computer graphics software. One of the more familiar names in this area is Pat Hanratty, whose MCS Company developed the well-known AD 2000 (a later version is ANVIL 4000), a general-purpose software CAD package.

4.2   THE DESIGN PROCESS

Before examining the several facets of computer-aided design, let us first consider the general design process. The process of designing something is characterized by Shigley [15] as an iterative procedure, which consists of six identifiable steps or phases:

1. Recognition of need
2. Definition of problem
3. Synthesis
4. Analysis and optimization
5. Evaluation
6. Presentation

Recognition of need involves the realization by someone that a problem exists for which some corrective action should be taken. This might be the identification of some defect in a current machine design by an engineer or the perception of a new product marketing opportunity by a salesperson. Definition of the problem involves a thorough specification of the item to be designed. This specification includes physical and functional characteristics, cost, quality, and operating performance.

Synthesis and analysis are closely related and highly iterative in the design process. A certain component or subsystem of the overall system is conceptualized by the designer, subjected to analysis, improved through this analysis procedure, and redesigned. The process is repeated until the design has been optimized within the constraints imposed on the designer. The components and subsystems are synthesized into the final overall system in a similar iterative manner.

Evaluation is concerned with measuring the design against the specifications established in the problem definition phase. This evaluation often requires the fabrication and testing of a prototype model to assess operating performance, quality, reliability, and other criteria. The final phase in the design process is the presentation of the design. This includes documentation of the design by means of drawings, material specifications, assembly lists, and so on. Essentially, the documentation requires that a design data base be created. Figure 4.2 illustrates the basic steps in the design process, indicating its iterative nature.

Engineering design has traditionally been accomplished on drawing boards, with the design being documented in the form of a detailed engineering drawing. Mechanical design includes the drawing of the complete product as well as its

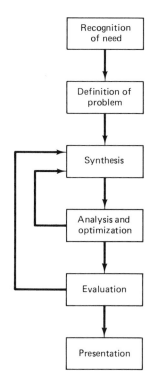

FIGURE 4.2 The general design process as defined by Shigley [13].

components and subassemblies, and the tools and fixtures required to manufacture the product. Electrical design is concerned with the preparation of circuit diagrams, specification of electronic components, and so on. Similar manual documentation is required in other engineering design fields (structural design, aircraft design, chemical engineering design, etc.). In each engineering discipline, the approach has traditionally been to synthesize a preliminary design manually and then to subject that design to some form of analysis. The analysis may involve sophisticated engineering calculations or it may involve a very subjective judgment of the aesthetc appeal possessed by the design. The analysis procedure identifies certain improvements that can be made in the design. As stated previously, the process is iterative. Each iteration yields an improvement in the design. The trouble with this iterative process is that it is time consuming. Many engineering labor hours are required to complete the design project.

4.3 THE APPLICATION OF COMPUTERS FOR DESIGN

The various design-related tasks which are performed by a modern computer-aided design system can be grouped into four functional areas:

1. Geometric modeling
2. Engineering analysis
3. Design review and evaluation
4. Automated drafting

These four areas correspond to the final four phases in Shigley's general design process, illustrated in Figure 4.3. Geometric modeling corresponds to the synthesis phase in which the physical design project takes form on the ICG system. Engineering analysis corresponds to phase 4, dealing with analysis and optimization. Design review and evaluation is the fifth step in the general design procedure. Automated drafting involves a procedure for converting the design image data residing in computer memory into a hard-copy document. It represents an important method for presentation (phase 6) of the design. The following four sections explore each of these four CAD functions.

Geometric modeling

In computer-aided design, geometric modeling is concerned with the computer-compatible mathematical description of the geometry of an object. The mathematical description allows the image of the object to be displayed and manipulated on a graphics terminal through signals from the CPU of the CAD system. The

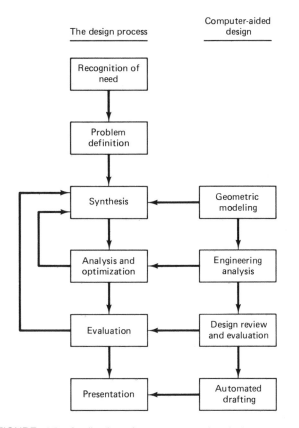

The design process   Computer-aided design

Recognition of need

Problem definition

Synthesis ← Geometric modeling

Analysis and optimization ← Engineering analysis

Evaluation ← Design review and evaluation

Presentation ← Automated drafting

FIGURE 4.3 Application of computers to the design process.

software that provides geometric modeling capabilities must be designed for efficient use both by the computer and the human designer.

To use geometric modeling, the designer constructs the graphical image of the object on the CRT screen of the ICG system by inputting three types of commands to the computer. The first type of command generates basic geometric elements such as points, lines, and circles. The second command type is used to accomplish scaling, rotation, or other transformations of these elements. The third type of command causes the various elements to be joined into the desired shape of the object being created on the ICG system. During this geometric modeling process, the computer converts the commands into a mathematical model, stores it in the computer data files, and displays it as an image on the CRT screen. The model can subsequently be called from the data files for review, analysis, or alteration.

There are several different methods of representing the object in geometric modeling. The basic form uses wire frames to represent the object. In this form,

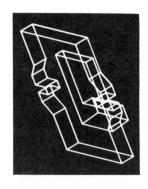

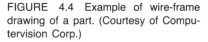
FIGURE 4.4 Example of wire-frame drawing of a part. (Courtesy of Computervision Corp.)

the object is displayed by interconnecting lines, as shown in Figure 4.4. Wire-frame geometric modeling is classified into three types, depending on the capabilites of the ICG system. The three types are:

1. *2D*.   Two-dimensional representation is used for a flat object.
2. *2½D*.   This goes somewhat beyond the 2D capability by permitting a three-dimensional object to be represented as long as it has no side-wall details.
3. *3D*.   This allows for full three-dimensional modeling of a more complex geometry.

Even three-dimensional wire-frame representations of an object are sometimes inadequate for complicated shapes. Wire-frame models can be enhanced by several different methods. Figure 4.5 shows the same object shown in the previous figure but with two possible improvements. The first uses dashed lines to portray the rear edges of the object, those which would be invisible from the front. The second enhancement removes the hidden lines completely, thus providing a less cluttered

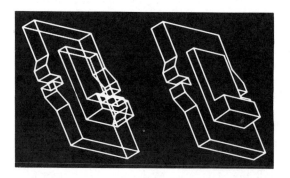

FIGURE 4.5 Same workpart as shown in Figure 4.4 but with (a) dashed lines to show rear edges of part, and (b) hidden-line removal. (Courtesy of Computervision Corp.)

FIGURE 4.6 Solid model of yoke part as displayed on a computer graphics system. (Courtesy of Computervision Corp.)

picture of the object for the viewer. Some CAD systems have an automatic "hidden-line removal feature," while other systems require the user to identify the lines that are to be removed from view. Another enhancement of the wire-frame model involves providing a surface representation which makes the object appear solid to the viewer. However, the object is still stored in the computer as a wire-frame model.

The most advanced method of geometric modeling is solid modeling in three dimensions. This method, illustrated in Figure 4.6, typically uses solid geometry shapes called primitives to construct the object. We shall discuss the difference between wire-frame and solid models in Chapter 6.

Another feature of some CAD systems is color graphics capability. By means of color, it is possible to display more information on the graphics screen. Colored images help to clarify components in an assembly, or highlight dimensions, or a host of other purposes. The benefits of color graphics are discussed in Chapter 5.

Engineering analysis

In the formulation of nearly any engineering design project, some type of analysis is required. The analysis may involve stress—strain calculations, heat-transfer computations, or the use of differential equations to describe the dynamic behavior of the system being designed. The computer can be used to aid in this analysis work. It is often necessary that specific programs be developed internally by the engineering analysis group to solve a particular design problem. In other situations, commercially available general-purpose programs can be used to perform the engineering analysis.

Turnkey CAD/CAM systems often include or can be interfaced to engineering analysis software which can be called to operate on the current design model.

We discuss two important examples of this type:

Analysis of mass properties
Finite-element analysis

The analysis of mass properties is the analysis feature of a CAD system that has probably the widest application. It provides properties of a solid object being analyzed, such as the surface area, weight, volume, center of gravity, and moment of inertia. For a plane surface (or a cross section of a solid object) the corresponding computations include the perimeter, area, and inertia properties.

Probably the most powerful analysis feature of a CAD system is the finite-element method. With this technique, the object is divided into a large number of finite elements (usually rectangular or triangular shapes) which form an interconnecting network of concentrated nodes. By using a computer with significant computational capabilities, the entire object can be analyzed for stress−strain, heat transfer, and other characteristics by calculating the behavior of each node. By determining the interrelating behaviors of all the nodes in the system, the behavior of the entire object can be assessed.

Some CAD systems have the capability to define automatically the nodes and the network structure for the given object. The user simply defines certain parameters for the finite-element model, and the CAD system proceeds with the computations.

The output of the finite-element analysis is often best presented by the system in graphical format on the CRT screen for easy visualization by the user. For example, in stress−strain analysis of an object, the output may be shown in the form of a deflected shape superimposed over the unstressed object. This is illustrated in Figure 4.7. Color graphics can also be used to accentuate the comparison

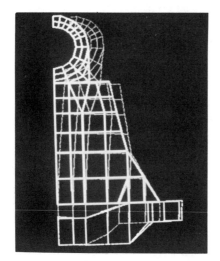

FIGURE 4.7 Finite-element modeling for stress–strain analysis. Graphics display shows strained part superimposed on unstrained part for comparison. (Courtesy of Applicon Inc.)

before and after deflection of the object. This is illustrated in Figure 5.6 for the same image as that shown in Figure 4.7. If the finite-element analysis indicates behavior of the design which is undesirable, the designer can modify the shape and recompute the finite-element analysis for the revised design.

Design review and evaluation

Checking the accuracy of the design can be accomplished conveniently on the graphics terminal. Semiautomatic dimensioning and tolerancing routines which assign size specifications to surfaces indicated by the user help to reduce the possibility of dimensioning errors. The designer can zoom in on part design details and magnify the image on the graphics screen for close scrutiny.

A procedure called layering is often helpful in design review. For example, a good application of layering involves overlaying the geometric image of the final shape of the machined part on top of the image of the rough casting. This ensures that sufficient material is available on the casting to accomplish the final machined dimensions. This procedure can be performed in stages to check each successive step in the processing of the part.

Another related procedure for design review is interference checking. This involves the analysis of an assembled structure in which there is a risk that the components of the assembly may occupy the same space. This risk occurs in the design of large chemical plants, air-separation cold boxes, and other complicated piping structures.

One of the most interesting evaluation features available on some computer-aided design systems is kinematics. The available kinematics packages provide the capability to animate the motion of simple designed mechanisms such as hinged components and linkages. This capability enhances the designer's visualization of the operation of the mechanism and helps to ensure against interference with other components. Without graphical kinematics on a CAD system, designers must often resort to the use of pin-and-cardboard models to represent the mechanism. Commercial software packages are available to perform kinematic analysis. Among these are programs such as ADAMS (Automatic Dynamic Analysis of Mechanical Systems), developed at the University of Michigan. This type of program can be very useful to the designer in constructing the required mechanism to accomplish a specified motion and/or force.

Automated drafting

Automated drafting involves the creation of hard-copy engineering drawings directly from the CAD data base. In some early computer-aided design departments, automation of the drafting process represented the principal justification for investing in the CAD system. Indeed, CAD systems can increase productivity in the drafting function by roughly five times over manual drafting.

Some of the graphics features of computer-aided design systems lend them-

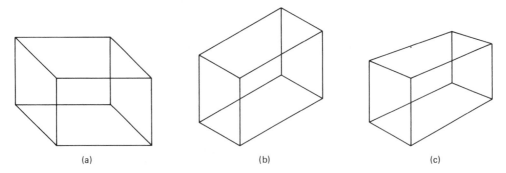

FIGURE 4.8   Three views of a wire frame block: (a) oblique, (b) isometric, (c) perspective.

selves especially well to the drafting process. These features include automatic dimensioning, generation of crosshatched areas, scaling of the drawing, and the capability to develop sectional views and enlarged views of particular part details. The ability to rotate the part or to perform other transformations of the image (e.g., oblique, isometric, or perspective views), as illustrated in Figure 4.8, can be of significant assistance in drafting. Most CAD systems are capable of generating as many as six views of the part. Engineering drawings can be made to adhere to company drafting standards by programming the standards into the CAD system. Figure 4.9 shows an engineering drawing with four views displayed. This drawing was produced automatically by a CAD system. Note how much the isometric view promotes a higher level of understanding of the object for the user than the three orthographic views.

We discuss the various pieces of equipment for creating the hard-copy drawing in Section 5.5.

Parts classification and coding

In addition to the four CAD functions described above, another feature of the CAD data base is that it can be used to develop a parts classification and coding system. Parts classification and coding involves the grouping of similar part designs into classes, and relating the similarities by means of a coding scheme. Designers can use the classification and coding system to retrieve existing part designs rather than always redesigning new parts. There are several uses of such systems in manufacturing also, and we postpone further discussion of this subject until Chapter 12.

4.4   CREATING THE MANUFACTURING DATA BASE

Section 4.3 has described the many ways in which computer-aided design can increase the productivity of the design department in the company. Another important reason for using a CAD system is that it offers the opportunity to

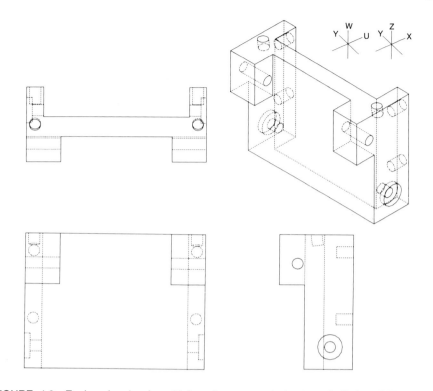

FIGURE 4.9 Engineering drawing with four views generated automatically by a CAD system. (From W. Fitzgerald, F. Gracer, and R. Wolfe, "GRIN: Interactive Graphics for Modeling Solids," *IBM Journal of Research and Development,* Vol. 25, No. 4, July, 1981. Copyright 1981 by International Business Machines Corporation; reprinted with permission.)

develop the data base needed to manufacture the product. In the conventional manufacturing cycle practiced for so many years in industry, engineering drawings were prepared by design draftsmen and then used by manufacturing engineers to develop the process plan (i.e., the "route sheets"). The activities involved in designing the product were separated from the activities associated with process planning. Essentially, a two-step procedure was employed. This was both time consuming and involved duplication of effort by design and manufacturing personnel. In an integrated CAD/CAM system, a direct link is established between product design and manufacturing. It is the goal of CAD/CAM not only to automate certain phases of design and certain phases of manufacturing, but also to automate the transition from design to manufacturing. Computer-based systems have been developed which create much of the data and documentation required to plan and manage the manufacturing operations for the product.

The manufacturing data base is an integrated CAD/CAM data base. It includes all the data on the product generated during design (geometry data, bill of materials and parts lists, material specifications, etc.) as well as additional data

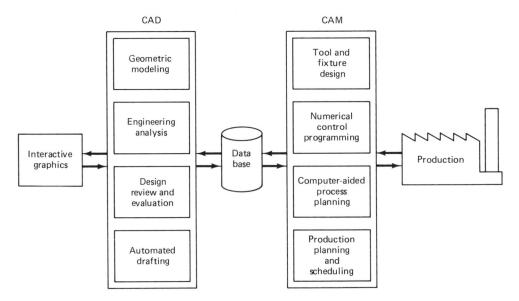

FIGURE 4.10 Desirable relationship of CAD/CAM data base to CAD and CAM.

required for manufacturing, much of which is based on the product design. Figure 4.10 shows how the CAD/CAM data base is related to design and manufacturing in a typical production-oriented company.

4.5 BENEFITS OF COMPUTER-AIDED DESIGN

There are many benefits of computer-aided design, only some of which can be easily measured. Some of the benefits are intangible, reflected in improved work quality, more pertinent and usable information, and improved control, all of which are difficult to quantify. Other benefits are tangible, but the savings from them show up far downstream in the production process, so that it is difficult to assign a dollar figure to them in the design phase. Some of the benefits that derive from implementing CAD/CAM can be directly measured. Table 4.1 provides a checklist of potential benefits of an integrated CAD/CAM system. In the subsections that follow, we elaborate on some of these advantages.

Productivity improvement in design

Increased productivity translates into a more competitive position for the firm because it will reduce staff requirements on a given project. This leads to lower costs in addition to improving response time on projects with tight schedules.

Surveying some of the larger CAD/CAM vendors, one finds that the productivity improvement ratio for a designer/draftsman is usually given as a range, typi-

TABLE 4.1   Checklist of Potential Benefits That May Result from Implementing CAD
as Part of an Integrated CAD/CAM System

---

1. Improved engineering productivity
2. Shorter lead times
3. Reduced engineering personnel requirements
4. Customer modifications are easier to make
5. Faster response to requests for quotations
6. Avoidance of subcontracting to meet schedules
7. Minimized transcription errors
8. Improved accuracy of design
9. In analysis, easier recognition of component interactions
10. Provides better functional analysis to reduce prototype testing
11. Assistance in preparation of documentation
12. Designs have more standardization
13. Better designs provided
14. Improved productivity in tool design
15. Better knowledge of costs provided
16. Reduced training time for routine drafting tasks and NC part programming
17. Fewer errors in NC part programming
18. Provides the potential for using more existing parts and tooling
19. Helps ensure designs are appropriate to existing manufacturing techniques
20. Saves materials and machining time by optimization algorithms
21. Provides operational results on the status of work in progress
22. Makes the management of design personnel on projects more effective
23. Assistance in inspection of complicated parts
24. Better communication interfaces and greater understanding among engineers, designers, drafters, management, and different project groups

---

cally from a low end of 3:1 to a high end in excess of 10:1 (often far in excess of that figure). There are individual cases in which productivity has been increased by a factor of 100, but it would be inaccurate to represent that figure as typical.

Productivity improvement in computer-aided design as compared to the traditional design process is dependent on such factors as:

Complexity of the engineering drawing
Level of detail required in the drawing
Degree of repetitiveness in the designed parts
Degree of symmetry in the parts
Extensiveness of library of commonly used entities

As each of these factors is increased, the productivity advantage of CAD will tend to increase.

Shorter lead times

Interactive computer-aided design is inherently faster than the traditional design process. It also speeds up the task of preparing reports and lists (e.g., the assembly lists) which are normally accomplished manually. Accordingly, it is possible with a CAD system to produce a finished set of component drawings and the associated reports in a relatively short time. Shorter lead times in design translate into shorter elapsed time between receipt of a customer order and delivery of the final product. The enhanced productivity of designers working with CAD systems will tend to reduce the prominence of design, engineering analysis, and drafting as critical time elements in the overall manufacturing lead time.

Design analysis

The design analysis routines available in a CAD system help to consolidate the design process into a more logical work pattern. Rather than having a back-and-forth exchange between design and analysis groups, the same person can perform the analysis while remaining at a CAD workstation. This helps to improve the concentration of designers, since they are interacting with their designs in a real-time sense. Because of this analysis capability, designs can be created which are closer to optimum. There is a time saving to be derived from the computerized analysis routines, both in designer time and in elapsed time. This saving results from the rapid response of the design analysis and from the time no longer lost while the design finds its way from the designer's drawing board to the design analyst's queue and back again.

An example of the success of this is drawn from the experience of the General Electric Company with the T700 engine [6]. In designing a jet engine, weight is an important design consideration. During the design of the engine, weights of each component for each design alternative must be determined. This had in the past been done manually by dividing each part into simple geometrical shapes to conveniently compute the volumes and weights. Through the use of CAD and its mass properties analysis function, the mass properties were obtained in 25% of the time formerly taken. The result of these calculations is illustrated in Fig. 4.11.

Since alterations in preliminary designs are generally easier to make and analyze with a CAD graphics system, more design alternatives can be explored and compared in the available development time. Consequently, it is reasonable to believe that a better design will result from the computer-aided design procedure.

Fewer design errors

Interactive CAD systems provide an intrinsic capability for avoiding design, drafting, and documentation errors. Data entry, transposition, and extension errors that occur quite naturally during manual data compilation for preparation of a bill of materials are virtually eliminated. One key reason for such accuracy is simply that

```
PARTNO    =    6034T00G02F — MASS P
DATE      =    10-6-76
XREF      =    0.0000 IN.


PLANE SECTION — ASSUME GDF IS IN INCHES

LGTH      =    12.3239 IN.
AREA      =    1.3261 IN. ** 2
CGX       =    28.9144 IN. FROM X-REF
CGY       =    2.5214 IN. FROM X-AXIS
          =    8.9281 IN. ** 4 ABOUT X-AXIS
          =    0.4978 IN. ** 4 ABOUT CGY
AMY       =    1109.2196 IN. ** 4 ABOUT X-REF
AMYC      =    0.5887 IN. ** 4 ABOUT CGX


ROTATED SOLID — DENSITY = 0.1610 LB. PER CU. IN.

SURF      =    188.7164 IN. ** 2 (ALL EXPOSED SURFACES)
VOL       =    21.0083 IN. ** 3
CGX       =    28.9863 IN. FROM X-REF.
WGHT      =    3.3823 LBS. (1535.5782 GRAMS)
MMX       =    25.3450 LB. IN. ** 2 ABOUT THE X-AXIS
MMY       =    2857.5319 LB. IN. ** 2 ABOUT THE X-REF
MMYC      =    13.7622 LB. IN. ** 2 ABOUT CGX
```

FIGURE 4.11 Weights of component parts as calculated by CAD system. (Reprinted by permission from Grimmer [6]).

no manual handling of information is required once the initial drawing has been developed. Errors are further avoided because interactive CAD systems perform time-consuming repetitive duties such as multiple symbol placement, and sorts by area and by like item, at high speeds with consistent and accurate results. Still more errors can be avoided because a CAD system, with its interactive capabilities, can be programmed to question input that may be erroneous. For example, the system might question a tolerance of 0.00002 in. It is likely that the user specified too many zeros. The success of this checking would depend on the ability of the CAD system designers to determine what input is likely to be incorrect and hence, what to question.

Greater accuracy in design calculations

There is also a high level of dimensional control, far beyond the levels of accuracy attainable manually. Mathematical accuracy is often to 14 significant decimal places. The accuracy delivered by interactive CAD systems in three-dimensional curved space designs is so far beyond that provided by manual calculation methods that there is no real comparison.

Computer-based accuracy pays off in many ways. Parts are labeled by the same recognizable nomenclature and number throughout all drawings. In some CAD systems, a change entered on a single item can appear throughout the entire documentation package, effecting the change on all drawings which utilize that part. The accuracy also shows up in the form of more accurate material and cost estimates and tighter procurement scheduling. These items are especially important in such cases as long-lead-time material purchases.

Standardization of design, drafting, and documentation procedures

The single data base and operating system is common to all workstations in the CAD system. Consequently, the system provides a natural standard for design/drafting procedures. With interactive computer-aided design, drawings are "standardized" as they are drawn; there is no confusion as to proper procedures because the entire format is "built into" the system program.

Drawings are more understandable

Interactive CAD is equally adept at creating and maintaining isometrics and oblique drawings as well as the simpler orthographics. All drawings can be generated and updated with equal ease. Thus an up-to-date version of any drawing type can always be made available.

In general, ease of visualization of a drawing relates directly to the projection used. Orthographic views are less comprehensible than isometrics. An isometric view is usually less understandable than a perspective view. Most actual construction drawings are "line drawings." The addition of shading increases comprehension. Different colors further enhance understanding. Finally, animation of the images on the CRT screen allows for even greater visualization capability. The various relationships are illustrated in Figure 4.12.

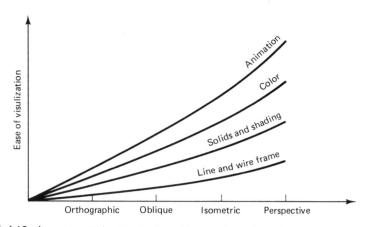

FIGURE 4.12   Improvement in visualization of images for various drawing types and computer graphics features.

Improved procedures for engineering changes

Control and implementation of engineering changes is significantly improved with computer-aided design. Original drawings and reports are stored in the data base of the CAD system. This makes them more accessible than documents kept in a drawing vault. They can be quickly checked against new information. Since data storage is extremely compact, historical information from previous drawings can be easily retained in the system's data base, for easy comparison with current design/drafting needs.

Benefits in manufacturing

The benefits of computer-aided design carry over into manufacturing. As indicated previously, the same CAD/CAM data base is used for manufacturing planning and control, as well as for design. These manufacturing benefits are found in the following areas:

Tool and fixture design for manufacturing
Numerical control part programming
Computer-aided process planning
Assembly lists (generated by CAD) for production
Computer-aided inspection
Robotics planning
Group technology
Shorter manufacturing lead times through better scheduling

These benefits are derived largely from the CAD/CAM data base, whose initial framework is established during computer-aided design. We will discuss the many facets of computer-aided manufacturing in later chapters. In the remainder of this chapter, let us explore several applications that utilize computer graphics technology to solve various problems in engineering and related fields.

4.6 SOME EXAMPLES

The following examples, with accompanying figures, were selected to demonstrate the capabilities and advantages of interactive computer graphics and computer-aided design. All the figures in this section were generated by computer, either in the form of a line drawing or as an image on the graphics screen. The reader should consider in all the examples the significant amount of human time and effort that would be required if the same illustrations were accomplished by hand, without the aid of a computer graphics system. We will provide a minimum of description in the examples. For the most part, the associated figures for each example speak for themselves.

EXAMPLE 4.1

Computer-aided design systems can be used to solve design and engineering problems in architectural and construction applications. Figure 4.13 provides two illustrations of these applications. The top drawing shows a three-dimensional view of a building section which possesses an unusual curved layout. The top view and certain other projections of the building would represent a fairly straightforward drawing problem in manual drafting, but it would be difficult to manually construct the three dimensional view of the building. The CAD

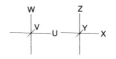

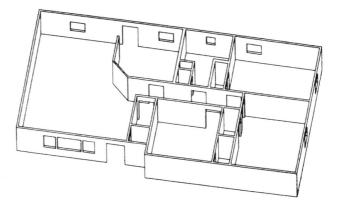

FIGURE 4.13 Examples of architectural application using a computer-aided design system.(From W. Fitzgerald, F. Gracer, and R. Wolfe, "GRIN: Interactive Graphics for Modeling Solids," *IBM Journal of Research and Development,* Vol. 25, No. 4, July, 1981. Copyright 1981 by International Business Machines Corporation; reprinted with permission.)

system, in this case with solids modeling capability, can generate the image of the building from any desired vantage point.

The lower drawing shows a three-dimensional floor plan of a house which adds a certain realism and perspective not achieved in a two-dimensional view. Both of these line drawings were generated using the GDP/GRIN system developed by IBM Corporation [5].

EXAMPLE 4.2

In the design of many products with complicated mechanisms or potential interferences between components, it is important to be able to test the product for these characteristics. This has traditionally been done by constructing a physical model or prototype of the product. With an interactive computer graphics system capable of motion simulation and interference checking, these problems can be analyzed without the need for a physical model. Figure 4.14 illustrates this capability. A bulldozer design is shown in Figure 4.14(a) (component identification and arrows were added to the computer drawing). The analysis was carried out using a solids modeling system with dynamic simulation capability [5]. The objective of the analysis was to find the range of motion (lift and tilt) which could be applied to the bulldozer blade without causing interference. Rotation commands were given to lift the blade over a range of elevations. At each position of the blade, interference analysis was performed to determine if any components were touching or overlapping. Figure 4.14(b) shows the blade in one of its extreme rotational orientations.

EXAMPLE 4.3

CAD/CAM systems have proven very useful in the design of integrated circuit devices [1]. Figure 4.15 shows a computer-generated design drawing of a LSI (large scale integrated)

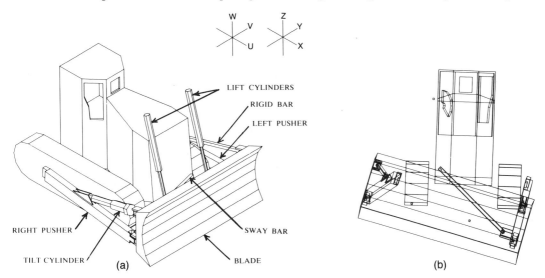

FIGURE 4.14 Bulldozer blade design problem; results of blade lift and tilt motion. (From W. Fitzgerald, F. Gracer, and R. Wolfe, "GRIN: Interactive Graphics for Modeling Solids," *IBM Journal of Research and Development,* Vol. 25, No. 4, July, 1981. Copyright 1981 by International Business Machines Corporation; reprinted with permission.)

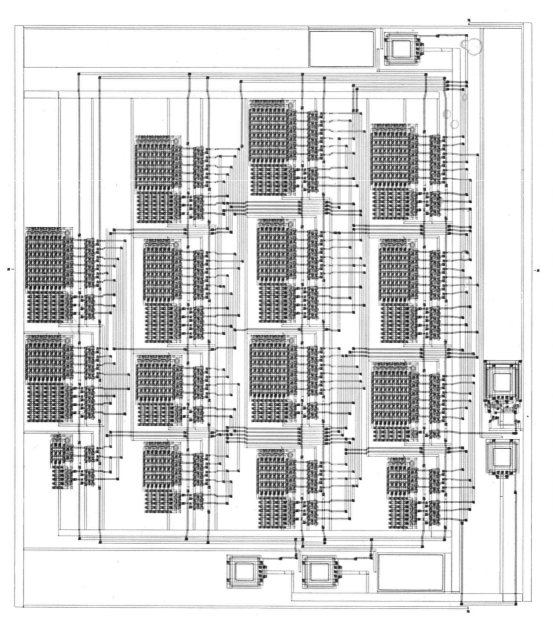

FIGURE 4.15 VLSI circut chip design. (Reprinted from Ayres [1] by permission.)

chip. This type of drawing can be developed from logic relationships which have been specified by the designer to define the function of the circuit. The drawing shows both the internal circuitry of the chip and the interface connectors at the periphery of the chip. CAD/CAM systems are useful in LSI design for generating the artwork required during fabrication of the chip. The complexity of this figure demonstrates the considerable trouble that would be involved in accomplishing the design manual methods.

EXAMPLE 4.4

CAD/CAM systems are also useful in the manufacture of mechanical parts and products. Figures 4.16 and 4.17 illustrate two examples. Figure 4.16 shows two views of the same sheet metal part. The left-hand view shows the final part after it has been blanked and formed into the desired form. One of the problems encountered in pressworking of this kind is the problem of shearing the flat sheet metal to the proper size and shape so that it can be bent to the specified final dimensions. This is not only a problem of geometry, but also the flat size must be made slightly undersized to allow for stretching of the metal during the bending operations. CAD/CAM software has been written to perform the necessary computations to determine the correct dimensions of the flat part in sheet metalworking. For the formed part in this example, the unbent flat is shown in the right-hand view.

Figure 4.17 shows a process drawing of a machined part that would be turned on a numerically controlled lathe. The hublike shape of the part in the left-hand view of the figure would be cut out of a round disk by means of a sequence of turning passes, as indicated by the series of horizontal lines. Numerically controlled machine tools operate according to a set of programmed instructions which control the machine's actions in a step-by-step manner. The left-hand side of Figure 4.17 shows the types of programming instructions that would be used to control the lathe. CAD/CAM systems can be utilized to aid the programmer in preparing the part programs for numerical control. In some cases, the programming can be done almost automatically by the system with very little interaction by the user. We discuss the programming of numerical control machines in some detail in Chapter 8. In particular, we examine the use of CAD/CAM packages to perform numerical control part programming in Section 8.8.

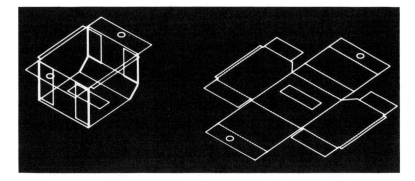

FIGURE 4.16 Sheet metal part in bent final shape (left) and flat blank (right). (Courtesy of Computervision Corp.)

```
PARTNO MS.7-07
MACHIN/GAAM01
CUTTER/1
OUTTOL/0
FROM/        0.00000,      0.00000,      0.00000
SPINDL/CLW, 300
COOLNT/ON
GOTO/       -5.50000,      8.50000,      0.00000
INTOL/       0.00500
FEDRAT/30, IPM
GOTO/       -5.50000,      7.73225,      0.00000
FEDRAT/0.015, IPR
GOTO/      -10.94875,      7.73225,      0.00000
GOTO/      -10.94875,      8.03125,      0.00000
FEDRAT/30, IPM
GOTO/      -10.89875,      8.08125,      0.00000
GOTO/       -7.94875,      8.08125,      0.00000
GOTO/       -7.94875,      7.43325,      0.00000
FEDRAT/0.15, IPR
GOTO/      -10.07645,      7.43325,      0.00000
GOTO/      -10.94875,      7.43325,      0.00000
GOTO/      -10.94875,      7.73225,      0.00000
FEDRAT/30, IPM
GOTO/      -10.89875,      7.78225,      0.00000
GOTO/       -7.94875,      7.78225,      0.00000
GOTO/       -7.94875,      7.13425,      0.00000
FEDRAT/0.015, IPR
GOTO/       -9.88141,      7.13425,      0.00000
GOTO/       -9.88263,      7.13425,      0.00000
AUTOPS
INDIRV/     -0.02921,      0.08577,      0.00000
TLON, GOFWD/(CIRCLE/ -10.41729,   7.00000,      0.00000, $
   0.551251, ON, (LINE/ -10.41729, $
   7.00000, 0.00000,    -10.07645,   7.43325,      0.00001
FEDRAT/30, IPM
GOTO/      -10.02645,      7.48325,      0.00000
GOTO/       -7.94875,      7.48325,      0.00000
GOTO/       -7.94875,      6.83525,      0.00000
```

FIGURE 4.17 Machined part (left) and numerical control part program (right). (Courtesy of Computervision Corp.)

EXAMPLE 4.5

Figure 4.18 illustrates an interesting application of interactive computer graphics which falls outside the usual scope of computer-aided design and manufacturing. The figure shows a three-dimensional drawing of a child's facial features, constructed from a high-resolution computed tomography (CT) scan. Tomography involves the use of X-ray techniques in which the shadows in front of and behind the sections of the subject under study do not show. The use of CT scans allows the three-dimensional bony and soft tissue surfaces to be reconstructed by computer software into the image illustrated in the figure. When applied to facial abnormalities in children, high-resolution CT scan techniques have provided an improved understanding of aberrant craniofacial anatomy in the medical community. These techniques have also been helpful in the planning of surgical operations and in improving postoperative care and evaluation.

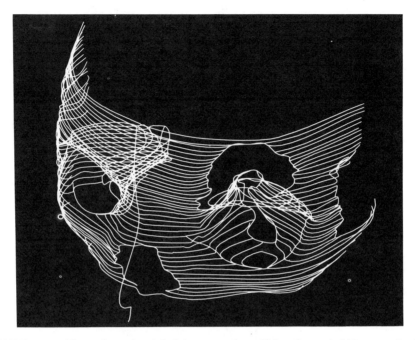

FIGURE 4.18 Three-dimensional facial construction utilizing Computed Tomography (CT) scan. (Courtesy of McDonnell-Douglas Automation Company.)

These examples illustrate the significant opportunities and benefits offered by the use of interactive computer graphics and CAD/CAM systems. The following two chapters examine the hardware and software technologies of computer-aided design which make these applications possible.

REFERENCES

[1] AYRES, R., *VLSI: Silicon Compilation and the Art of Automatic Microchip Design*, Prentice-Hall, Inc., Englewood Cliffs, N.J., 1983, Chapter 8.

[2] BYLINSKI, G., "A New Industrial Revolution Is on the Way," *Fortune*, October 5, 1981, pp. 106–114.

[3] CHASEN, S. H., "Historical Highlights of Interactive Computer Graphics," *Mechanical Engineering*, November, 1981, pp. 32–41.

[4] CLAYTON, R. J., "CAD/CAM Integration: 2 + 2 = 5," *CAD/CAM Technology*, Spring, 1982, pp. 21–26.

[5] FITZGERALD, W., GRACER, F., AND WOLFE, R., "GRIN: Interactive Graphics for Modeling Solids," *IBM Journal of Research and Development,* July, 1981, pp. 281–294.

[6] GRIMMER, G., "Design and Manufacture of a Jet Engine," in *The CAD/CAM Handbook,* C. Machover and R. Blauth, Eds., Computervision Corp., Bedford, Mass., 1980, pp. 159–172.

[7] GROOVER, M. P., *Automation, Production Systems, and Computer-Aided Manufacturing,* Prentice-Hall, Inc., Englewood Cliffs, N.J., 1980, Chapter 10.

[8] INGLESBY, T., "CAD/CAM: Should We or Shouldn't We?" *Assembly Engineering,* March, 1982, pp. 48–50.

[9] KROUSE, J. K., "CAD/CAM—Bridging the Gap from Design to Production," *Machine Design,* June 12, 1980, pp. 117–125.

[10] LERRO, J. P., JR., "CAD/CAM System: More than an Automated Drafting Tool," *Design News,* November 17, 1980.

[11] LERRO, J. P., JR., "CAD/CAM System: Start of the Productivity Revolution," *Design News,* November 16, 1981, pp. 46–65.

[12] MACHOVER C., AND BLAUTH, R. E., Eds., *The CAD/CAM Handbook,* Computervision Corp., Bedford, Mass., 1980.

[13] MYERS, W., "Interactive Graphics: Flying High," *Computer,* July, 1979, pp. 8–11.

[14] SCHAFFER, G., "Computer Graphics Goes to Work," Special Report 724, *American Machinist,* July, 1980, pp. 149–164.

[15] SHIGLEY, J. E., *Mechanical Engineering Design,* 3rd ed., McGraw-Hill Book Company, New York, 1977.

[16] SMYTH, S. J., "CAD/CAM Data Handling from Conceptual Design through Product Support," *Journal of Aircraft,* October, 1980, pp. 753–760.

chapter 5

Hardware in Computer-Aided Design

5.1 INTRODUCTION

Chapter 4 provided a general survey of computer-aided design. In the present chapter we examine the various hardware components that make up a modern CAD system.

Hardware components for computer-aided design are available in a variety of sizes, configurations, and capabilities. Hence it is possible to select a CAD system that meets the particular computational and graphics requirements of the user firm. Engineering firms that are not involved in production would choose a system exclusively for drafting and design-related functions. Manufacturing firms would choose a system to be part of a company-wide CAD/CAM system. Of course, the CAD hardware is of little value without the supporting software for the system, and we shall discuss the software for computer-aided design in the following chapter.

As indicated in Chapter 4, a modern computer-aided design system is based on interactive computer graphics (ICG). However, the scope of computer-aided design includes other computer systems as well. For example, computerized design has also been accomplished in a batch mode, rather than interactively. Batch design means that data are supplied to the system (a deck of computer cards is traditionally used for this purpose) and then the system proceeds to develop the details of the design. The disadvantage of the batch operation is that there is a time lag between when the data are submitted and when the answer is received back as output. With interactive graphics, the system provides an immediate response to inputs by the user. The user and the system are in direct communication with each other, the user entering commands and responding to questions generated by the system.

Computer-aided design also includes nongraphic applications of the computer in design work. These consist of engineering results which are best displayed in other than graphical form. Nongraphic hardware (e.g., line printers) can be employed to create rough images on a piece of paper by appropriate combinations of characters and symbols. However, the resulting pictures, while they may create interesting wall posters, are not suitable for design purposes.

The hardware we discuss in this chapter is restricted to CAD systems that utilize interactive computer graphics. Typically, a stand-alone CAD system would include the following hardware components:

> One or more design workstations. These would consist of:
>> A graphics terminal
>> Operator input devices
> One or more plotters and other output devices
> Central processing unit (CPU)
> Secondary storage

These hardware components would be arranged in a configuration as illustrated in Figure 5.1. The following sections discuss these various hardware components and the alternatives and options that can be obtained in each category.

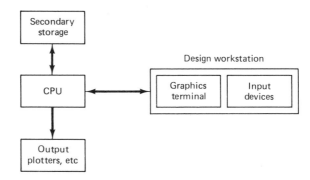

FIGURE 5.1 Typical configuration of hardware components in a stand-alone CAD system. There would likely be more than one design workstation.

5.2   THE DESIGN WORKSTATION

The CAD workstation is the system interface with the outside world. It represents a significant factor in determining how convenient and efficient it is for a designer to use the CAD system. The workstation must accomplish five functions [3]:

1. It must interface with the central processing unit.
2. It must generate a steady graphic image for the user.
3. It must provide digital descriptions of the graphic image.
4. It must translate computer commands into operating functions.
5. It must facilitate communication between the user and the system.

The use of interactive graphics has been found to be the best approach to accomplish these functions. A typical interactive graphics workstation would consist of the following hardware components:

A graphics terminal
Operator input devices

A graphics design workstation showing these components is illustrated in Figure 5.2.

FIGURE 5.2   Interactive graphics design workstation showing graphics terminal and two input devices: alphanumeric keyboard and electronic tablet and pen. (Courtesy of Applicon Inc.)

5.3   THE GRAPHICS TERMINAL

There are various technological approaches which have been applied to the development of graphics terminals. The technology continues to evolve as CAD system manufactures attempt to improve their products and reduce their costs. In this section we present a discussion of the current technology in interactive computer graphics terminals.

Image generation in computer graphics

Nearly all computer graphics terminals available today use the cathode ray tube (CRT) as the display device. Television sets use a form of the same device as the picture tube. The operation of the CRT is illustrated in Figure 5.3. A heated cathode emits a high-speed electron beam onto a phosphor-coated glass screen. The electrons energize the phosphor coating, causing it to glow at the points where the beam makes contact. By focusing the electron beam, changing its intensity, and controlling its point of contact against the phosphor coating through the use of a deflector system, the beam can be made to generate a picture on the CRT screen.

There are two basic techniques used in current computer graphics terminals for generating the image on the CRT screen. They are:

1.   Stroke writing
2.   Raster scan

Other names for the stroke-writing technique include line drawing, random position, vector writing, stroke writing, and directed beam. Other names for the raster scan technique include digital TV and scan graphics.

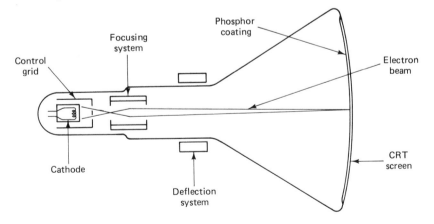

FIGURE 5.3   Diagram of cathode ray tube (CRT).

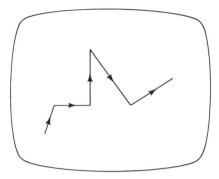

FIGURE 5.4 Stroke writing for generating images in computer graphics.

The stroke-writing system uses an electron beam which operates like a pencil to create a line image on the CRT screen. The image is constructed out of a sequence of straight-line segments. Each line segment is drawn on the screen by directing the beam to move from one point on the screen to the next, where each point is defined by its x and y coordinates. The process is portrayed in Figure 5.4. Although the procedure results in images composed of only straight lines, smooth curves can be approximated by making the connecting line segments short enough.

In the raster scan approach, the viewing screen is divided into a large number of discrete phosphor picture elements, called pixels. The matrix of pixels constitutes the raster. The number of separate pixels in the raster display might typically range from 256×256 (a total of over 65,000) to 1024×1024 (a total of over 1,000,000 points). Each pixel on the screen can be made to glow with a different brightness. Color screens provide for the pixels to have different colors as well as brightness. During operation, an electron beam creates the image by sweeping along a horizontal line on the screen from left to right and energizing the pixels in that line during the sweep. When the sweep of one line is completed, the electron beam moves to the next line below and proceeds in a fixed pattern as indicated in Figure 5.5. After sweeping the entire screen the process is repeated at a rate of 30 to 60 entire scans of the screen per second.

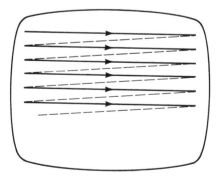

FIGURE 5.5 Raster scan approach for generating images in computer graphics.

Graphics terminals for computer-aided design

The two approaches described above are used in the overwhelming majority of current-day CAD graphics terminals. There are also a variety of other technical factors which result in different types of graphics terminals. These factors include the type of phosphor coating on the screen, whether color is required, the pixel density, and the amount of computer memory available to generate the picture. We will discuss three types of graphics terminals, which seem to be the most important today in commercially available CAD systems. The three types are:

1. Directed-beam refresh
2. Direct-view storage tube (DVST)
3. Raster scan (digital TV)

The following paragraphs describe the three basic types. We then discuss some of the possible enhancements, such as color and animation. A more thorough treatment of these graphics hardware technologies is given in Ref. [3], [4], and [7].

DIRECTED-BEAM REFRESH. The directed-beam refresh terminal utilizes the stroke-writing approach to generate the image on the CRT screen. The term "refresh" in the name refers to the fact that the image must be regenerated many times per second in order to avoid noticeable flicker of the image. The phosphor elements on the screen surface are capable of maintaining their brightness for only a short time (sometimes measured in microseconds). In order for the image to be continued, these picture tubes must be refreshed by causing the directed beam to retrace the image repeatedly. On densely filled screens (very detailed line images or many characters of text), it is difficult to avoid flickering of the image with this process. On the other hand, there are several advantages associated with the directed-beam refresh systems. Because the image is being continually refreshed, selective erasure and alteration of the image is readily accomplished. It is also possible to provide animation of the image with a refresh tube.

The directed-beam refresh system is the oldest of the modern graphics display technologies. Other names sometimes used to identify this system include vector refresh and stroke-writing refresh. Early refresh tubes were very expensive. but the steadily decreasing cost of solid-state circuitry has brought the price of these graphics systems down to a level which is competitive with other types.

DIRECT-VIEW STORAGE TUBE (DVST). DVST terminals also use the stroke-writing approach to generate the image on the CRT screen. The term "storage tube" refers to the ability of the screen to retain the image which has been projected against it, thus avoiding the need to rewrite the image constantly. What makes this possible is the use of an electron flood gun directed at the phosphor coated screen which keeps the phosphor elements illuminated once they have been energized by the stroke-writing electron beam. The resulting image on the

CRT screen is flicker-free. Lines may be readily added to the image without concern over their effect on image density or refresh rates. However, the penalty associated with the storage tube is that individual lines cannot be selectively removed from the image.

Storage tubes have historically been the lowest-cost terminals and are capable of displaying large amounts of data, either graphical or textual. Because of these features, there are probably more storage tube terminals in service in industry at the time of this writing than any other graphics display terminal. The principal disadvantage of a storage CRT is that selective erasure is not possible. Instead, if the user wants to change the picture, the change will not be manifested on the screen until the entire picture is regenerated. Other disadvantages include its lack of color capability, the inability to use a light pen as a data entry device (we discuss the light pen in Section 5.4), and its lack of animation capability.

RASTER SCAN TERMINALS. Raster scan terminals operate by causing an electron beam to trace a zigzag pattern across the viewing screen, as described earlier. The operation is similar to that of a commercial television set. The difference is that a TV set uses analog signals originally generated by a video camera to construct the image on the CRT screen, while the raster scan ICG terminal uses digital signals generated by a computer. For this reason, the raster scan terminals used in computer graphics are sometimes called digital TVs.

The introduction of the raster scan graphics terminal using a refresh tube had been limited by the cost of computer memory. For example, the simplest and lowest-cost terminal in this category uses only two beam intensity levels, on or off. This means that each pixel in the viewing screen is either illuminated or dark. A picture tube with 256 lines of resolution and 256 addressable points per line to form the image would require 256×256 or over 65,000 bits of storage. Each bit of memory contains the on/off status of the corresponding pixel on the CRT screen. This memory is called the frame buffer or refresh buffer. The picture quality can be improved in two ways: by increasing the pixel density or adding a gray scale (or color). Increasing pixel density for the same size screen means adding more lines of resolution and more addressable points per line. A 1024×1024 raster screen would require more than 1 million bits of storage in the frame buffer. A gray scale is accomplished by expanding the number of intensity levels which can be displayed on each pixel. This requires additional bits for each pixel to store the intensity level. Two bits are required for four levels, three bits for eight levels, and so forth. Five or six bits would be needed to achieve an approximation of a continuous gray scale. For a color display, three times as many bits are required to get various intensity levels for each of the three primary colors: red, blue, and green. (We discuss color in the following section.) A raster scan graphics terminal with high resolution and gray scale can require a very large capacity refresh buffer. Until recent developments in memory technology, the cost of this storage capacity was prohibitive for a terminal with good picture quality. The capability to achieve color and animation was not possible except for very low resolution levels.

TABLE 5.1   Comparison of Graphics Terminal Features

	Directed-beam refresh	DVST	Raster scan
Image generation	Stroke writing	Stroke writing	Raster scan
Picture quality	Excellent	Excellent	Moderate to good
Data content	Limited	High	High
Selective erase	Yes	No	Yes
Gray scale	Yes	No	Yes
Color capability	Moderate	No	Yes
Animation capability	Yes	No	Moderate

Source: Adapted from Ref. [6].

It is now possible to manufacture digital TV systems for interactive computer graphics at prices which are competitive with the other two types. The advantages of the present raster scan terminals include the feasibility to use low-cost TV monitors, color capability, and the capability for animation of the image. These features, plus the continuing improvements being made in raster scan technology, make it the fastest-growing segment of the graphics display market.

Many of the important characteristics of the three types of graphics terminals are summarized in Table 5.1.

Color and animation in computer graphics

The capabilities for multicolored images and animated pictures in computer graphics are largely dependent on hardware considerations. Table 5.1 shows the relative capabilities for the three types of commercial graphics terminals for color and animation. This section briefly explores these two features.

Since the late 1970s, there has been a growing appreciation for the opportunities and advantages provided by color graphics. Figures 5.6 through 5.10 are color plates taken from computer graphics screens. Figure 5.6 shows the improved clarity of information provided by color in displaying the results of a finite element analysis. Figure 4.7 shows the same image in black and white. Figure 5.7 shows a wire-frame model of a jet engine with different sections of the assembly displayed in various colors. Figure 5.8 illustrates a shaded model of an industrial building, showing parking lot, landscaping, and other details. The capabilities of interactive computer graphics would permit us to zoom in on certain sections of the structure, including interior layout of the plant. Figures 5.9 and 5.10 show two examples of solid modeling. (We discuss the topic of solid modeling in Chapter 6.) Figure 5.9 shows the same yoke as in Figure 4.6 but with several components assembled to it, all displayed in various colors. Figure 5.10 illustrates a welded frame assembly for a water-cooled power frame. Cable and piping are shown in different colors. A solid computer model such as this would be helpful for interference checking in various types of process plants. As indicated by these figures, the use of color in

FIGURE 5.6 Finite-element analysis results displayed in color. (Courtesy of Applicon Inc.)

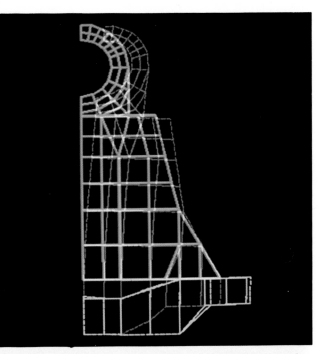

FIGURE 5.7 Wire-frame model of a jet engine. (Courtesy of Computervision Corp.)

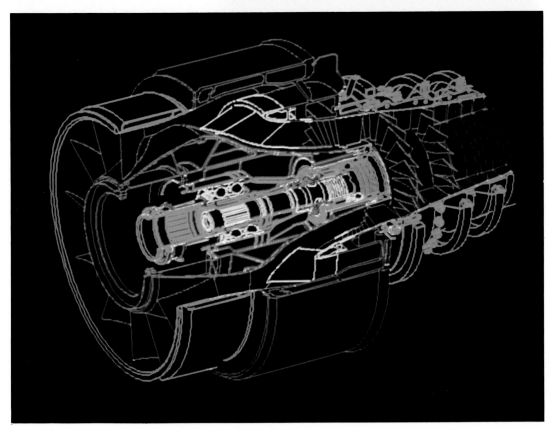

FIGURE 5.8 Color display of industrial plant. (Courtesy of Computervision Corp.)

FIGURE 5.9 Color solid model of yoke assembly. (Courtesy of Computervision Corp.)

FIGURE 5.10 Color graphics for display of pipes and cables in steel frame. (From W. Fitzgerald, F. Gracer, and R. Wolfe, "GRIN: Interactive Graphics for Modeling Solids," *IBM Journal of Research and Development,* Vol. 25, No. 4, July, 1981. Copyright 1981 by International Business Machines Corporation; reprinted with permission.)

the graphics display allows significantly more information to be clearly communicated to the viewer.

The typical color CRT uses three electron beams and a triad of color dots on the phosphor screen to provide each of the three colors, red, green, and blue. By combining the three colors at different intensity levels, a variety of colors can be created on the screen. It is more difficult to fabricate a stroke-writing tube which is precise enough for color because of the technical problem of getting the three beams to converge properly against the screen.

The raster scan approach has superior color graphics capabilities because of the developments which have been made over the years in the color television industry. Color raster scan terminals with 1024×1024 resolution are commercially available for computer graphics. The problem in the raster terminals is the memory requirements of the refresh buffer. Each pixel on the viewing screen may require up to 24 bits of memory in the refresh buffer in order to display the full range of color tones. When multiplied by the number of pixels in the display screen, this translates into a very large storage buffer.

The capability for animation in computer graphics is limited to display methods in which the image can be quickly redrawn. This limitation excludes the storage tube terminals. Both the directed-beam refresh and the raster scan systems are capable of animation. However, this capability is not automatically acquired with these systems. It must be accomplished by means of a powerful and fast CPU interfaced to the graphics terminal to process the large volumes of data required for animated images. In computer-aided design, animation would be a powerful feature in applications where kinematic simulation is required. The analysis of linkage mechanisms and other mechanical behavior would be examples. In computer-aided manufacturing, the planning of a robotic work cycle would be improved through the use of an animated image of the robot simulating the motion of the arm during the cycle. The popular video games marketed by Atari and other manufacturers for use with home TV sets are primitive examples of animation in computer graphics. Animation in these TV games is made possible by sacrificing the quality of the picture. This keeps the price of these games within an affordable range.

5.4 OPERATOR INPUT DEVICES

Operator input devices are provided at the graphics workstation to facilitate convenient communication between the user and the system. Workstations generally have several types of input devices to allow the operator to select the various preprogrammed input functions. These functions permit the operator to create or modify an image on the CRT screen or to enter alphanumeric data into the system. This results in a complete part on the CRT screen as well as a complete geometric description of the part in the CAD data base.

Different CAG system vendors offer different types of operator input devices. These devices can be divided into three general categories:

1.   Cursor control devices
2.   Digitizers
3.   Alphanumeric and other keyboard terminals

Of the three, cursor control devices and digitizers are both used for graphical interaction with the system. Keyboard terminals are used as input devices for commands and numerical data.

There are two basic types of graphical interaction accomplished by means of cursor control and digitizing:

Creating and positioning new items on the CRT screen

Pointing at or otherwise identifying locations on the screen, usually associated with existing images

Ideally, a graphical input device should lend itself to both of these functions. However, this is difficult to accomplish with a single unit and that is why most workstations have several different input devices.

Cursor control

The cursor normally takes the form of a bright spot on the CRT screen that indicates where lettering or drawing will occur. The computer is capable of reading the current position of the cursor. Hence the user's capability to control the cursor position allows locational data to be entered into the CAD system data base. A typical example would be for the user to locate the cursor to identify the starting point of a line. Another, more sophisticated case, would be for the user to position the cursor to select an item from a menu of functions displayed on the screen. For instance, the screen might be divided into two sections, one of which is an array of blocks which correspond to operator input functions. The user simply moves the cursor to the desired block to execute the particular function.

There are a variety of cursor control devices which have been employed in CAD systems. These include:

Thumbwheels
Direction keys on a keyboard terminal
Joysticks
Tracker ball
Light pen
Electronic tablet/pen

The first four items in the list provide control over the cursor without any direct physical contact of the screen by the user. The last two devices in the list require the user to control the cursor by touching the screen (or some other flat surface which is related to the screen) with a pen-type device.

The thumbwheel device uses two thumbwheels, one to control the horizontal position of the cursor, the other to control the vertical position. This type of device is often mounted as an integral part of the CRT terminal. The cursor in this arrangement is often represented by the intersection of a vertical line and a horizontal line displayed on the CRT screen. The two lines are like crosshairs in a gunsight which span the height and width of the screen.

Direction keys on the keyboard are another basic form of cursor control used not only for graphics terminals but also for CRT terminals without graphics capabilities. Four keys are used for each of the four directions in which the cursor can be moved (right or left, and up or down).

The joystick apparatus is pictured in Figure 5.11. It consists of a box with a vertical toggle stick that can be pushed in any direction to cause the cursor to be moved in that direction. The joystick gets its name from the control stick that was used in old airplanes.

The tracker ball is pictured in Figure 5.12. Its operation is similar to that of the joystick except that an operator-controlled ball is rotated to move the cursor in the desired direction on the screen.

The light pen is a pointing device in which the computer seeks to identify the

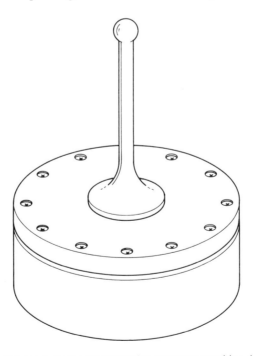

FIGURE 5.11 Joystick input device for interactive computer graphics. (Adapted with permission from Machover [6].)

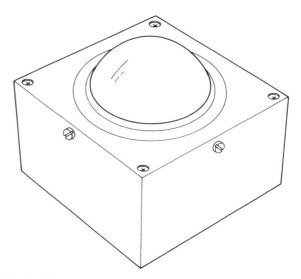

FIGURE 5.12 Tracker ball input device for interactive computer graphics. (Adapted with permission from Machover [6].)

position where the light pen is in contact with the screen. Contrary to what its name suggests, the light pen does not project light. Instead, it is a detector of light on the CRT screen and uses a photodiode, phototransistor, or some other form of light sensor. The light pen can be utilized with a refresh-type CRT but not with a storage tube. This is because the image on the refresh tube is being generated in time sequence. The time sequence is so short that the image appears continuous to the human eye. However, the computer is capable of discerning the time sequence and it coordinates this timing with the position of the pen against the screen. In essence, the system is performing as an optical tracking loop to locate the cursor or to execute some other input function. Figure 5.13 illustrates the use of the light pen.

The tablet and pen in computer graphics describes an electronically sensitive tablet used in conjunction with an electronic stylus. The tablet is a flat surface, separate from the CRT screen, on which the user draws with the penlike stylus to input instructions or to control the cursor. This form of input is illustrated in Figure 5.14.

It should be noted that thumbwheels, direction keys, joysticks, and tracker balls are generally limited in their functions to cursor control. The light pen and tablet/pen are typically used for other input functions as well as cursor control. Some of these functions are:

Selecting from a function menu

Drawing on the screen or making strokes on the screen or tablet which indicate what image is to be drawn

Selecting a portion of the screen for enlargement of an existing image

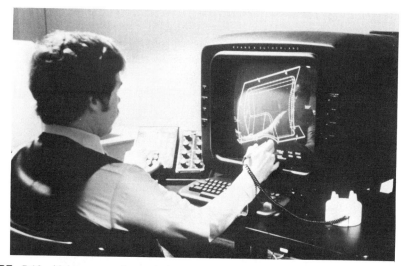

FIGURE 5.13 Light pen as input device for interactive computer graphics. (Courtesy of McDonnell-Douglas Automation Company.)

FIGURE 5.14 Electronic tablet and pen. (Courtesy of Computervision Corp.)

Digitizers

The digitizer is an operator input device which consists of a large, smooth board (the appearance is similar to a mechanical drawing board) and an electronic tracking device which can be moved over the surface to follow existing lines. It is a common technique in CAD systems for taking x, y coordinates from a paper drawing. The electronic tracking device contains a switch for the user to record the desired x and y coordinate positions. The coordinates can be entered into the computer memory or stored on an off-line storage medium such as magnetic tape. High-resolution digitizers, typically with a large board (e.g., 42 in. by 60 in.) can provide resolution and accuracy on the order of 0.001 in. It should be mentioned that the electronic tablet and pen, previously discussed as a cursor control device, can be considered to be a small, low-resolution digitizer.

Not all CAD systems would include a digitizer as part of its core of operator input devices. It would be inadequate, for example, in three-dimensional mechanical design work since the digitizer is limited to two dimensions. For two-dimensional drawings, drafters can readily adapt to the digitizer because it is similar to their drafting boards. It can be tilted, raised,or lowered to assume a comfortable position for the drafter.

The digitizer can be used to digitize line drawings. The user can input data from a rough schematic or large layout drawing and edit the drawing to the desired level of accuracy and detail. The digitizer can also be used to freehand a new design, with subsequent editing to finalize the drawing.

Keyboard terminals

Several forms of keyboard terminals are available as CAD input devices. The most familiar type is the alphanumeric terminal, which is available with nearly all interactive graphics systems. The alphanumeric terminal can be either a CRT or a hard-copy terminal, which prints on paper. For graphics, the CRT has the advantage because of its faster speed, the ability to easily edit, and the avoidance of large volumes of paper. On the other hand, a permanent record is sometimes desirable and this is most easily created with a hard-copy terminal. Many CAD systems use the graphics screen to display the alphanumeric data, but there is an advantage in having a separate CRT terminal so that the alphanumeric messages can be created without disturbing or overwriting the image on the graphics screen.

The alphanumeric terminal is used to enter commands, functions, and supplemental data to the CAD system. This information is displayed for verification on the CRT or typed on paper. The system also communicates back to the user in a similar manner. Menu listings, program listings, error messages, and so forth, can be displayed by the computer as part of the interactive procedure. An example of an alphanumeric keyboard is shown in Figure 5.2 with an electronic tablet/pen and a graphics CRT terminal.

Some CAD systems make use of special function keyboards, as pictured in Figure 5.15. These function keyboards are provided to eliminate extensive typing of commands, or calculate coordinate positions, and other functions. The number of function keys varies from about 8 to 80. The particular function corresponding with each button is generally under computer control so that the button function can be changed as the user proceeds from one phase of the design to the next. In this way the number of alternative functions can easily exceed the number of but-

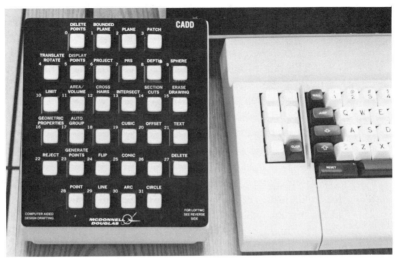

FIGURE 5.15 Special function keyboard for interactive computer graphics. (Courtesy of McDonnell-Douglas Automation Company.)

tons on the keyboard. Also, lighted buttons are used on the keyboards to indicate which functions are possible in the current phase of design activity. A menu of the various function alternatives is typically displayed on the CRT screen for the user to select the desired function.

5.5   PLOTTERS AND OTHER OUTPUT DEVICES

There are various types of output devices used in conjunction with a computer-aided design system. These output devices include:

Pen plotters
Hard-copy units
Electrostatic plotters
Computer-output-to-microfilm (COM) units

We discuss these devices in the following sections.

Pen plotters

The accuracy and quality of the hard-copy plot produced by a pen plotter is considerably greater than the apparent accuracy and quality of the corresponding image on the CRT screen. In the case of the CRT image, the quality of the picture is degraded because of lack of resolution and because of losses in the digital-to-analog conversion through the display generators. On the other hand, a high-precision pen plotter is capable of achieving a hard-copy drawing whose accuracy is nearly consistent with the digital definitions in the CAD data base.

The pen plotter uses a mechanical ink pen (either wet ink or ballpoint) to write on paper through relative movement of the pen and paper. There are two basic types of pen plotters currently in use:

Drum plotters
Flat-bed plotters

The drum plotter, pictured in Figure 5.16, is generally the least expensive. It uses a round drum, usually mounted horizontally, and a slide which can be moved along a track mounted axially with respect to the drum. The paper is attached to the drum and the pen is mounted on the slide. The relative motion between pen and paper is achieved by coordinating the rotation of the drum with the motion of the slide. The drum plotter is fast and it can make drawings of virtually unlimited length. The width, however, is limited by the length of the drum. These lengths typically range between 8½ in. (216 mm) and 42 in. (1067 mm).

The flat-bed plotter, illustrated in Figure 5.17, is more expensive. It uses a flat drawing surface to which the paper is attached. On some models, the surface is horizontal, while other models use a drawing surface which is mounted in a nearly vertical orientation to conserve floor space. This type is shown in Figure 5.18. Parallel tracks are located on two sides of the flat surface. A bridge is driven along these tracks to provide the *x*-coordinate motion. Attached to the bridge is another track, on which rides a writing head. Movement of the writing head relative to the

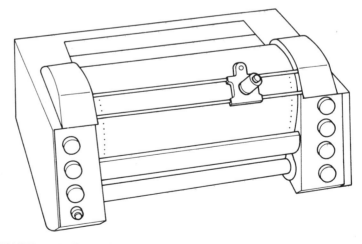

FIGURE 5.16   Drum plotter. (Reprinted with permission from Machover [6].)

FIGURE 5.17 Flat-bed x-y plotter in near-vertical postion. (Courtesy of Gerber Scientific Instrument Company.)

FIGURE 5.18 Belt-bed plotter in near-vertical position. (Courtesy of Applicon Inc.)

bridge produces the *y*-coordinate motion. The writing head carries the pen or pencil, which can be raised or lowered to provide contact with the paper as desired. The size of these automated drafting tables can range up to roughly 5 ft (1.5 m) by 20 ft (6.1 m) with plotting accuracies approaching ±0.001 in. (±0.025 mm).

The pen plotter accepts digitized data either on-line from the computer or off-line in the form of magnetic tape or punched tape. On modern pen plotters, a microprocessor is often used as the control unit. This allows certain shapes, such as circles and ellipses, to be programmed in the form of simple instructions to the plotter. In this way the digital data for a complicated shape can be made more compact and efficient.

Many plotters work with several pens of different colors to achieve multicolor plots. Also, in some models, the pen may be replaced by a highly focused, high-intensity light and the conventional drafting paper by a photosensitive paper. This arrangement would be used for certain artwork applications. Another option available on a flat-bed plotter is to combine the plotter function with the operation of a digitizer (previously discussed as an operator input device in Section 5.4). Such a device is called a digitizer-plotter.

Hard-copy unit

A hard-copy unit is a machine that can make copies from the same image data displayed on the CRT screen. The image on the screen can be duplicated in a matter of seconds. The copies can be used as records of intermediate steps in the design process or when rough hard copies of the screen are needed quickly. The hard copies produced from these units are not suitable as final drawings because the accuracy and quality of the reproduction is not nearly as good as the output of a pen plotter.

Most hard-copy units are dry silver copiers that use light-sensitive paper exposed through a narrow CRT window inside the copier. The window is typically 8½ in. (216 mm), corresponding to the width of the paper, by about ½ in. (12 mm) wide. The paper is exposed by moving it past the window and coordinating the CRT beam to gradually transfer the image. A heated roller inside the copier is used to develop the exposed paper. The size of the paper is usually limited on these hard-copy units to 8½ by 11 in. Another drawback is that the dry silver copies will darken with time when they are left exposed to normal light.

Electrostatic plotters

Hard-copy units are relatively fast but their accuracy and resolution are poor. Pen plotters are highly accurate but plotting time can take many minutes (up to a half-hour or longer for complicated drawings). The electrostatic plotter offers a compromise between these two types in terms of speed and accuracy. It is almost as fast as the hard-copy unit and almost as accurate as the pen plotter.

The electrostatic copier consists of a series of wire styli mounted on a bar which spans the width of the charge-sensitive paper. The styli have a density of up to 200 per linear inch. The paper is gradually moved past the bar and certain styli are activated to place dots on the paper. By coordinating the generation of the dots with the paper travel, the image is progressively transferred from the data base into hard-copy form. The dots overlap each other slightly to achieve continuity. For example, a series of adjacent dots gives the appearance of a continuous line.

A limitation of the electrostatic plotter is that the data must be in the raster format (i.e., in the same format used to drive the raster-type CRT) in order to be readily converted into hard copy using the electrostatic method. If the data are not in raster format, some type of conversion is required to change them into the required format. The conversion mechanism is usually based on a combination of software and hardware.

An advantage of the electrostatic plotter which is shared with the drum-type pen plotter is that the length of the paper is virtually unlimited. Typical plotting widths might be up to 6 ft (1.83 m). Another advantage is that the electrostatic plotter can be utilized as a high-speed line printer, capable of up to 1200 lines of text per minute.

Computer-output-to-microfilm (COM) units

COM units reproduce the drawings on microfilm rather than as full-size engineering drawings. It is an expensive piece of equipment with a current price of around $300,000. However, for the large corporation able to afford a COM unit, there are several important advantages. One advantage is storage capability. A large engineering department may have tens of thousands of engineering drawings to be stored. Reducing the size of each drawing to microfilm achieves a significant storage benefit. If a full-size hard-copy drawing is ever required, the microfilm can be easily retrieved to be photographically enlarged to full size. Another advantage is speed. COM units produce a microfilm copy much faster than a pen plotter, perhaps several hundred times faster for a complicated line drawing. Computer-output-to-microfilm is also faster than electrostatic plotters.

Disadvantages of the COM process are that the user cannot write notes on the microfilm as is possible with a paper copy. Also, enlargements of the microfilm onto paper, although adequate, are not of as high quality as the output from a pen plotter.

5.6 THE CENTRAL PROCESSING UNIT

The CPU operates as the central ''brain'' of the computer-aided design system. It is typically a minicomputer. It executes all the mathematical computations needed to accomplish graphics and other functions, and it directs the various activities within the system. These activities include:

Managing the design workstations (operator inputs, editing, etc.)

Directing plotters in the generation of engineering drawings

Copying data currently on disk onto magnetic tapes for semiactive storage

Reading magnetic tapes containing drawing data for possible revision or other use

Transmitting data to and from other larger computers

The computer in a CAD system must be capable of a wide variety of both graphical and nongraphical functions.

Most CAD systems use CPUs purchased from commercial computer companies such as Digital Equipment Corporation, Data General, and Hewlett-Packard. Computervision Corporation is a notable exception to this rule. Its CAD systems use CPUs designed and built within the company.

At the time of this writing, most of the CAD minicomputers use a 16-bit word. However, there is a very definite trend toward the use of 32-bit CPUs in commercial CAD systems. There are several advantages which advocate the use of 32-bit processors. These include greater speed, greater accuracy, and more addressable memory. These advantages permit programs of greater complexity to be executed efficiently on the system. Many FORTRAN programs, for example, are written to be run on 32-bit processors. Primary storage capacity for a stand-alone CAD system can range up to several million characters of memory. More typical values for today's systems would be around 256,000 characters for today's 16-bit systems.

There are several factors related to computer and storage technology which are appropriate to mention in this discussion. First, the cost of computers and computer memory continues to decrease as the technology develops. This will permit future CAD systems to use more powerful computers and greater memory capacity.

Second, CAD systems would normally use secondary storage units to reduce the cost of more expensive main memory. We discuss secondary storage for computer-aided design in Section 5.7.

Third, many CAD systems have their CPUs connected to larger mainframe computers (often called the host computer) to gain access to greater computational and memory capacity. This arrangement is illustrated in Figure 5.19. The host computer is called in to execute complex engineering and numerical analysis which would overburden the CAD system CPU. When the analysis is completed by the large computer, the results are downloaded to the graphics system for display and possible hard-copy output. This procedure allows each computer (the larger host and the smaller graphics CPU) to function in what it does best. Some computer-aided design systems use "intelligent" terminals in a distributed computing system. Each terminal contains a microprocessor which functions as a small CPU for that terminal. This CPU can handle many of the user input commands, and would only require access to the host computer for plotting, output, and access

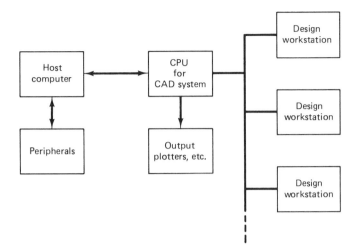

FIGURE 5.19   Configuration of host-satellite CAD system.

to secondary storage. In this configuration the host computer could service perhaps a dozen intelligent terminals since it would be relieved of many of the computational requirements of a conventional CAD system. Rao et al. [8] report the results of a study performed at McDonnell-Douglas Automation Company (McAuto) which compared the speed and performance of a distributed graphics system (using intelligent terminals) to those of the conventional centralized configuration (using "dumb" terminals). After adjusting for other differences in hardware between the two systems, they found that the distributed system had more than a 40% improvement in operating cost and performance. Of course, a distributed system would probably have a higher initial cost because of the more expensive intelligent terminals.

5.7   SECONDARY STORAGE

In addition to primary computer storage, secondary storage capacity is provided in a computer-aided design system. This usually takes the form of magnetic disk and magnetic tape. The purpose in using secondary storage is to reduce the cost of main computer memory. The secondary storage can be used for engineering drawing files, CAD software which can be transferred to main memory as needed, and temporary files for CPU output which will be downloaded to individual graphics terminals, plotters, or other output devices.

Magnetic disks are typically used for the CAD system software and the CAD data base (engineering drawings). Disks have the advantage of fairly rapid retrieval, owing to their random access configuration. Because of this feature, the

CPU can load and swap programs and files between primary and secondary memory as needed.

Magnetic tape would be used for storing programs and files which are less frequently used by the system. Storage on magnetic tape is less expensive than on disks; however, the access time is significantly longer because data are stored sequentially (refer to Table 2.2). Magnetic tape would be suitable for disk backup, permanent archival files, and data transfer to output devices or other computers.

REFERENCES

[1]   BESANT, C. B., *Computer-Aided Design and Manufacture*, Ellis Horwood Ltd., Chichester, England, 1980.

[2]   CHASEN, S. H., AND DOW, J. W., *The Guide for Evaluation and Implementation of CAD/ CAM Systems*, CAD/CAM Decisions, Atlanta, Ga., 1979

[3]   FOLEY, J. D., AND VAN DAM, A., *Fundamentals of Interactive Computer Graphics*, Addision-Wesley Publishing Co., Inc., Reading, Mass., 1982.

[4]   HOBBS, L. C., "Computer Graphics Display Hardware," *IEEE Computer Graphics and Applications*, January, 1981, pp. 25−39.

[5]   LERRO, J. P., JR., "CAD/CAM System: More than an Automated Drafting Tool," *Design News*, November 17, 1980.

[6]   MACHOVER, C., "What Are the Tools?" in *The CAD/CAM Handbook* (C. Machover and R. E. Blauth, Eds.), Computervision Corp., Bedford, Mass., 1980, pp. 17−45.

[7]   NEWMAN, W., AND SPROUL, R., *Principles of Interactive Computer Graphics*, 2nd ed., McGraw-Hill Book Company, New York, 1979.

[8]   RAO, J. R., WINTERS, W. F., AND SCHMIDT, L. D., "Performance Evaluation of a Test Distributed Graphics System," technical paper presented at IEEE Computer Society's Third International Computer Software and Applications Conference, Chicago, November, 1979.

[9]   RYAN, D. L., *Computer-Aided Graphics and Design*, Marcel Dekker, Inc., New York, 1979.

[10] SCHAFFER, G., "Computer Graphics Goes to Work," Special Report 724, *American Machinist*, July, 1980, pp. 149−164.

[11] ZIMMERS, E. W., JR, *Computer-Aided Design Module*, General Electric CAD/CAM Seminar, Lehigh University, Bethlehem, Pa., 1982.

[12] ZIMMERS, E. W., JR. AND PLEBANI, L. J., "Using a Turnkey Graphics System in Computer Integrated Manufacturing," *Industrial Engineering*, November, 1981, pp. 98−104.

PROBLEMS

5.1.   The hardware components of a computer-aided design system are to be specified for the two cases cited below. The particular CAD/CAM vendor need not be identified, but the number of terminals, types of input devices, and types of output devices should be specified. Also, will the CPU be stand-alone or connected to a larger host computer? It is important to maintain a high benefit/cost ratio in the selection of hardware. Explain and justify your specifications. The two cases are:

(a) An engineering firm with seven mechanical designers and 12 drafters. The firm designs mechanical equipment for client companies, which fabricate the equipment. It is important that the engineering drawings be clear and of high quality.

(b) A manufacturing firm which produces products in high volume for the public. The design engineering staff consists of 10 mechanical and electrical engineers.

5.2. For each of the following situations, identify which of the three types of graphics terminals would be most appropriate. Explain your selection by reference to the comparison features listed in Table 5.1.

(a) In the first situation, there is the requirement to display large amounts of textual output.

(b) In the second case, it is necessary to perform kinematic analysis of automobile door hinges.

(c) In the third application, it is desired to perform stress analysis and to display the before and after strain effects at the same time.

(d) In the fourth case, the quality and resolution of the image must be very high. There will be a large amount of graphical data displayed on the screen.

chapter 6

Computer Graphics Software and Data Base

6.1 INTRODUCTION

The CAD hardware discussed in Chapter 5 would be useless without the software to support it. This chapter discusses some of the issues and methods related to the software and accompanying data base for interactive computer graphics and computer-aided design. A more complete coverage of this topic is available in several of the references listed at the end of the chapter. References [4] and [8] are comprehensive in their coverage of both hardware and software. References [2] and [10] are concerned more with the mathematics required in computer graphics.

The graphics software is the collection of programs written to make it convenient for a user to operate the computer graphics system. It includes programs to generate images on the CRT screen, to manipulate the images, and to accomplish various types of interaction between the user and the system. In addition to the graphics software, there may be additional programs for implementing certain specialized functions related to CAD/CAM. These include design analysis programs (e.g., finite-element analysis and kinematic simulation) and manufacturing planning programs (e.g., automated process planning and numerical control part programming). This chapter deals mainly with the graphics software.

The graphics software for a particular computer graphics system is very much a function of the type of hardware used in the system. The software must be written specifically for the type of CRT and the types of input devices used in the system. The details of the software for a stroke-writing CRT would be different than for a raster scan CRT. The differences between a storage tube and a refresh tube would also influence the graphics software. Although these differences in software may be invisible to the user to some extent, they are important considerations in the design of an interactive computer graphics system.

Newman and Sproull [8] list six "ground rules" that should be considered in designing graphics software:

1. *Simplicity*. The graphics software should be easy to use.
2. *Consistency*. The package should operate in a consistent and predictable way to the user.
3. *Completeness*. There should be no inconvenient omissions in the set of graphics functions.
4. *Robustness*. The graphics system should be tolerant of minor instances of misuse by the operator.
5. *Performance*. Within limitations imposed by the system hardware, the performance should be exploited as much as possible by software. Graphics programs should be efficient and speed of response should be fast and consistent.
6. *Economy*. Graphics programs should not be so large or expensive as to make their use prohibitive.

6.2 THE SOFTWARE CONFIGURATION OF A GRAPHICS SYSTEM

In the operation of the graphics system by the user, a variety of activities take place, which can be divided into three categories:

1. Interact with the graphics terminal to create and alter images on the screen.
2. Construct a model of something physical out of the images on the screen. The models are sometimes called application models.
3. Enter the model into computer memory and/or secondary storage.

In working with the graphics system the user performs these various activities in combination rather than sequentially. The user constructs a physical model and inputs it to memory by interactively describing images to the system. This is done without any thought about whether the activity falls into category 1, 2, or 3.

The reason for separating these activities in this fashion is that they correspond to the general configuration of the software package used with the interactive computer graphics (ICG) system. The graphics software can be divided into three modules according to a conceptual model suggested by Foley and Van Dam [4]:

1. The graphics package (Foley and Van Dam called this the graphics system)
2. The application program
3. The application data base

This software configuration is illustrated in Figure 6.1. The central module is the application program. It controls the storage of data into and retrieves data out of the application data base. The application program is driven by the user through the graphics package.

The application program is implemented by the user to construct the model of a physical entity whose image is to be viewed on the graphics screen. Application programs are written for particular problem areas. Problem areas in engineering design would include architecture, construction, mechanical components, electronics, chemical engineering, and aerospace engineering. Problem areas other than design would include flight simulators, graphical display of data, mathematical analysis, and even artwork. In each case, the application software is developed to deal with images and conventions which are appropriate for that field.

The graphics package is the software support between the user and the graphics terminal. It manages the graphical interaction between the user and the system. It also serves as the interface between the user and the application software. The graphics package consists of input subroutines and output subroutines. The input routines accept input commands and data from the user and forward them to the application program. The output subroutines control the display terminal (or other output device) and converts the application models into two-dimensional or three-dimensional graphical pictures.

The third module in the ICG software is the data base. The data base contains mathematical, numerical, and logical definitions of the application models,

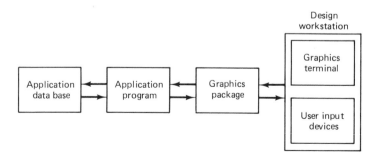

FIGURE 6.1 Model of graphics software configuration.

such as electronic circuits, mechanical components, automobile bodies, and so forth. It also includes alphanumeric information associated with the models, such as bills of materials, mass properties, and other data. The contents of the data base can be readily displayed on the CRT or plotted out in hard-copy form. Section 6.6 presents a discussion of the data base for computer graphics.

6.3   FUNCTIONS OF A GRAPHICS PACKAGE

To fulfill its role in the software configuration, the graphics package must perform a variety of different functions. These functions can be grouped into function sets. Each set accomplishes a certain kind of interaction between the user and the system. Some of the common function sets are:

>   Generation of graphic elements
>
>   Transformations
>
>   Display control and windowing functions
>
>   Segmenting functions
>
>   User input functions

We examine some of these functions in more detail in subsequent sections of this chapter. What we present in the sections below is a brief description of each.

Generation of graphic elements

A graphic element in computer graphics is a basic image entity such as a dot (or point), line segment, circle, and so forth. The collection of elements in the system could also include alphanumeric characters and special symbols. There is often a special hardware component in the graphics system associated with the display of many of the elements. This speeds up the process of generating the element. The user can construct the application model out of a collection of elements available on the system.

The term "primitive" is often used in reference to graphic elements. We shall reserve the use of this term to three-dimensional graphics construction. Accordingly, a primitive is a three-dimensional graphic element such as a sphere, cube, or cylinder. In three-dimensional wire-frame models and solid modeling, primitives are used as building blocks to construct the three-dimensional model of the particular object of interest to the user.

Transformations

Transformations are used to change the image on the display screen and to reposition the item in the data base. Transformations are applied to the graphic elements in order to aid the user in constructing an application model. These transformations include enlargement and reduction of the image by a process called scaling, repositioning the image or translation, and rotation. We discuss two- and three-dimensional transformations in Section 6.5.

Display control and windowing functions

This function set provides the user with the ability to view the image from the desired angle and at the desired magnification. In effect, it makes use of various transformations to display the application model the way the user wants it shown. This is sometimes referred to as windowing because the graphics screen is like a window being used to observe the graphics model. The notion is that the window can be placed wherever desired in order to look at the object being modeled.

Another aspect of display control is hidden-line removal. In most graphics systems, the image is made up of lines used to represent a particular object. Hidden-line removal is the procedure by which the image is divided into its visible and invisible (or hidden) lines. In some systems, the user must identify which lines (or portions of lines) are invisible so that they can be removed from the image to make it more understandable. In other systems, the graphics package is sufficiently sophisticated to remove the hidden lines from the picture automatically.

Segmenting functions

Segmenting functions provide users with the capability to selectively replace, delete, or otherwise modify portions of the image. The term ''segment'' refers to a particular portion of the image which has been identified for purposes of modifying it. The segment may define a single element or logical grouping of elements that can be modified as a unit.

Storage-type CRT tubes are unsuited to segmenting functions. To delete or modify a portion of the image on a storage tube requires erasing the entire picture and redrawing it with the changes incorporated. Raster scan refresh tubes are ideally suited to segmenting functions because the screen is automatically redrawn 30 or more times per second anyway. The image is regenerated each cycle from a display file, a file used for storage that is part of the hardware in the raster scan CRT. The segment can readily be defined as a portion of that display file by giving it a name. The contents of that portion of the file would then be deleted or altered to execute the particular segmenting function.

User input functions

User input functions constitute a critical set of functions in the graphics package because they permit the operator to enter commands or data to the system. The entry is accomplished by means of operator input devices, and from Chapter 5 we note that there are a large variety of these input devices. The user input functions must, of course, be written specifically for the particular compliment of input devices used on the system. The extent to which the user input functions are well designed has a significant effect on how "friendly" the system is to the user, that is, how easy it is to work on the system.

The input functions should be written to maximize the benefits of the interactive feature of ICG. The software design compromise is to find the optimum balance between providing enough functions to conveniently cover all data entry situations without inundating the user with so many commands that they cannot be remembered. One of the goals that are sought after by software designers in computer graphics is to simplify the user interface enough that a designer with little or no programming experience can function effectively on the system.

6.4 CONSTRUCTING THE GEOMETRY

The use of graphics elements

The graphics system accomplishes the definition of the model by constructing it out of graphic elements. These elements are called by the user during the construction process and added, one by one, to create the model. There are several aspects about this construction process which should be discussed.

First, as each new element is being called but before it is added to the model, the user can specify its size, its position, and its orientation. These specifications are necessary to form the model to the proper shape and scale. For this purpose, the various transformations mentioned previously are utilized.

A second aspect of the geometric construction process is that graphics elements can be subtracted as well as added. Another way of saying this is that the model can be formed out of negative elements as well as positive elements. Figure 6.2 illustrates this construction feature for a two-dimensional object, C. The object is drawn by subtracting circle B from rectangle A.

A third feature available during model building is the capability to group several elements together into units which are sometimes called cells. A cell, in this context, refers to a combination of elements which can be called to use any-

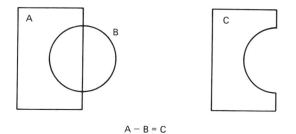

A − B = C

FIGURE 6.2 Example of two-dimensional model construction by subtraction of circle B from rectangle A.

where in the model. For example, if a bolt is to be used several places in the construction of a mechanical assembly model, the bolt can be formed as a cell and added anywhere to the model. The use of graphic cells is a convenient and powerful feature in geometric model construction.

Defining the graphic elements

The user has a variety of different ways to call a particular graphic element and position it on the geometric model. Table 6.1 lists several ways of defining points, lines, arcs, and other components of geometry through interaction with the ICG system. These components are maintained in the data base in mathematical form and referenced to a three-dimensional coordinate system. For example, a point would be defined simply by its x, y, and z coordinates. A polygon would be defined as an ordered set of points representing the corners of the polygon. A circle would be defined by its center and radius. Mathematically, a circle can be defined in the x-y plane by the equation

$$(x - m)^2 + (y - n)^2 = r^2 \qquad (6.1)$$

This specifies that the radius of the circle is r and the x and y of the center are m and n. In each case, the mathematical definition can be converted into its corresponding edges and surfaces for filing in the data base and display on the CRT screen.

Editing the geometry

A computer-aided design system provides editing capabilities to make corrections and adjustments in the geometric model. When developing the model, the user must be able to delete, move, copy, and rotate components of the model. We have previously discussed some of these adjustments in our discussion of the functions of a graphics package in Section 6.3. The editing procedure involves selecting the desired portion of the model (usually by means of one of the segmenting functions), and executing the appropriate command (often involving one of the transformation functions).

The method of selecting the segment of the model to be modified varies from system to system. With cursor control, a common method is for a rectangle to be formed on the CRT screen around the model segment. The rectangle is defined by entering the upper left and lower right corners of the rectangle. Another method involving a light pen is to place the pen over the component to be selected. With the electronic pen and tablet, the method might be to stroke a line across the portion of the model which is to be altered.

The computer must somehow indicate to the user which portion of the model has been selected. The reason for this is verification that the portion selected by the computer is what the user intended. Various techniques are used by different ICG

TABLE 6.1   Methods of Defining Elements in Interactive Computer Graphics

Points

Methods of defining points in computer graphics include:

1. Pointing to the location on the screen by means of cursor control
2. Entering the coordinates via the alphanumeric keyboard
3. Entering the offset (distance in x, y, and z) from a previously defined point
4. The intersection of two ~~points~~ LINES
5. Locating points at fixed intervals along an element

Lines

Methods of defining lines include:

1. Using two previously defined points
2. Using one point and specifying the angle of the line with the horizontal
3. Using a point and making the line either normal or tangent to a curve
4. Using a point and making the line either parallel or perpendicular to another line
5. Making the line tangent to two curves
6. Making the line tangent to a curve and parallel or perpendicular to a line

Arcs and Circles

Methods of defining arcs and circles include:

1. Specifying the center and the radius
2. Specifying the center and a point on the circle
3. Making the curve pass through three previously defined points
4. Making the curve tangent to three lines
5. Specifying the radius and making the curve tangent to two lines or curves

Conics

Conics, including ellipses, parabolas, and hyperbolas, can be defined in any plane by methods which include:

1. Specifying five points on the element
2. Specifying three points and a tangency condition

Curves

Mathematical splines are used to fit a curve through given data. For example, in a cubic spline, third-order polynomial segments are fitted between each pair of adjacent data points. Other curve-generating techniques used in computer graphics include Bezier curves and B-spline methods. Both of these methods use a blending procedure which smooths the effect of the data points. The resulting curve does not pass through all the points. In these cases the data points would be entered to the graphics system and the type of curve-fitting technique would be specified for determining the curve.

TABLE 6.1   (*cont.*)

Surfaces

The methods described for generating curves can also be used for determining the mathematical definition of a surface. Automobile manufacturers use these methods to represent the sculptured surfaces of the sheet metal car body. Some of the methods for generating surfaces include:

1. Using a surface of revolution formed by rotating any lines and/or curves around a specific axis.
2. Using the intersection line or surface of two intersecting surfaces. For example, this could be used to generate cross sections of parts, by slicing a plane through the part at the desired orientation.

---

systems to identify the segment. These include: placing a mark on the segment, making the segment brighter than the rest of the image, and making the segment blink.

A selection of common editing capabilities available in commercial CAD systems is presented in Table 6.2.

TABLE 6.2   Some Common Editing Features Available on a CAD System

---

1. Move an item to another location. This involves the translation of the item from one location to another.
2. Duplicate an item at another location. The copy function is similar to the move function except that it preserves a copy of the item at its original location.
3. Rotate an item. This is the rotation transformation, in which the item is rotated through a specified angle from its original orientation.
4. Mirror an item. This creates a mirror image of the item about a specified plane.
5. Delete an item. This function causes the selected segment of the model to be removed from the screen and from the data base.
6. Remove an item from the display (without deleting it from the data base). This removes the particular segment from the current image on the screen. However, it is not removed from the data base. Therefore, repainting the screen from the data base will cause the segment to reappear.
7. Trim a line or other component. This function would remove the portion of the line that extends beyond a certain point.
8. Create a cell out of graphic elements. This feature provides the capability to construct a cell out of selected elements. The cell can then be added to the model in any orientation as needed.
9. Scale an item. A selected component can be scaled by a specified factor in x, y, and z directions. The entire size of the model can be scaled, or it can be scaled in only one or two directions.

6.5 TRANSFORMATIONS

Many of the editing features involve transformations of the graphics elements or cells composed of elements or even the entire model. In this section we discuss the mathematics of these transformations. Two-dimensional transformations are considered first to illustrate concepts. Then we deal with three dimensions.

Two-dimensional transformations

To locate a point in a two-axis cartesian system, the x and y coordinates are specified. These coordinates can be treated together as a 1×2 matrix: (x,y). For example, the matrix $(2, 5)$ would be interpreted to be a point which is 2 units from the origin in the x-direction and 5 units from the origin in the y-direction.

This method of representation can be conveniently extended to define a line as a 2×2 matrix by giving the x and y coordinates of the two end points of the line. The notation would be

$$L = \begin{bmatrix} x_1 & y_1 \\ x_2 & y_2 \end{bmatrix} \tag{6.2}$$

Using the rules of matrix algebra, a point or line (or other geometric element represented in matrix notation) can be operated on by a transformation matrix to yield a new element.

There are several common transformations used in computer graphics. We will discuss three transformations: translation, scaling, and rotation.

TRANSLATION. Translation involves moving the element from one location to another. In the case of a point, the operation would be

$$x' = x + m, \qquad y' = y + n \tag{6.3}$$

where x',y' = coordinates of the translated point

$\qquad x,y$ = coordinates of the original point

$\qquad m,n$ = movements in the x and y directions, respectively

In matrix notation this can be represented as

$$(x', y') = (x,y) + T \tag{6.4}$$

where

$$T = (m,n), \text{ the translation matrix} \tag{6.5}$$

Any geometric element can be translated in space by applying Eq. (6.4) to each point that defines the element. For a line, the transformation matrix would be applied to its two end points.

SCALING. Scaling of an element is used to enlarge it or reduce its size. The scaling need not necessarily be done equally in the x and y directions. For example, a circle could be transformed into an ellipse by using unequal x and y scaling factors.

The points of an element can be scaled by the scaling matrix as follows:

$$(x',y') = (x,y)S \tag{6.6}$$

where

$$S = \begin{bmatrix} m & 0 \\ 0 & n \end{bmatrix} \qquad \text{the scaling matrix} \tag{6.7}$$

This would produce an alteration in the size of the element by the factor m in the x-direction and by the factor n in the y direction. It also has the effect of repositioning the element with respect to the cartesian system origin. If the scaling factors are less than 1, the size of the element is reduced and it is moved closer to the origin. If the scaling factors are larger than 1, the element is enlarged and removed farther from the origin.

ROTATION. In this transformation, the points of an object are rotated about the origin by an angle 0. For a positive angle, this rotation is in the counterclockwise direction. This accomplishes rotation of the object by the same angle, but it also moves the object. In matrix notation, the procedure would be as follows:

$$(x',y') = (x,y)R \tag{6.8}$$

where

$$R = \begin{bmatrix} \cos 0 & \sin 0 \\ -\sin 0 & \cos 0 \end{bmatrix} \qquad \text{the rotation matrix} \tag{6.9}$$

EXAMPLE 6.1

As an illustration of these transformations in two dimensions, consider the line defined by

$$L = \begin{bmatrix} 1 & 1 \\ 2 & 4 \end{bmatrix}$$

Let us suppose that it is desired to translate the line in space by 2 units in the x direction and 3 units in the y direction. This would involve adding 2 to the current x value and 3 to the current y value of the end points defining the line. That is,

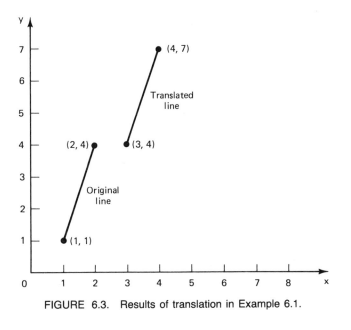

FIGURE 6.3.   Results of translation in Example 6.1.

$$\begin{bmatrix} 1 & 1 \\ 2 & 4 \end{bmatrix} + \begin{bmatrix} 2, & 3 \\ 2, & 3 \end{bmatrix} = \begin{bmatrix} 3 & 4 \\ 4 & 7 \end{bmatrix}$$

The new line would have end points at (3, 4) and (4, 7). The effect of the transformation is illustrated in Figure 6.3.

EXAMPLE 6.2

For the same original line as in Example 6.1, let us apply the scaling factor of 2 to the line. The scaling matrix for the 2×2 line definition would therefore be

$$T = \begin{bmatrix} 2 & 0 \\ 0 & 2 \end{bmatrix}$$

The resulting line would be determined by Eq. (6.6) as follows:

$$\begin{bmatrix} 1 & 1 \\ 2 & 4 \end{bmatrix} \cdot \begin{bmatrix} 2 & 0 \\ 0 & 2 \end{bmatrix} = \begin{bmatrix} 2 & 2 \\ 4 & 8 \end{bmatrix}$$

The new line is pictured in Figure 6.4.

EXAMPLE 6.3

We will again use our same line and rotate the line about the origin by 30°. Equation (6.8) would be used to determine the transformed line where the rotation matrix would be:

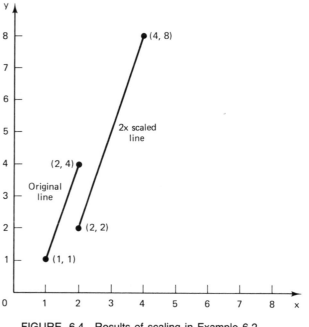

FIGURE 6.4   Results of scaling in Example 6.2.

$$R = \begin{bmatrix} \cos 30 & \sin 30 \\ -\sin 30 & \cos 30 \end{bmatrix} = \begin{bmatrix} 0.866 & 0.500 \\ -0.500 & 0.866 \end{bmatrix}$$

The new line would be defined as:

$$\begin{bmatrix} 1 & 1 \\ 2 & 4 \end{bmatrix} \cdot \begin{bmatrix} 0.866 & 0.500 \\ -0.50 & 0.866 \end{bmatrix} = \begin{bmatrix} 0.366 & 1.366 \\ -0.268 & 4.464 \end{bmatrix}$$

The effect of applying the rotation matrix to the line is shown in Figure 6.5.

Three-dimensional transformations

Transformations by matrix methods can be extended to three-dimensional space. We consider the same three general categories defined in the preceding section. The same general procedures are applied to use these transformations that were defined for the three cases by Eqs. (6.4), (6.6), and (6.8).

TRANSLATION. The translation matrix for a point defined in three dimensions would be

$$T = (m, \ n, \ p) \tag{6.10}$$

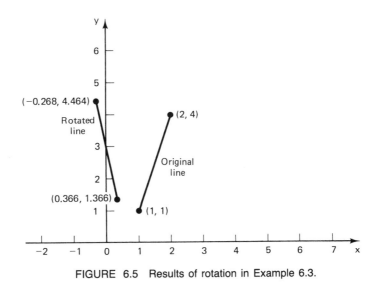

FIGURE 6.5   Results of rotation in Example 6.3.

and would be applied by adding the increments m, n, and p to the respective coordinates of each of the points defining the three-dimensional geometry element.

SCALING. The scaling transformation is given by

$$S = \begin{bmatrix} m & 0 & 0 \\ 0 & n & 0 \\ 0 & 0 & p \end{bmatrix} \qquad (6.11)$$

For equal values of m, n, and p, the scaling is linear.

ROTATION. Rotation in three dimensions can be defined for each of the axes. Rotation about the z axis by an angle θ is accomplished by the matrix

$$R_z = \begin{bmatrix} \cos\theta & -\sin\theta & 0 \\ \sin\theta & \cos\theta & 0 \\ 0 & 0 & 1 \end{bmatrix} \qquad (6.12)$$

Rotation about the y axis by the angle θ is accomplished similarly.

$$R_y = \begin{bmatrix} \cos\theta & 0 & \sin\theta \\ 0 & 1 & 0 \\ -\sin\theta & 0 & \cos\theta \end{bmatrix} \qquad (6.13)$$

Rotation about the x axis by the angle θ is done with an analogous transformation matrix.

$$R_x = \begin{bmatrix} 1 & 0 & 0 \\ 0 & \cos\theta & -\sin\theta \\ 0 & \sin\theta & \cos\theta \end{bmatrix} \quad (6.14)$$

Concatenation

The previous single transformations can be combined as a sequence of transformations. This is called concatenation, and the combined transformations are called concatenated transformations.

During the editing process when a graphic model is being developed, the use of concatenated transformations is quite common. It would be unusual that only a single transformation would be needed to accomplish a desired manipulation of the image. Two examples of where combinations of transformations would be required would be:

Rotation of the element about an arbitrary point in the element

Magnifying the element but maintaining the location of one of its points in the same location

In the first case, the sequence of transformations would be: translation to the origin, then rotation about the origin, then translation back to the original location. In the second case, the element would be scaled (magnified) followed by a translation to locate the desired point as needed.

The objective of concatenation is to accomplish a series of image manipulations as a single transformation. This allows the concatenated transformation to be defined more concisely and the computation can generally be accomplished more efficiently.

Determining the concatenation of a sequence of single transformations can be fairly straightforward if the transformations are expressed in matrix form as we have done. For example, if we wanted to scale a point by the factor of 2 in a two-dimensional system and then rotate it about the origin by 45°, the concatenation would simply be the product of the two transformation matrices. It is important that the order of matrix multiplication be the same as the order in which the transformations are to be carried out. Concatenation of a series of transformations becomes more complicated when a translation is involved, and we will not consider this case.

EXAMPLE 6.4

Let us consider the example cited in the text in which a point was to be scaled by a factor of 2 and rotated by 45°. Suppose that the point under consideration was (3, 1). This might be one of several points defining a geometric element. For purposes of illustration let us first accomplish the two transformations sequentially. First, consider the scaling.

$$(x', y') = (x, y)S$$

$$(x', y') = (3, 1) \begin{bmatrix} 2 & 0 \\ 0 & 2 \end{bmatrix} = (6, 2)$$

Next, the rotation can be performed.

$$(x'', y'') = (x', y')R$$

$$(x'', y'') = (6, 2) \begin{bmatrix} \cos 45 & \sin 45 \\ -\sin 45 & \cos 45 \end{bmatrix}$$

$$= (6, 2) \begin{bmatrix} 0.7071 & 0.7071 \\ -0.7071 & 0.7071 \end{bmatrix} = (2.828, 5.657)$$

The same result can be accomplished by concatenating the two separate transformation matrices. The product of the two matrices would be

$$SR = \begin{bmatrix} 2 & 0 \\ 0 & 2 \end{bmatrix} \cdot \begin{bmatrix} 0.7071 & 0.7071 \\ -0.7071 & 0.7071 \end{bmatrix}$$

$$= \begin{bmatrix} 1.414 & 1.414 \\ -1.414 & 1.414 \end{bmatrix}$$

Now, applying this concatenated transformation matrix to the original point, we have

$$(x'', y'') = (3, 1) \begin{bmatrix} 1.414 & 1.414 \\ -1.414 & 1.414 \end{bmatrix}$$

$$= (2.828, 5.657)$$

6.6 DATA BASE STRUCTURE AND CONTENT

In Section 6.2 the application data base was identified as one of the three modules in the graphics software. Nearly all of the functions of a CAD system depend on its data base. The computer-aided design data base contains the application models, designs, drawings, assemblies, and alphanumeric information such as bills of materials and text. The data base also includes much of the interactive graphics software, such as system commands, function menus, and plotter output routines. The data base resides in computer memory (primary storage) and secondary storage. Since particular parts of the data base can be exchanged readily between primary and secondary storage as needed, we shall not be concerned in this discussion with how the data base is physically stored.

Our principal concern is with the contents of the data base and with its organization. Foley and Van Dam [4] define the basic ingredients of the application model which must be carried in the data base. The following model structure is patterned after their suggested data base organization:

1. Basic graphic elements (points and other elements)
2. Geometry (shape) of the model components and their layout in space
3. Topology or structure of the models—how the various components are connected to form the model
4. Application-specific data, such as material properties
5. Application-specific analysis programs, such as finite-element analysis programs

The list represents a building-block approach to model formulation, with items in category 1 being the most elementary ingredients. They are combined to form the components in category 2, which in turn are used to construct category 3, and so forth. The model structure consists of both data and procedures to connect, describe, and analyze the model.

The model data base can be organized in various ways. This depends on the type of model (mechanical, electrical, etc.) and the preference of the CAD system designer. Some systems lean toward more complete descriptions of the model stored explicitly as data. This requires more storage space. Other systems are designed to store a minimum of data but with more complete procedures so that the model can be recomputed when needed. This saves on storage space, but requires more computation time.

One possible data structure involves storing the coordinates of the geometry, together with other information which might be required to more completely define the model or to use certain application programs (e.g., engineering analysis or numerical control part programming). There are disadvantages to this type of data structure. For example, consider the definition of a cylinder. The definition might consist of a line segment, parallel to the y axis, and rotated about that axis to form the cylinder. The data record for this component would consist of the points for the line segment and an axis of rotation. The computer can generate an image of the cylinder for display purposes, but it does not have all of the solid cylinder data elements stored in the record. It would be difficult for the computer system to determine that the cylinder was a solid object. The characteristic of being solid might be important in a subsequent analysis for interference checking when the cylinder was assembled along with other solid components.

Alternative forms for the data base include the graph-based model. The graph-based model, illustrated in Figure 6.6 for a tetrahedron, is composed of a series of points and lines which establish relationships among the points, edges, and surfaces of the geometric element. Only the points (vertices) are contained as spatial data in the record. However, the relationships that connect edges to vertices, faces to edges, and the solid to faces are also recorded. This turns out to be a compact way to define a solid.

Boolean operations can be used to construct the geometric model. The process is illustrated in Figure 6.7 and is sometimes referred to by the term Boolean modeling. The solid model in part (a) of the figure is formed by the intersection of

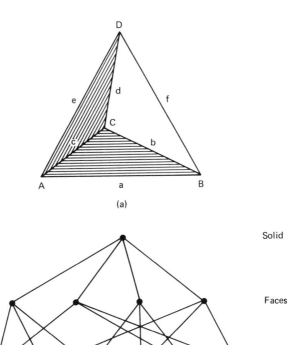

(a)

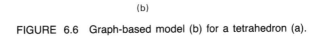

(b)

FIGURE 6.6 Graph-based model (b) for a tetrahedron (a).

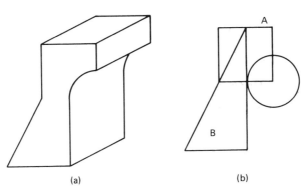

(a)   (b)

FIGURE 6.7 Boolean operation $\overline{C}(A + B)$ performed on elements in (b) to form solid in (a).

121

the compliment of the cylinder C with the union of rectangular solid A and triangular prism B. Putting this more concisely, the object in (a) is equal to

$$\overline{C}(A + B)$$

Part (b) of the figure shows the three elements A, B, and C in cross-sectional view. A more complex geometric model, created by similar Boolean operations, is shown in Figure 6.8.

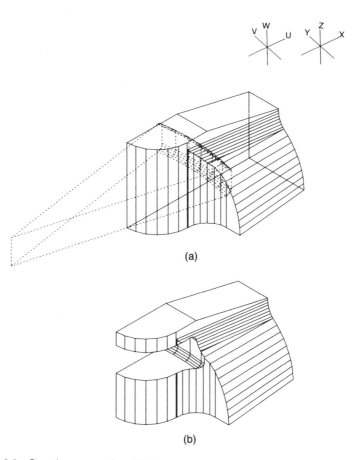

(a)

(b)

FIGURE 6.8   Complex geometric solid shape showing removal of material in (a) to form object in (b).(From W. Fitzgerald, F. Gracer, and R. Wolfe, "GRIN: Interactive Graphics for Modeling Solids," *IBM Journal of Research and Development,* Vol. 25, No. 4, July, 1981. Copyright 1981 by International Business Machines Corporation; reprinted with permission.)

6.7   WIRE-FRAME VERSUS SOLID MODELING

The importance of three-dimensional geometry

Early CAD systems were basically automated drafting board systems which displayed a two-dimensional representation of the object being designed. Operators (e.g., the designer or drafter) could use these graphics systems to develop the line drawing the way they wanted it and then obtain a very high quality paper plot of the drawing. By using these systems, the drafting process could be accomplished in less time, and the productivity of the designers could be improved.

However, there was a fundamental shortcoming of these early systems. Although they were able to reproduce high-quality engineering drawings efficiently and quickly, these systems stored in their data files a two-dimensional record of the drawings. The drawings were usually of three-dimensional objects and it was left to the human beings who read these drawings to interpret the three-dimensional shape from the two-dimensional representation. The early CAD systems were not capable of interpreting the three-dimensionality of the object. It was left to the user of the system to make certain that the two-dimensional representation was correct (e.g., hidden lines removed or dashed, etc.), as stored in the data files.

More recent computer-aided design systems possess the capability to define objects in three dimensions. This is a powerful feature because it allows the designer to develop a full three-dimensional model of an object in the computer rather than a two-dimensional illustration. The computer can then generate the orthogonal views, perspective drawings, and close-ups of details in the object.

The importance of this three-dimensional capability in interactive computer graphics should not be underestimated. To illustrate the limitations of a two-dimensional line drawing, consider Figure 6.9. It could represent any of a number of possible geometric shapes. Even a three-dimensional perspective drawing, as illustrated in Figure 6.10, does not always uniquely define the solid shape of the object. It is important that the graphics system work with three-dimensional shapes in developing the model of an object.

Wire-Frame models

Most current-day graphics systems use a form of modeling called wire-frame modeling. In the construction of the wire-frame model, the edges of the objects are shown as lines. Figure 4.4 illustrated this form of representation. For objects in which there are curved surfaces, contour lines can be added, as shown in Figure 6.11, to indicate the contour. The image assumes the appearance of a frame constructed out of wire—hence the name ''wire-frame'' model.

There are limitations to the models which use the wire-frame approach to form the image. These limitations are, of course, especially pronounced in the case

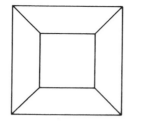

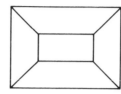
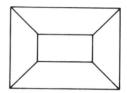

FIGURE 6.9 Orthographic views of three-dimensional object without hidden-line removal.

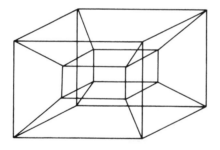

FIGURE 6.10 Perspective view of three-dimensional object of Figure 6.9 without hidden line removal.

of three-dimensional objects. In many cases, wire-frame models are quite adequate for two-dimensional representation. The most conspicuous limitation is that all of the lines that define the edges (and contoured surfaces) of the model are shown in the image. Many three-dimensional wire-frame systems in use today do not possess an automatic hidden-line removal feature. Consequently, the lines that indicate the edges at the rear of the model show right through the foreground surfaces. This can cause the image to be somewhat confusing to the viewer, and in some cases the image might be interpretable in several different ways (e.g., Figure 6.10). This interpretation problem can be alleviated to some extent through human intervention in removing the hidden background lines in the image.

There are also limitations with the wire-frame models in the way many CAD systems define the model in their data bases. For example, there might be ambiguity in the case of a surface definition as to which side of the surface is solid. This

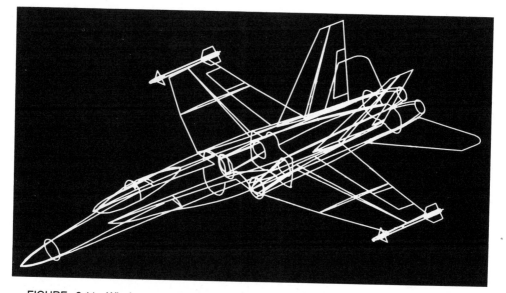

FIGURE 6.11 Wireframe model of F/A-18 fighter aircraft, showing primary control curves. (Photo courtesy of McDonnell Douglas Automation Company.)

type of limitation prevents the computer system from achieving a comprehensive and unambiguous definition of the object.

Solid models

An improvement over wire-frame models, both in terms of realism to the user and definition to the computer, is the solid modeling approach. In this approach, the models are displayed as solid objects to the viewer, with very little risk of misinterpretation. When color is added to the image, the resulting picture becomes strikingly realistic. Two examples of three-dimensional color solid modeling were shown in Figures 5.9 and 5.10. It is anticipated that graphics systems with this capability will find a wide range of applications outside computer-aided design and manufacturing. These applications will include color illustrations in magazines and technical publications, animation in movie films, and training simulators (e.g., aircraft pilot training).

There are two factors which promote future widespread use of solid modelers (i.e., graphics systems with the capability for solid modeling). The first is the increasing awareness among users of the limitations of wire-frame systems. As powerful as today's wire-frame-based CAD systems have become, solid model systems represent a dramatic improvement in graphics technology. The second reason is the continuing development of computer hardware and software which make solid modeling possible. Solid modelers require a great deal of computational power, in terms of both speed and memory, in order to operate. The advent of powerful, low-cost minicomputers has supplied the needed capacity to meet this requirement. Developments in software will provide application programs which take advantage of the opportunities offered by solid modelers. Among the possibilities are more highly automated model building and design systems, more complete three-dimensional engineering analysis of the models, including interference checking, automated manufacturing planning, and more realistic production simulation models.

Two basic approaches to the problem of solid modeling have been developed:

1. Constructive solid geometry (CSG or C-rep), also called the building-block approach
2. Boundary representation (B-rep)

The CSG systems allow the user to build the model out of solid graphic primitives, such as rectangular blocks, cubes, spheres, cylinders, and pyramids. This building-block approach is similar to the methods described in Section 6.4 except that a solid three-dimensional representation of the object is produced. The most common method of structuring the solid model in the graphics data base is to use Boolean operations, described in the preceding section and pictured in Figure 6.7.

The boundary representation approach requires the user to draw the outline or boundary of the object on the CRT screen using an electronic tablet and pen or analogous procedure. The user would sketch the various views of the object (front, side, and top, more views if needed), drawing interconnecting lines among the views to establish their relationship. Various transformations and other specialized editing procedures are used to refine the model to the desired shape. The general scheme is illustrated in Figure 6.12.

The two approaches have their relative advantages and disadvantages. The C-rep systems usually have a significant procedural advantage in the initial formulation of the model. It is relatively easy to construct a precise solid model out of regular solid primitives by adding, subtracting, and intersecting the components. The building-block approach also results in a more compact file of the model in the data base.

On the other hand, B-rep systems have their relative advantages. One of them becomes evident when unusual shapes are encountered that would not be included within the available repertoire of the CSG systems. This kind of situation is exemplified by aircraft fuselage and wing shapes and by automobile body styling. Such shapes would be quite difficult to develop with the building-block approach, but the boundary representation method is very feasible for this sort of problem.

Another point of comparison between the two approaches is the difference in the way the model is stored in the data base for the two systems. The CSG approach stores the model by a combination of data and logical procedures (the

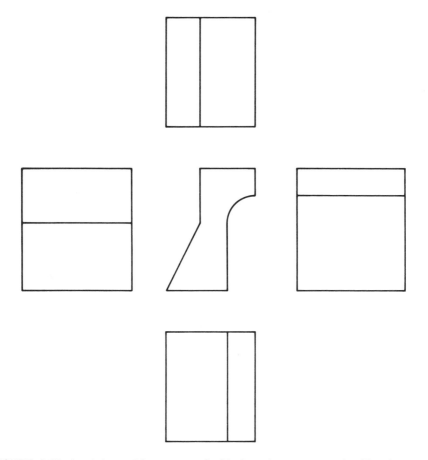

FIGURE 6.12   Input views of the types required for boundary representation (B-rep) approach for solid modeling of object from Figure 6.7.

Boolean model). This generally requires less storage but more computation to reproduce the model and its image. By contrast, the B-rep system stores an explicit definition of the model boundaries. This requires more storage space but does not necessitate nearly the same computation effort to reconstruct the image. A related benefit of the B-rep systems is that it is relatively simple to convert back and forth between a boundary representation and a corresponding wire-frame model. The reason is that the model's boundary definition is similar to the wire-frame definition, which facilitates conversion of one form to the other. This makes the newer solid B-rep systems compatible with existing CAD systems out in the field.

Because of the relative benefits and weaknesses of the two approaches, hybrid systems have been developed which combine the CSG and B-rep approaches. With these systems, users have the capability to construct the geometric model by either approach, whichever is more appropriate to the particular problem.

6.8   OTHER CAD FEATURES AND CAD/CAM INTEGRATION

Most computer-aided design systems currently available offer extensive capabilities for developing engineering drawings. These capabilities include:

1. *Automatic crosshatching of surfaces* in drawing from wire-frame models.

2. *Capability to write text on the drawings.*   Control can typically be exercised over such parameters as:

    Size and style of the lettering

    Horizontal and vertical justification of the text

3. *Semiautomatic dimensioning.*   The dimensions can be obtained from the data base. Dimensioning features typically include:

    Linear or angular conventions, depending on which is more appropriate

    English and/or metric (International System)

    Dimensions displayed in decimal or fractional notation

    Various types of tolerance displays

4. *Automatic generation of bill of materials*   (assembly and materials listings).

All of these features are helpful in reducing the time required to complete the design and drafting process.

In most CAD systems, an interface is generally provided to various high-level programming languages such as FORTRAN. This interface allows the development of application programs for analysis purposes. It also provides the opportunity to utilize programs and routines written at other facilities. This interface allows a wide variety of powerful design analysis and manufacturing planning procedures to be applied to the product development.

Experience has demonstrated that the benefits obtained from an integrated CAD/CAM data base are far greater than those which can be realized from applying CAD and CAM as separate technologies. There is a significant overlap in the data base required for design and that which is required for manufacturing. Bridging the gap that has traditionally existed between design and manufacturing is a critical objective of CAD/CAM.

Much of the remainder of this book will be concerned with topics in computer-aided manufacturing. However, it is important to recognize that in most of the topics discussed, the degree to which computerized automation can be achieved, and the success of the application, relies heavily on the generation of the data base during computer-aided design.

REFERENCES

[1] BOYSE, J. W., AND GILCHRIST, J. E., "GM Solid: Interactive Modeling for Design and Analysis of Solids," *IEEE Computer Graphics*, March, 1982, pp. 27–40.

[2] CHASEN, S. H., *Geometric Principles and Procedures for Computer Graphic Applications*, Prentice-Hall, Inc., Englewood Cliffs, N.J., 1978.

[3] FITZGERALD, W., GRACER, F., AND WOLFE, R., "GRIN: Interactive Graphics for Modeling Solids," *IBM Journal of Research and Development*, July, 1981, pp. 281–294.

[4] FOLEY, J. D., AND VAN DAM, A., *Fundamentals of Interactive Computer Graphics*, Addison-Wesley Publishing Co., Inc., Reading, Mass., 1982.

[5] GILOI, W. K., *Interactive Computer Graphics*, Prentice-Hall, Inc., Englewood Cliffs, N.J., 1978.

[6] KINNUCAN, P., "Solid Modelers Make the Scene," *High Technology*, July/August, 1982, pp. 38–44.

[7] MYERS, W., "An Industrial Perspective on Solid Modeling," *IEEE Computer Graphics*, March, 1982, pp. 86–97.

[8] NEWMAN, W. M., AND SPROULL, R. F., *Principles of Interactive Computer Graphics*, 2nd ed., McGraw-Hill Book Company, New York, 1979.

[9] REQUICHA, A. A. G., AND VOELCKER, H. B., "Solid Modeling: A Historical Summary and Contemporary Assessment," *IEEE Computer Graphics*, March, 1982, pp. 9–24.

[10] ROGERS, D. F., AND ADAMS, J. A., *Mathematical Elements in Computer Graphics*, McGraw-Hill Book Company, New York, 1976.

[11] ZIMMERS, E. W., JR., *Computer-Aided Design Module,* General Electric CAD/CAM Seminar, Lehigh University, Bethlehem, Pa., 1982.

PROBLEMS

6.1. Figures 6.9 and 6.10 show a wire-frame drawing of a three-dimensional solid object. However, there are several possible interpretations of this drawing. Make a list of as many interpretations as you can think of.

6.2. Make a sketch of two or three solid primitives (such as a rectangular solid) and show how these various primitives can be scaled, reoriented, added, subtracted, and intersected to create the geometry of the object in Figure 6.10.

6.3. A line is defined by its end points (0, 0) and (2, 3) in a two-dimensional graphics system. Express the line in matrix notation and perform the following transformations on this line:

(a) Scale the line by a factor of 2.0.

(b) Scale the original line by a factor of 3.0 in the x direction and 2.0 in the y direction.

(c) Translate the original line by 2.0 units in the x direction and 2.0 units in the y direction.

(d) Rotate the original line by 45° about the origin.

(e) Plot the original line and each of the four transformations on a piece of graph paper.

6.4. A triangle is defined in a two-dimensional ICG system by its vertices (0, 2), (0, 3), and (1, 2). Perform the following transformations on this triangle:

(a) Translate the triangle in space by 2 units in the x direction and 5 units in the y direction.

(b) Scale the original triangle by a factor of 1.5.

(c) Scale the original triangle by a factor of 1.5 in the x direction and 3.0 in the y direction.

(d) Rotate the original triangle by 45° about the origin.

6.5. A line is defined in two-dimensional space by its end points (1, 2) and (6, 4). Express this in matrix notation and perform the following transformations in succession on this line:

(a) Rotate the line by 90° about the origin.

(b) Scale the line by a factor of 1/2.

(c) Show the sequence of transformations on a piece of graph paper.

6.6 Determine the concatenation of the transformations performed in Problem 6.5. It should be expressed in matrix notation.

6.7. A cube is defined in three-dimensional space with edges which are one unit in length. The corners of the cube are located at (0, 0, 0), (0, 0, 1), (0, 1, 0,), (0, 1, 1), (1, 0, 0), (1, 0, 1), (1, 1, 0), (1, 1, 1). Determine the locations of the corners if the cube is first translated by 2.0 units in the x direction and then scaled by a factor of 3.0.

6.8. Using the same cube from Problem 6.7, carry out the two transformations indicated in that problem in the reverse order. Are the resulting cube and its position the same as in Problem 6.7?

6.9. Rotate the cube from Problem 6.7 about the y axis by an angle of 45°.

6.10. A line in two-dimensional space has end points defined by (1, 1) and (1, 3). It is desired to move this line by a series of transformations so that its end points will be at (0, 1) and (0, 5).

(a) Describe the sequence of transformations required to accomplish the movement of the line as specified.

(b) For each transformation in the sequence, write the transformation matrix.

6.11. A point in two dimensions is located at (3, 4). It is desired to relocate the point by means of rotation and scaling transformations only (no translation) to a new position defined by (0, 8).

(a) Describe the sequence of transformations required to accomplish the movement of the line as specified.

(b) Write the transformation matrix for each step in the sequence.

(c) Write the concatenated transformation matrix for the sequence.

6.12. In order to perform the translation transformation in two dimensions, the transformation must be represented by a 3×3 matrix. In order to be compatible, the representation of a point would be given as $(x, y, 1)$. This $n + 1$ component vector for n-

dimensional space is called a homogeneous coordinate representation. Thus, for translating a point, the homogeneous representation would be

$$(x', y', 1) = (x, y, 1) \begin{bmatrix} 1 & 0 & 0 \\ 0 & 1 & 0 \\ n & m & 1 \end{bmatrix}$$

where n and m in the 3×3 translation matrix represent the displacement in the x and y directions, respectively.

(a) Use the equation above to translate the point at $(1, 1)$ in the x direction by 2 units and in the y direction by 5 units.

(b) Develop the analogous equation for the translation of a line, and translate the line whose matrix is

$$\begin{bmatrix} 1 & 1 \\ 2 & 3 \end{bmatrix}$$

by 3 units in the x direction.

PART III

**Numerical Control,
The Beginnings of CAM**

chapter 7

Conventional Numerical Control

7.1 INTRODUCTION

Many of the achievements in computer-aided design and manufacturing have a common origin in numerical control (abbreviated NC). The conceptual framework established during the development of numerical control is still undergoing further refinement and enhancement in today's CAD/CAM technology. It is appropriate that we devote a major part of this book to the subject of NC.

This chapter defines the basic concepts and applications of conventional numerical control. Modern NC systems rely heavily on computer technology. Chapter 8 deals with the topic of NC part programming, a procedure that depends greatly on computer-aided methods. Chapter 9 is concerned with computer control in NC systems.

Numerical control defined

Numerical control can be defined as a form of programmable automation in which the process is controlled by numbers, letters, and symbols. In NC, the numbers form a program of instructions designed for a particular workpart or job. When the job changes, the program of instructions is changed. This capability to change the program for each new job is what gives NC its flexibility. It is much easier to write new programs than to make major changes in the production equipment.

NC technology has been applied to a wide variety of operations, including drafting, assembly, inspection, sheet metal presswork, and spot welding. However, numerical control finds its principal applications in metal machining processes. The machined workparts are designed in various sizes and shapes, and most machined parts produced in industry today are made in small to medium-size batches. To produce each part, a sequence of drilling operations may be required, or a series of turning or milling operations. The suitability of NC for these kinds of jobs is the reason for the tremendous growth of numerical control in the metalworking industry over the last 25 years.

Historical background

Conventional NC is based largely on the pioneering work of a man named John T. Parsons. In the late 1940s, Parsons conceived a method of using punched cards containing coordinate position data to control a machine tool. The machine was directed to move in small increments, thus generating the desired surface of an airfoil. In 1948, Parsons demonstrated his concept to the U.S. Air Force, which subsequently sponsored a series of research projects at the Servomechanisms Laboratory of the Massachusetts Institute of Technology.

The initial work at MIT involved the development of a prototype NC milling machine, by retrofitting a conventional tracer mill with position servomechanisms for the three machine tool axes. The first demonstration of the NC prototype was held in 1952. By 1953, the potential usefulness of the NC concept had been proven.

Shortly thereafter, the machine tool builders began initiating their own development projects to introduce commercial NC units. Also, certain user industries, especially airframe builders, worked to devise numerical control machines to satisfy their own particular needs. The Air Force continued its encouragement of NC development by sponsoring additional research at MIT to design a part programming language that could be used for controlling the NC machines. This work resulted in the APT language, which stands for Automatically Programmed Tools. The objective of the APT research was to provide a means by which the part programmer could communicate the machining instructions to the machine tool in simple English-like statements. Although the APT language has been criticized as being too large for many computers, it nevertheless represents a major accomplishment. APT is still widely used in industry today, and most other modern part programming languages are based on APT concepts.

7.2   BASIC COMPONENTS OF AN NC SYSTEM

An operational numerical control system consists of the following three basic components:

1.   Program of instructions
2.   Controller unit, also called a machine control unit (MCU)
3.   Machine tool or other controlled process

The general relationship among the three components is illustrated in Figure 7.1. The program of instructions serves as the input to the controller unit, which in turn commands the machine tool or other process to be controlled. We will discuss the three components in the sections below.

Program of instructions

The program of instructions is the detailed step-by-step set of directions which tell the machine tool what to do. It is coded in numerical or symbolic form on some type of input medium that can be interpreted by the controller unit. The most common input medium today is 1-in.-wide punched tape. Over the years, other forms of input media have been used, including punched cards, magnetic tape, and even 35-mm motion picture film.

There are two other methods of input to the NC system which should be mentioned. The first is by manual entry of instructional data to the controller unit. This method is called manual data input, abbreviated MDI, and is appropriate only for relatively simple jobs where the order will not be repeated. We will discuss MDI more thoroughly in Chapter 8. The second other method of input is by means

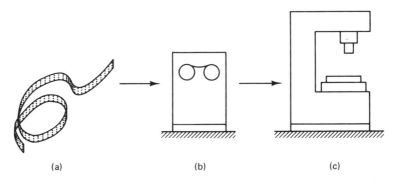

FIGURE 7.1   Three basic components of a numerical control system: (a) program of instruction; (b) controller unit; (c) machine tool.

of a direct link with a computer. This is called *direct numerical control*, or DNC, and we discuss this in Chapter 9.

The program of instructions is prepared by someone called a part programmer. The programmer's job is to provide a set of detailed instructions by which the sequence of processing steps is to be performed. For a machining operation, the processing steps involve the relative movement between the cutting tool and the workpiece.

Controller unit

The second basic component of the NC system is the controller unit. This consists of the electronics and hardware that read and interpret the program of instructions and convert it into mechanical actions of the machine tool. The typical elements of a conventional NC controller unit include the tape reader, a data buffer, signal output channels to the machine tool, feedback channels from the machine tool, and the sequence controls to coordinate the overall operation of the foregoing elements. It should be noted that nearly all modern NC systems today are sold with a microcomputer as the controller unit. This type of NC is called computer numerical control (CNC). We discuss CNC in Chapter 9.

The tape reader is an electromechanical device for winding and reading the punched tape containing the program of instructions. The data contained on the tape are read into the data buffer. The purpose of this device is to store the input instructions in logical blocks of information. A block of information usually represents one complete step in the sequence of processing elements. For example, one block may be the data required to move the machine table to a certain position and drill a hole at that location.

The signal output channels are connected to the servomotors and other controls in the machine tool. Through these channels, the instructions are sent to the machine tool from the controller unit. To make certain that the instructions have been properly executed by the machine, feedback data are sent back to the controller via the feedback channels. The most important function of this return loop is to assure that the table and workpart have been properly located with respect to the tool.

Sequence controls coordinate the activities of the other elements of the controller unit. The tape reader is actuated to read data into the buffer from the tape, signals are sent to and from the machine tool, and so on. These types of operations must be synchronized and this is the function of the sequence controls.

Another element of the NC system, which may be physically part of the controller unit or part of the machine tool, is the control panel. The control panel or control console contains the dials and switches by which the machine operator runs the NC system. It may also contain data displays to provide information to the operator. Although the NC system is an automatic system, the human operator is still needed to turn the machine on and off, to change tools (some NC systems have automatic tool changers), to load and unload the machine, and to perform various other duties. To be able to discharge these duties, the operator must be able to control the system, and this is done through the control panel.

Machine tool or other controlled process

The third basic component of an NC system is the machine tool or other controlled process. It is the part of the NC system which performs useful work. In the most common example of an NC system, one designed to perform machining operations, the machine tool consists of the worktable and spindle as well as the motors and controls necessary to drive them. It also includes the cutting tools, work fixtures, and other auxiliary equipment needed in the machining operation. Figure 7.2 illustrates an NC machine tool.

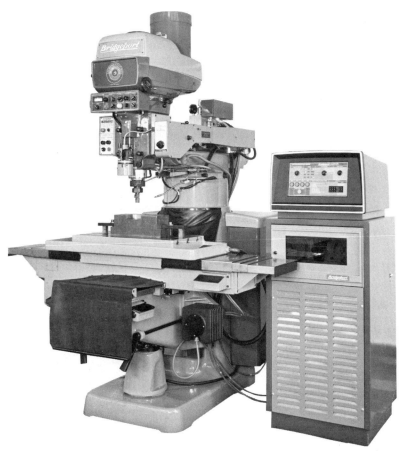

FIGURE 7.2 NC system showing machine tool and controller. (Courtesy of Bridgeport Machines Division of Textron Inc.)

NC machines range in complexity from simple tape-controlled drill presses to highly sophisticated and versatile machining centers. The NC machining center was first introduced in the late 1950s. It is a multifunction machine which incorporates several timesaving features into a single piece of automated production equipment. First, a machining center is capable of performing a variety of different operations: drilling, tapping, reaming, milling, and boring. Second, it has the capacity to change tools automatically under tape command. A variety of machining operations means that a variety of cutting tools are required. The tools are kept in a tool drum or other holding device. When the tape calls a particular tool, the drum rotates to position the tool for insertion into the spindle. The automatic tool changer then grasps the tool and places it into the spindle chuck. A third capability of the NC machining center is workpiece positioning. The machine table can orient the job so that it can be machined on several surfaces, as required. Finally, a fourth feature possessed by some machining centers is the presence of two tables or pallets on which the workpiece can be fixtured. While the machining sequence is being performed on one workpart, the operator can be unloading the previously completed piece, and loading the next one. This improves machine tool utilization because the machine does not have to stand idle during loading and unloading of the workparts. An example of an NC machining center is shown in Figure 7.3.

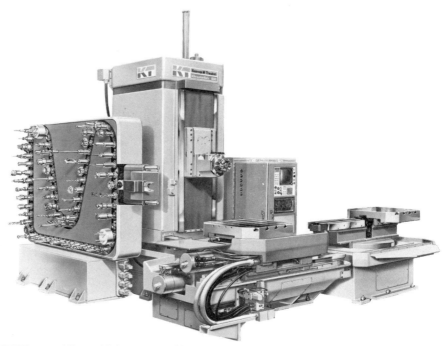

FIGURE 7.3 NC machining center with computer control. (Courtesy of Kearney & Trecker Corp.)

7.3 THE NC PROCEDURE

To utilize numerical control in manufacturing, the following steps must be accomplished.

1. *Process planning.* The engineering drawing of the workpart must be interpreted in terms of the manufacturing processes to be used. This step is referred to as process planning and it is concerned with the preparation of a route sheet. The route sheet is a listing of the sequence of operations which must be performed on the workpart. It is called a route sheet because it also lists the machines through which the part must be routed in order to accomplish the sequence of operations. We assume that some of the operations will be performed on one or more NC machines.

2. *Part programming.* A part programmer plans the process for the portions of the job to be accomplished by NC. Part programmers are knowledgeable about the machining process and they have been trained to program for numerical control. They are responsible for planning the sequence of machining steps to be performed by NC and to document these in a special format. There are two ways to program for NC:

Manual part programming

Computer-assisted part programming

In manual part programming, the machining instructions are prepared on a form called a part program manuscript. The manuscript is a listing of the relative cutter/workpiece positions which must be followed to machine the part. In computer-assisted part programming, much of the tedious computational work required in manual part programming is transferred to the computer. This is especially appropriate for complex workpiece geometries and jobs with many machining steps. Use of the computer in these situations results in significant savings in part programming time.

3. *Tape preparation.* A punched tape is prepared from the part programmer's NC process plan. In manual part programming, the punched tape is prepared directly from the part program manuscript on a typewriterlike device equipped with tape punching capability. In computer-assisted part programming, the computer interprets the list of part programming instructions, performs the necessary calculations to convert this into a detailed set of machine tool motion commands, and then controls a tape punch device to prepare the tape for the specific NC machine.

4. *Tape verification.* After the punched tape has been prepared, a method is usually provided for checking the accuracy of the tape. Sometimes the tape is checked by running it through a computer program which plots the various tool movements (or table movements) on paper. In this way, major errors in the tape can be discovered. The "acid test" of the tape involves trying it out on the

machine tool to make the part. A foam or plastic material is sometimes used for this tryout. Programming errors are not uncommon, and it may require about three attempts before the tape is correct and ready to use.

5. *Production.* The final step in the NC procedure is to use the NC tape in production. This involves ordering the raw workparts, specifying and preparing the tooling and any special fixturing that may be required, and setting up the NC machine tool for the job. The machine tool operator's function during production is to load the raw workpart in the machine and establish the starting position of the cutting tool relative to the workpiece. The NC system then takes over and machines the part according to the instructions on tape. When the part is completed, the operator removes it from the machine and loads the next part.

7.4 NC COORDINATE SYSTEMS

In order for the part programmer to plan the sequence of positions and movements of the cutting tool relative to the workpiece, it is necessary to establish a standard axis system by which the relative positions can be specified. Using an NC drill press as an example, the drill spindle is in a fixed vertical position, and the table is moved and controlled relative to the spindle. However, to make things easier for the programmer, we adopt the viewpoint that the workpiece is stationary while the drill bit is moved relative to it. Accordingly, the coordinate system of axes is established with respect to the machine table.

Two axes, x and y, are defined in the plane of the table, as shown in Figure 7.4. The z axis is perpendicular to this plane and movement in the z direction is controlled by the vertical motion of the spindle. The positive and negative directions of motion of tool relative to table along these axes are as shown in Figure 7.4. NC drill presses are classified as either two-axis or three-axis machines, depending on whether or not they have the capability to control the z axis.

A numerical control milling machine and similar machine tools (boring mill, for example) use an axis system similar to that of the drill press. However, in addition to the three linear axes, these machines may possess the capacity to con-

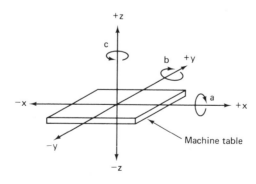

FIGURE 7.4 NC machine tool axis system for milling and drilling operations.

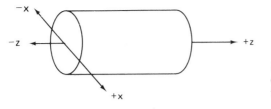

FIGURE 7.5 NC machine tool axis system for turning operation.

trol one or more rotational axes. Three rotational axes are defined in NC: the a, b, and c axes. These axes specify angles about the x, y, and z axes, respectively. To distinguish positive from negative angular motions, the "right-hand rule" can be used. Using the right hand with the thumb pointing in the positive linear axis direction (x, y, or z), the fingers of the hand are curled to point in the positive rotational direction.

For turning operations, two axes are normally all that are required to command the movement of the tool relative to the rotating workpiece. The z axis is the axis of rotation of the workpart, and x axis defines the radial location of the cutting tool. This arrangement is illustrated in Figure 7.5.

The purpose of the coordinate system is to provide a means of locating the tool in relation to the workpiece. Depending on the NC machine, the part programmer may have several different options available for specifying this location.

Fixed zero and floating zero

The programmer must determine the position of the tool relative to the origin (zero point) of the coordinate system. NC machines have either of two methods for specifying the zero point. The first possibility is for the machine to have a *fixed zero*. In this case, the origin is always located at the same position on the machine table. Usually, that position is the southwest corner (lower left-hand corner) of the table and all tool locations will be defined by positive x and y coordinates.

The second and more common feature on modern NC machines allows the machine operator to set the zero point at any position on the machine table. This feature is called *floating zero*. The part programmer is the one who decides where the zero point should be located. The decision is based on part programming convenience. For example, the workpart may be symmetrical and the zero point should be established at the center of symmetry. The location of the zero point is communicated to the machine operator. At the beginning of the job, the operator moves the tool under manual control to some "target point" on the table. The target point is some convenient place on the workpiece or table for the operator to position the tool. For example, it might be a predrilled hole in the workpiece. The target point has been referenced to the zero point by the part programmer. In fact, the programmer may have selected the target point as the zero point for tool positioning. When the tool has been positioned at the target point, the machine opera-

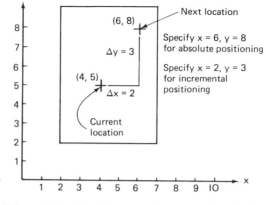

FIGURE 7.6   Absolute versus incremental positioning.

tor presses a "zero" button on the machine tool console, which tells the machine where the origin is located for subsequent tool movements.

Absolute positioning and incremental positioning

Another option sometimes available to the part programmer is to use either an absolute system of tool positioning or an incremental system. *Absolute positioning* means that the tool locations are always defined in relation to the zero point. If a hole is to be drilled at a spot that is 8 in. above the x axis and 6 in. to the right of the y axis, the coordinate location of the hole would be specified as $x = +6.000$ and $y = +8.000$. By contrast, *incremental positioning* means that the next tool location must be defined with reference to the previous tool location. If in our drilling example, suppose that the previous hole had been drilled at an absolute position of $x = +4.000$ and $y = +5.000$. Accordingly, the incremental position instructions would be specified as $x = +2.000$ and $y = +3.000$ in order to move the drill to the desired spot. Figure 7.6 illustrates the difference between absolute and incremental positioning.

7.5   NC MOTION CONTROL SYSTEMS

In order to accomplish the machining process, the cutting tool and workpiece must be moved relative to each other. In NC, there are three basic types of motion control systems:

1.   Point-to-point
2.   Straight cut
3.   Contouring

Point-to-point systems represent the lowest level of motion control between the tool and workpiece. Contouring represents the highest level of control.

Point-to-point NC

Point-to-point (PTP) is also sometimes called a positioning system. In PTP, the objective of the machine tool control system is to move the cutting tool to a predefined location. The speed or path by which this movement is accomplished is not important in point-to-point NC. Once the tool reaches the desired location, the machining operation is performed at that position.

NC drill presses are a good example of PTP systems. The spindle must first be positioned at a particular location on the workpiece. This is done under PTP control. Then the drilling of the hole is performed at the location, and so forth. Since no cutting is performed between holes, there is no need for controlling the relative motion of the tool and workpiece between hole locations. Figure 7.7 illustrates the point-to-point type of control.

Positioning systems are the simplest machine tool control systems and are therefore the least expensive of the three types. However, for certain processes, such as drilling operations and spot welding, PTP is perfectly suited to the task and any higher level of control would be unnecessary.

Straight-cut NC

Straight-cut control systems are capable of moving the cutting tool parallel to one of the major axes at a controlled rate suitable for machining. It is therefore appropriate for performing milling operations to fabricate workpieces of rectangular configurations. With this type of NC system it is not possible to combine movements in more than a single axis direction. Therefore, angular cuts on the workpiece would not be possible. An example of a straight-cut operation is shown in Figure 7.8. An NC machine capable of straight cut movements is also capable of PTP movements.

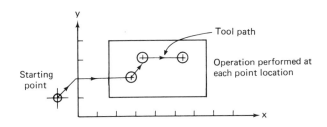

FIGURE 7.7   Point-to-point (positioning) NC system.

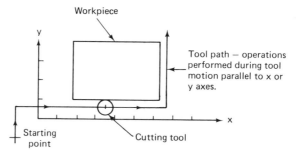

FIGURE 7.8 Straight-cut system.

Contouring NC

Contouring is the most complex, the most flexible, and the most expensive type of machine tool control. It is capable of performing both PTP and straight-cut operations. In addition, the distinguishing feature of contouring NC systems is their capacity for simultaneous control of more than one axis movement of the machine tool. The path of the cutter is continuously controlled to generate the desired geometry of the workpiece. For this reason, contouring systems are also called continuous-path NC systems. Straight or plane surfaces at any orientation, circular paths, conical shapes, or most any other mathematically definable form are possible under contouring control. Figure 7.9 illustrates the versatility of continuous-path NC. Milling and turning operations are common examples of the use of contouring control.

In order to machine a curved path in a numerical control contouring system, the direction of the feed rate must continuously be changed so as to follow the path. This is accomplished by breaking the curved path into very short straight-line segments that approximate the curve. Then the tool is commanded to machine each segment in succession. What results is a machined outline that closely approaches

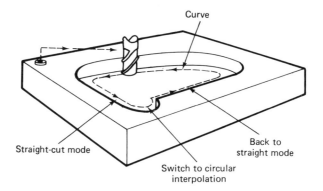

FIGURE 7.9 Contouring (continuous path) NC system for two-dimensional operations. (Reprinted from Olesten [9].)

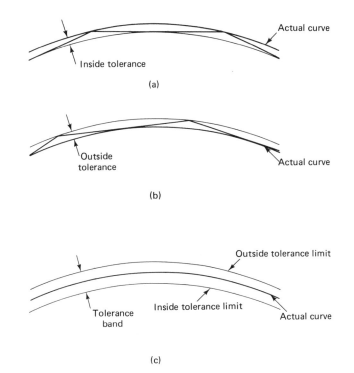

FIGURE 7.10 Approximation of a curved path in NC by a series of straight-line segments. The accuracy of the approximation is controlled by the "tolerance" between the actual curve and the maximum deviation of the straight-line segments. In (a), the tolerance is defined on the inside of the curve. It is also possible to define the tolerance on the outside of the curve, as in (b). Finally, the tolerance can be specified on both inside and outside, as shown in (c).

the desired shape. The maximum error between the two can be controlled by the length of the individual line segments, as illustrated in Figure 7.10.

7.6 APPLICATIONS OF NUMERICAL CONTROL

Numerical control systems are widely used in industry today, especially in the metalworking industry. By far the most common application of NC is for metal cutting machine tools. Within this category, numerically controlled equipment has been built to perform virtually the entire range of material removal processes, including:

Milling
Drilling and related processes
Boring

Turning
Grinding
Sawing

Within the machining category, NC machine tools are appropriate for certain jobs and inappropriate for others. Following are the general characteristics of production jobs in metal machining for which numerical control would be most appropriate:

1. Parts are processed frequently and in small lot sizes.
2. The part geometry is complex.
3. Many operations must be performed on the part in its processing.
4. Much metal needs to be removed.
5. Engineering design changes are likely.
6. Close tolerances must be held on the workpart.
7. It is an expensive part where mistakes in processing would be costly.
8. The parts require 100% inspection.

It has been estimated that most manufactured parts are produced in lot sizes of 50 or fewer. Small-lot and batch production jobs represent the ideal situations for the application of NC. This is made possible by the capability to program the NC machine and to save that program for subsequent use in future orders. If the NC programs are long and complicated (complex part geometry, many operations, much metal removed), this makes NC all the more appropriate when compared to manual methods of production. If engineering design changes or shifts in the production schedule are likely, the use of tape control provides the flexibility needed to adapt to these changes. Finally, if quality and inspection are important issues (close tolerances, high part cost, 100% inspection required), NC would be most suitable, owing to its high accuracy and repeatability.

In order to justify that a job be processed by numerical control methods, it is not necessary that the job possess every one of these attributes. However, the more of these characteristics that are present, the more likely it is that the part is a good candidate for NC.

Figures 7.11 and 7.12 illustrate several NC machine tools that perform machining operations.

In addition to metal machining, numerical control has been applied to a variety of other operations. The following, although not a complete list, will give the reader an idea of the wide range of potential applications of NC:

Pressworking machine tools
Welding machines
Inspection machines

FIGURE 7.11 Four-axis NC chucking center for turning operations. (Courtesy of Cincinnati Milacron.)

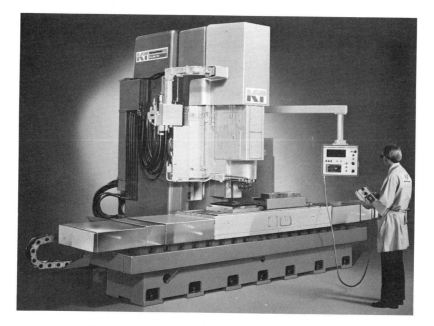

FIGURE 7.12 NC milling machine. (Courtesy of Kearney & Trecker Corp.)

Automatic drafting

Assembly machines

Tube bending

Flame cutting

Plasma arc cutting

Laser beam processes

Automated knitting machines

Cloth cutting

Automatic riveting

Wire-wrap machines

Figures 7.13 and 7.14 illustrate two of the NC machines that perform these non-machining applications.

7.7 ECONOMICS OF NUMERICAL CONTROL

The great variety of numerical control applications were explored in the preceding section. We also examined the general characteristics of production jobs for which NC seems to be particularly well suited. When properly applied, numerical control provides the user with a significant number of economic advantages. In this section we present the advantages and disadvantages of NC compared to conventional manual methods of production.

Advantages of NC

Following are the advantages of numerical control when it is utilized in the types of production jobs described in Section 7.6:

1. *Reduced nonproductive time.* Numerical control has little or no effect on the basic metal cutting (or other manufacturing) process. However, NC can increase the proportion of time the machine is engaged in the actual process. It accomplishes this by means of fewer setups, less time in setting up, reduced workpiece handling time, automatic tool changes on some machines, and so on.

In a University of Michigan survey reported by Smith and Evans [10], a comparison was made between the machining cycle times for conventional machine tools versus the cycle times for NC machines. NC cycle times, as a percentage of their conventional counterparts, ranged from 35% for five-axis machining centers to 65% for presswork punching [10, p. 185]. The advantage for numerical control tends to increase with the more complex processes.

2. *Reduced fixturing.* NC requires fixtures which are simpler and less costly to fabricate because the positioning is done by the NC tape rather than the jig or fixture.

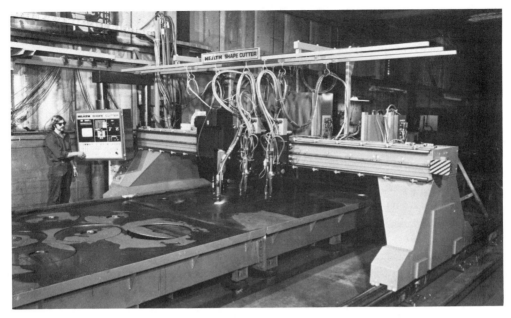

FIGURE 7.13 NC plasma arc cutting machine (Courtesy of Heath Corp.)

FIGURE 7.14 NC punch press for sheet metal operations. (Courtesy of Warner & Swasey Company, Wiedemann Division.)

3. *Reduced manufacturing lead time.* Because jobs can be set up more quickly with NC and fewer setups are generally required with NC, the lead time to deliver a job to the customer is reduced. According to the University of Michigan survey [10, p. 241], average lead-time reductions range between 26 and 44%, depending on type of machine tool.

4. *Greater manufacturing flexibility.* With numerical control it is less difficult to adapt to engineering design changes, alterations of the production schedule, changeovers in jobs for rush orders,and so on.

5. *Improved quality control.* NC is ideal for complicated workparts where the chances of human mistakes are high. Numerical control produces parts with greater accuracy, reduced scrap, and lower inspection requirements. Smith and Evans report that the average scrap decrease from NC ranges from 31 to 44% [10, p. 212], depending on type of machine tool, and that the annual cost savings from reduced inspection needs of NC are about 38% [10, p. 214].

6. *Reduced inventory.* Owing to fewer setups and shorter lead times with numerical control, the amount of inventory carried by the company is reduced.

7. *Reduced floor space requirements.* Since one NC machining center can often accomplish the production of several conventional machines, the amount of floor space required in an NC shop is usually less than in a conventional shop.

Disadvantages of NC

Along with the advantages of NC, there are several features about NC which must be considered disadvantages:

1. *Higher investment cost.* Numerical control machine tools represent a more sophisticated and complex technology. This technology costs more to buy than its non-NC counterpart. The higher cost requires manufacturing managements to use these machines more aggressively than ordinary equipment. High machine utilization is essential in order to get reasonable returns on investment. Machine shops must operate their NC machines two or three shifts per day to achieve this high machine utilization.

2. *Higher maintenance cost.* Because NC is a more complex technology and because NC machines are used harder, the maintenance problem becomes more acute. Although the reliability of NC systems has been improved over the years, maintenance costs for NC machines will generally be higher than for conventional machine tools. According to the companies responding to the University of Michigan survey, the average percentage increase in NC maintenance cost compared to non-NC ranged from 48% for milling machines to 63% for machining centers [10, p. 206].

3. *Finding and/or training NC personnel.* Certain aspects of numerical control shop operations require a higher skill level than conventional operations. Part programmers and NC maintenance personnel are two skill areas where available personnel are in short supply. The problems of finding, hiring, and training these people must be considered a disadvantage to the NC shop.

7.8 SUMMARY

In this chapter we have attempted to cover the fundamentals of numerical control technology and how NC is used in manufacturing. Numerical control continues to be a developing technology. Each year, NC machines with new features and refinements are offered to industrial customers who are eager to improve productivity. The common theme in the majority of these innovations involves the expanding use of the computer. This trend began in the late 1950s with the development of the APT language. It gained momentum in the late 1960s, when NC machines were first controlled directly by computers. Further innovations have been made in both the programming and control of NC machines during the 1970s and early 1980s as computer technology has advanced to provide us with smaller, less expensive, yet more powerful computers. In the following two chapters, we examine some of these developments in numerical control. Chapter 8 is concerned with NC part programming. We will discuss the basics of manual part programming, but most of the chapter focuses on computer–assisted part programming, with particular emphasis on the APT language. Since the introduction of APT, many innovations have been made in NC languages and NC programming. Among these are the use of interactive graphics and voice communication for part programming.

Chapter 9 deals with the control of NC machines and how computers are used to perform this control function. We trace the development of NC controls up to the current-day computerized NC systems. Included within this chapter is a discussion of direct numerical control (DNC) and adaptive control. We conclude Chapter 9 by exploring some of the new developments and trends in NC technology.

REFERENCES

[1] BEERCHECK, R. C., "Machine Tools: Cutting Edge of Technology," *Machine Design*, January 25, 1979, pp. 18–47.

[2] CHILDS, J. J., *Numerical Control Part Programming*, Industrial Press, Inc., New York, 1973.

[3] DALLAS, D. B. (Ed.), *Tool and Manufacturing Engineers Handbook*, 3rd ed., Society of Manufacturing Engineers/McGraw-Hill Book Company, New York, 1976, Chapter 12.

[4] DEVRIES, M. F., "Two Case Studies of the Investment Justification for N/C Machinery Using the MAPI Method," in *Education Module*, Manufacturing Productivity Educational Committee, Purdue Research Foundation, West Lafayette, Ind., 1977.

[5] DROY, J., "Machining Centers Mean Production Profits," *Production Engineering*, June, 1981, pp. 50–54.

[6] GROOVER, M. P., *Automation, Production Systems, and Computer-Aided Manufacturing*, Prentice-Hall, Inc., Englewood Cliffs, N.J., 1980.

[7] HEGLAND, D. E., "Numerical Control—Your Best Investment in Productivity," *Production Engineering*, March, 1981, pp. 42−47.

[8] JENKINS, L. J., GAY, J. M., MULDOON, T. F., SMITH, D., HUNT, R. C., and HARRINGTON, J., JR., "Getting More Out of NC." *American Machinist*, October, 1981, pp. 185−192.

[9] OLESTEN, N. O., *Numerical Control*, John Wiley & Sons, Inc. (Wiley-Interscience), New York, 1970.

[10] SMITH, D. N., and EVANS, L., *Management Standards for Computer and Numerical Controls*, University of Michigan, Ann Arbor, 1977.

chapter 8

NC Part Programming

8.1 INTRODUCTION

Numerical control part programming is the procedure by which the sequence of processing steps to be performed on the NC machine is planned and documented. It involves the preparation of a punched tape (or other input medium) used to transmit the processing instructions to the machine tool. As indicated in Section 7.3, there are two methods of part programming: manual part programming and computer-assisted part programming. In this chapter we describe both of these methods, with emphasis on the latter.

It is appropriate to begin the discussion of NC part programming by examining the way in which the punched tape is coded. Coding of the punched tape is concerned with the basic symbols used to communicate a complex set of instructions to the NC machine tool. In numerical control, the punched tape must be generated whether the part programming is done manually or with the assistance of some computer package. With either method of part programming, the tape is the net result of the programming effort. In Sections 8.2 and 8.3 our attention will be focused on the punched tape and the structure of the basic language used by the NC system.

8.2 THE PUNCHED TAPE IN NC

The part program is converted into a sequence of machine tool actions by means of the input medium, which contains the program, and the controller unit, which interprets the input medium. The controller unit and the input medium must be compatible. That is, the input medium uses coded symbols which represent the part program, and the controller unit must be capable of reading those symbols. The most common input medium is punched tape. The tape has been standardized so that tape punchers are manufactured to prepare the NC tapes, and tape readers (part of the controller unit) can be manufactured to read the tapes. The punched tape used for NC is 1 in. wide. It is standardized as shown in Figure 8.1 by the Electronics Industries Association (EIA), which has been responsible for many of the important standards in the NC industry.

There are two basic methods of preparing the punched tape. The first method is associated with manual part programming and involves the use of a typewriter-like device. Figure 8.2 illustrates a modern version of this kind of equipment. The operator types directly from the part programmer's handwritten list of coded instructions. This produces a typed copy of the program as well as the punched tape. The second method is used with computer-assisted part programming. By this approach, the tape is prepared directly by the computer using a device called a tape punch.

By either method of preparation, the punched tape is ready for use. During production on a conventional NC machine, the tape is fed through the tape reader

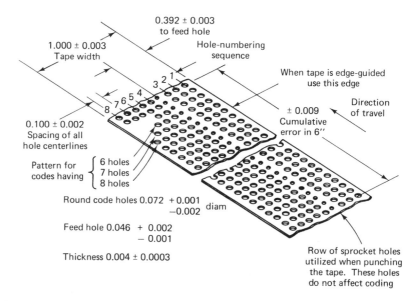

FIGURE 8.1 Numerical control punched tape format as standardized by Electronic Industries Association. (Reprinted fron Childs [2].)

FIGURE 8.2   Equipment used to prepare punched tape for NC. (Courtesy of Numeridex Corp.)

once for each workpiece. It is advanced through the tape reader one instruction at a time. While the machine tool is performing one instruction, the next instruction is being read into the controller unit's data buffer. This makes the operation of the NC system more efficient. After the last instruction has been read into the controller, the tape is rewound back to the start of the program to be ready for the next workpart.

8.3   TAPE CODING AND FORMAT

NC tape coding

As shown in Figure 8.1, there are eight regular columns of holes running in the lengthwise direction of the tape. There is also a ninth column of holes between the third and fourth regular columns. However, these are smaller and are used as sprocket holes for feeding the tape.

Figure 8.1 shows a hole present in nearly every position of the tape. However, the coding of the tape is provided by either the presence or absence of a hole in the various positions. Because there are two possible conditions for each position—either the presence or absence of a hole—this coding system is called the binary code. It uses the base 2 number system, which can represent any number in the more familiar base 10 or decimal system. The NC tape coding system is used to code not only numbers, but also alphabetical letters and other symbols. Eight columns provide more than enough binary digits to define any of the required symbols.

How instructions are formed

A binary digit is called a bit. It has a value of 0 or 1 depending on the absence or presence of a hole in a certain row and column position on the tape. (Columns of hole positions run lengthwise along the tape. Row positions run across the tape.)

Out of a row of bits, a character is made. A character is a combination of bits, which represents a letter, number, or other symbol. A word is a collection of characters used to form part of an instruction. Typical NC words are *x* position, *y* position, cutting speed, and so on. Out of a collection of words, a block is formed. A block of words is a complete NC instruction. Using an NC drilling operation as an example, a block might contain information on the *x* and *y* coordinates of the hole location, the speed and feed at which the cut should be run, and perhaps even a specification of the cutting tool.

To separate blocks, an end-of-block (EOB) symbol is used (in the EIA standard, this is a hole in column 8). The tape reader feeds the data from the tape into the buffer in blocks. That is, it reads in a complete instruction at a time.

NC words

Following is a list of the different types of words in the formation of a block. Not every NC machine uses all the words. Also, the manner in which the words are expressed will differ between machines. By convention, the words in a block are given in the following order:

SEQUENCE NUMBER (n-words):   This is used to identify the block.

PREPARATORY WORD (g-words):   This word is used to prepare the controller for instructions that are to follow. For example, the word g02 is used to prepare the NC controller unit for circular interpolation along an arc in the clockwise direction. The preparatory word is needed so that the controller can correctly interpret the data that follow it in the block.

COORDINATES (x-, y-, and z-words):   These give the coordinate positions of the tool. In a two-axis system, only two of the words would be used. In a four- or five-axis machine, additional a-words and/or b-words would specify the angular positions.

Although different NC systems use different formats for expressing a coordinate, we will adopt the convention of expressing it in the familiar decimal form: For example, x+7.235 or y−0.500. Some formats do not use the decimal point in writing the coordinate. The + sign to define a positive coordinate location is optional. The negative sign is, of course, mandatory.

FEED RATE (f-word):   This specifies the feed in a machining operation. Units are inches per minute (ipm) by convention.

CUTTING SPEED (s-word):   This specifies the cutting speed of the process, the rate at which the spindle rotates.

TOOL SELECTION (t-word):   This word would be needed only for machines with a tool turret or automatic tool changer. The t-word specifies which tool is to be used in the operation. For example, t05 might be the designation of a ½-in. drill bit in turret position 5 on an NC turret drill.

MISCELLANEOUS FUNCTION (m-word):   The m-word is used to specify certain miscellaneous or auxiliary functions which may be available on the machine tool. Of course, the machine must possess the function that is being called. An example would be m03 to start the spindle rotation. The miscellaneous function is the last word in the block. To identify the end of the instruction, an end-of-block (EOB) symbol is punched on the tape.

8.4 MANUAL PART PROGRAMMING

To prepare a part program using the manual method, the programmer writes the machining instructions on a special form called a part programming manuscript. The instructions must be prepared in a very precise manner because the typist prepares the NC tape directly from the manuscript. Manuscripts come in various forms, depending on the machine tool and tape format to be used. For example, the manuscript form for a two-axis point-to-point drilling machine would be different than one for a three-axis contouring machine. The manuscript is a listing of the relative tool and workpiece locations. It also includes other data, such as preparatory commands, miscellaneous instructions, and speed/feed specifications, all of which are needed to operate the machine under tape control.

Manual programming jobs can be divided into two categories: point-to-point jobs and contouring jobs. Except for complex workparts with many holes to be drilled, manual programming is ideally suited for point-to-point applications. On the other hand, except for the simplest milling and turning jobs, manual programming can become quite time consuming for applications requiring continuous-path control of the tool. Accordingly, we shall be concerned only with manual part programming for point-to-point operations. Contouring is much more appropriate for computer-assisted part programming.

The basic method of manual part programming for a point-to-point application is best demonstrated by means of an example.

EXAMPLE 8.1

Suppose that the part to be programmed is a drilling job. The engineering drawing for the part is presented in Figure 8.3. Three holes are to be drilled at a diameter of $^{31}\!/_{64}$ in. The close hole size tolerance requires reaming to 0.500 in. diameter. Recommended speeds and feeds[1] are as follows:

	Speed (rpm)	Feed (in./min)
0.484-in.-diameter drill	592	3.55
0.500-in.-diameter drill	382	3.82

The NC drill press operates as follows. Drill bits are manually changed by the machine operator, but speeds and feeds must be programmed on the tape. The machine has the floating-zero feature and absolute positioning.

---

[1]Recommended cutting speeds and feeds could be obtained from machinability data handbooks.

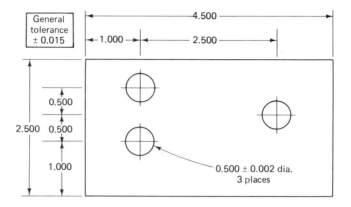

FIGURE 8.3   Part drawing for Example 8.1.

The first step in preparing the part program is to define the axis coordinates in relation to the workpart. We assume that the outline of the part has already been machined before the drilling operation. Therefore, the operator can use one of the corners of the part as the target point. Let us define the lower left-hand corner as the target point and the origin of our axis system. The coordinates are shown in Figure 8.4 for the example part. The x and y locations of each hole can be seen in the figure. The completed manuscript would appear as in Figure 8.5. The first line shows the x and y coordinates at the zero point. The machine operator would insert the tape and read this first block into the system. (A block of instruction corresponds generally to one line on the manuscript form.) The tool would then be positioned over the target point on the machine table. The operator would then press the zero buttons to set the machine.

The next line on the manuscript is RWS, which stands for rewind-stop. This signal is coded into the tape as holes in columns 1, 2, and 4. The symbol stops the tape after it has

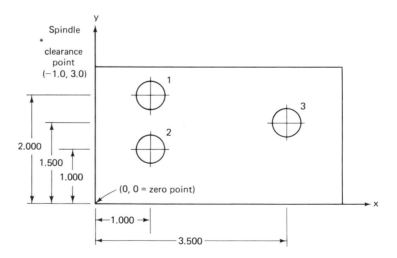

FIGURE 8.4   Coordinate system defined for part in Example 8.1.

NC part programming manuscript
Two axis point-to-point drill press
tab sequential format

Part No. _EXAMPLE 8.2_   Date ___4/4/79___

Part Name _HOLE PLATE_   Prepared by ___MPG___

Seq. No.	Tab EOB	x-COORD	Tab EOB	y-COORD	Tab EOB	Feed	Tab EOB	Speed	Tab EOB	m-WORD	Tab EOB	Comments
00	TAB	0.0	TAB	0.0	EOB							ZERO
RWS												
01	TAB	1.0	TAB	2.0	TAB	3.55	TAB	592	TAB	13	EOB	DRILL 1
02	TAB		TAB	1.0	EOB							DRILL 2
03	TAB	3.5	TAB	1.5	EOB							DRILL 3
04	TAB	-1.0	TAB	3.0	TAB		TAB		TAB	06	EOB	TOOL CHG
05	TAB	3.5	TAB	1.5	TAB	3.82	TAB	382	TAB	13	EOB	REAM 3
06	TAB	1.0	TAB	1.0	EOB							REAM 2
07	TAB		TAB	2.0	EOB							REAM 1
08	TAB	-1.0	TAB	3.0	TAB		TAB		TAB	06	EOB	TOOL CHG
09	TAB		TAB		TAB		TAB		TAB	30	EOB	REWIND & CHG PART

FIGURE 8.5   Part program manuscript for Example 8.1.

been rewound. The last line on the tape contains the m30 word, causing the tape to be rewound at the end of the machining cycle. Other m-words used in the program are m06, which stops the machine for an operator tool change, and m13, which turns on the spindle and coolant. Note in the last line that the tool has been repositioned away from the work area to allow for changing the workpiece.

8.5   COMPUTER-ASSISTED PART PROGRAMMING

The workpart of Example 8.1 was relatively simple. It was a suitable application for manual programming. Most parts machined on NC systems are considerably more complex. In the more complicated point-to-point jobs and in contouring applications, manual part programming becomes an extremely tedious task and subject to errors. In these instances it is much more appropriate to employ the high-speed digital computer to assist in the part programming process. Many part programming language systems have been developed to perform automatically most of the calculations which the programmer would otherwise be forced to do. This saves time and results in a more accurate and more efficient part program.

The part programmer's job

In computer-assisted part programming, the NC procedure for preparing the tape from the engineering drawing is followed as described in Section 7.3. The machining instructions are written in English-like statements of the NC programming language, which are then processed by the computer to prepare the tape. The computer automatically punches the tape in the proper tape format for the particular NC machine.

The part programmer's responsibility in computer-assisted part programming consists of two basic steps:

1.   Defining the workpart geometry

2.   Specifying the operation sequence and tool path

No matter how complicated the workpart may appear, it is composed of basic geometric elements. Using a relatively simple workpart to illustrate, consider the component shown in Figure 8.6. Although somewhat irregular in overall appearance, the outline of the part consists of intersecting straight lines and a partial circle. The holes in the part can be expressed in terms of the center location and radius of the hole. Nearly any component that can be conceived by a designer can be described by points, straight lines, planes, circles, cylinders, and other mathematically defined surfaces. It is the part programmer's task to enumerate the elements out of which the part is composed. Each geometric element must be identified and the dimensions and location of the element explicitly defined.

After defining the workpart geometry, the programmer must next construct the path that the cutter will follow to machine the part. This tool path specification involves a detailed step-by-step sequence of cutter moves. The moves are made along the geometry elements, which have previously been defined. The part programmer can use the various motion commands to direct the tool to machine along

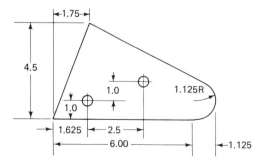

FIGURE 8.6 Sample workpart, like other parts, can be defined in terms of basic geometric elements, such as points, lines, and circles.

the workpart surfaces, to go to point locations, to drill holes at these locations, and so on. In addition to part geometry and tool motion statements, the programmer must also provide other instructions to operate the machine tool properly. We will consider the various categories of instructions for the APT language in Section 8.6.

The computer's job

The computer's job in computer-assisted part programming consists of the following steps:

1. Input translation
2. Arithmetic calculations
3. Cutter offset computation
4. Postprocessor

The sequence of these steps and their relationships to the part programmer and the machine tool are illustrated in Figure 8.7.

The part programmer enters the program written in the APT or other language. The input translation component converts the coded instructions contained in the program into computer-usable form, preparatory to further processing.

The arithmetic calculations unit of the system consists of a comprehensive set of subroutines for solving the mathematics required to generate the part surface. These subroutines are called by the various part programming language statements. The arithmetic unit is really the fundamental element in the part programming package. This unit frees the programmer from the time-consuming geometry and trigonometry calculations, to concentrate on the workpart processing.

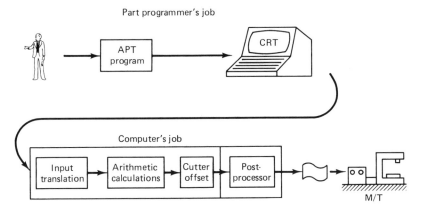

FIGURE 8.7 Steps in computer-assisted part programming.

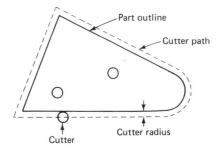

FIGURE 8.8 Cutter offset problem in part programming for contouring.

The second task of the part programmer is that of constructing the tool path. However, the actual tool path is different from the part outline because the tool path is defined as the path taken by the center of the cutter. It is at the periphery of the cutter that machining takes place. The purpose of the cutter offset computations is to offset the tool path from the desired part surface by the radius of the cutter. This means that the part programmer can define the exact part outline in the geometry statements. Thanks to the cutter offset calculation provided by the programming system, the programmer need not be concerned with this task. The cutter offset problem is illustrated in Figure 8.8.

As noted previously, NC machine tool systems are different. They have different features and capabilities. They use different NC tape formats. Nearly all of the part programming languages, including APT, are designed to be general-purpose languages, not limited to one or two machine tool types. Therefore, the final task of the computer in computer-assisted part programming is to take the general instructions and make them specific to a particular machine tool system. The unit that performs this task is called a postprocessor.

The postprocessor is a separate computer program that has been written to prepare the punched tape for a specific machine tool. The input to the postprocessor is the output from the other three components: a series of cutter locations and other instructions. The output of the postprocessor is the NC tape written in the correct format for the machine on which it is to be used.

NC part programming languages

An NC part programming language consists of a software package (computer program) plus the special rules, conventions, and vocabulary words for using that software. Its purpose is to make it convenient for a part programmer to communicate the necessary part geometry and tool motion information to the computer so that the desired part program can be prepared. The vocabulary words are typically mnemonic and English-like, to make the NC language easy to use.

There have probably been over 100 NC part programming languages developed since the initial MIT research on NC programming in the mid-1950s. Most of the languages were developed to meet particular needs and have not survived the test of time. Today, there are several dozen NC languages still in use. Refinements and enhancements to existing languages are continually being made. The following list provides a description of some of the important NC languages in current use.

APT (Automatically Programmed Tools).   The APT language was the product of the MIT developmental work on NC programming systems. Its development began in June, 1956, and it was first used in production around 1959. Today it is the most widely used language in the United States. Although first intended as a contouring language, modern versions of APT can be used for both positioning and continuous-path programming in up to five axes. Versions of APT for particular processes include APTURN (for lathe operations), APTMIL (for milling and drilling operations), and APTPOINT (for point-to-point operations).

ADAPT (ADaptation of APT).   Several part programming languages are based directly on the APT program. One of these is ADAPT, which was developed by IBM under Air Force contract. It was intended to provide many of the features of APT but to utilize a smaller computer. The full APT program requires a computing system that would have been considered large by the standards of the 1960s. This precluded its use by many small and medium-sized firms that did not have access to a large computer. ADAPT is not as powerful as APT, but it can be used to program for both positioning and contouring jobs.

EXAPT (EXtended subset of APT).   This was developed in Germany starting around 1964 and is based on the APT language. There are three versions: EXAPT I—designed for positioning (drilling and also straight-cut milling), EXAPT II—designed for turning, and EXAPT III—designed for limited contouring operations. One of the important features of EXAPT is that it attempts to compute optimum feeds and speeds automatically.

UNIAPT.   The UNIAPT package represents another attempt to adapt the APT language to use on smaller computers. The name derives from the developer, the United Computing Corp. of Carson, California. Their efforts have provided a limited version of APT to be implemented on minicomputers, thus allowing many smaller shops to possess computer-assisted programming capacity.

SPLIT (Sundstrand Processing Language Internally Translated). This is a proprietary system intended for Sundstrand's machine tools. It can handle up to five-axis positioning and possesses contouring capability as well. One of the unusual features of SPLIT is that the postprocessor is built into the program. Each machine tool uses its own SPLIT package, thus obviating the need for a special postprocessor.

COMPACT II.   This is a package available from Manufacturing Data Systems, Inc. (MDSI), a firm based in Ann Arbor, Michigan. The NC language is similar to SPLIT in many of its features. MDSI leases the COMPACT II system to its users on a time-sharing basis. The part programmer uses a remote terminal to feed the program into one of the MDSI computers, which in turn produces the NC tape. The COMPACT II language is one of the most widely used programming languages. MDSI has roughly 3000 client companies which use this system.

PROMPT.   This is an interactive part programming language offered by Weber N/C System, Inc., of Milwaukee, Wisconsin. It is designed for use with a variety of machine tools, including lathes, machining centers, flame cutters, and punch presses.

CINTURN II. This is a high-level language developed by Cincinnati Milacron to facilitate programming of turning operations.

The most widely used NC part programming language is APT, including its derivatives (ADAPT, EXAPT, UNIAPT, etc.). We present the basic elements of the APT language in Section 8.6.

Time-sharing versus in-house computer

There are two ways to implement a computer-assisted part programming system. The first is by means of a time-sharing service and the second is by means of a computer at the company's own plant.

Time-sharing services provide a relatively low cost means for a company to get started in computer-assisted NC part programming. The company accesses the NC language of the service over telephone lines. Several of the NC languages listed previously are available primarily through time-sharing service companies (COMPACT II and PROMPT are good examples). The cost advantage to the client company is that the billing is based on the use of the service. There is no large fixed cost required to purchase a lot of hardware. The remote terminal and the tape punching equipment are basically all that the company needs in order to use the time-sharing service. The billing for the service includes such items as CPU time (actual time spent using the computer's central processing unit), terminal-connect time, data storage, and so forth. There is also the telephone bill.

In the growth of NC within a particular plant, the company eventually reaches a point at which the cost of using the time-sharing service becomes very high. This point varies for different companies, but when the plant is paying monthly bills of several thousand dollars for the time-sharing service, it should probably begin to consider its own in-house programming system. This is accomplished either by purchasing a dedicated computer which will be used exclusively for NC part programming, or by loading one or more NC **language** packages on the company's existing data processing facilities. The cost of installing an in-house capability can be considerable, but if the volume of part programming work is high, the overall cost is lower than the cost of time sharing. Many companies have installed their own computer capability, but have also continued to retain time-sharing services for maximum flexibility.

8.6   THE APT LANGUAGE

In this section we present an introduction to the APT language for computer-assisted part programming. Our objectives are to demonstrate the English-like statements of this NC language and to show how they are used to command the cutting tool through its sequence of machining operations.

APT is not only an NC language; it is also the computer program that performs the calculations to generate cutter positions based on APT statements. We will not consider the internal workings of the computer program. Instead, we will concentrate on the language that the part programmmer must use.

APT is a three-dimensional system that can be used to control up to five axes. We will limit our discussion to the more familiar three axes, x, y, and z, and exclude rotational coordinates. APT can be used to control a variety of different machining operations. We will cover only drilling and milling applications. There are over 400 words in the APT vocabulary. Only a small (but important) fraction will be covered here.

There are four types of statements in the APT language:

1. *Geometry statements*. These define the geometric elements that comprise the workpart. They are also sometimes called definition statements.
2. *Motion statements*. These are used to describe the path taken by the cutting tool.
3. *Postprocessor statements*. These apply to the specific machine tool and control system. They are used to specify feeds and speeds and to actuate other features of the machine.
4. *Auxiliary statements*. These are miscellaneous statements used to identify the part, tool, tolerances, and so on.

Geometry statements

To program in APT, the workpart geometry must first be defined. The tool is subsequently directed to move to the various point locations and along surfaces of the workpart which have been defined by these geometry statements. The definition of the workpart elements must precede the motion statements.

The general form of an APT geometry statement is this:

$$symbol = geometry \ type/descriptive \ data \qquad (8.1)$$

An example of such a statement is

$$P1 = POINT/5.0, 4.0, 0.0 \qquad (8.2)$$

The statement is made up of three sections. The first is the symbol used to identify the geometric element. A symbol can be any combination of six or fewer alphabetic and numeric characters. At least one of the six must be an alphabetic character. Also, although it may seem obvious, the symbol cannot be one of the APT vocabulary words.

The second section of the geometry statement is an APT vocabulary word that identifies the type of geometry element. Besides POINT, other geometry elements in the APT vocabulary include LINE, PLANE, and CIRCLE.

The third section of the geometry statement comprises the descriptive data that define the element precisely, completely, and uniquely. These data may include quantitative dimensional and positional data, previously defined geometry elements, and other APT words.

The punctuation used in the APT geometry statement is illustrated in the example, Eq. (8.2). The statement is written as an equation, the symbol being equated to the surface type. A slash separates the surface type from the descriptive data. Commas are used to separate the words and numbers in the descriptive data.

There are a variety of ways to specify the different geometry elements. The appendix at the end of this chapter presents a dictionary of APT vocabulary words as well as a sampling of statements for defining the geometry elements we will be using: points, lines, circles, and planes. The reader may benefit from a few examples.

To specify a line, the easiest method is by two points through which the line passes:

$$L3 = LINE/P3, \ P4$$

The part programmer may find it convenient to define a new line parallel to another line which has previously been defined. One way of doing this would be

$$L4 = LINE/P5, \ PARLEL, \ L3$$

This states that the line L4 must pass through point P5 and be parallel (PARLEL) to line L3.

A plane can be defined by specifying three points through which it passes:

$$PL1 = PLANE/P1, \ P4, \ P5$$

Of course, the three points must not lie along a straight line. A plane can also be defined as being parallel to another plane, similar to the previous line parallelism statement.

$$PL2 = PLANE/P2, PARLEL, PL1$$

Plane PL2 is parallel to plane PL1 and passes through point P2.

A circle can be specified by its center and its radius.

$$C1 = CIRCLE/CENTER, P1, RADIUS, 5.0$$

The two APT descriptive words are used to identify the center and radius. The orientation of the circle perhaps seems undefined. By convention, it is a circle located in the x-y plane.

There are several ground rules that must be followed in formulating an APT geometry statement:

1. The coordinate data must be specified in the order x, y, z. For example, the statement

$$P1 = POINT/5.0, 4.0, 0.0$$

is interpreted by the APT program to mean a point $x = 5.0$, $y = 4.0$, and $z = 0.0$.

2. Any symbols used as descriptive data must have been previously defined. For example, in the statement

$$P2 = POINT/INTOF, L1, L2$$

the two lines L1 and L2 must have been previously defined. In setting up the list of geometry statements, the APT programmer must be sure to define symbols before using them in subsequent statements.

3. A symbol can be used to define only one geometry element. The same symbol cannot be used to define two different elements. For example, the following sequence would be incorrect:

$$P1 = POINT/1.0, 1.0, 1.0$$
$$P1 = POINT/2.0, 3.0, 4.0$$

4. Only one symbol can be used to define any given element. For example, the following two statements in the same program would render the program incorrect:

$$P1 = POINT/1.0, 1.0, 1.0$$
$$P2 = POINT/1.0, 1.0, 1.0$$

5. Lines defined in APT are considered to be of infinite length in both directions. Similarly, planes extend indefinitely and circles defined in APT are complete circles.

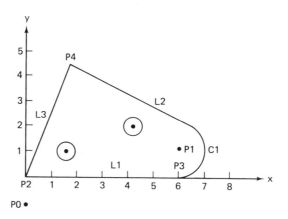

FIGURE 8.9 Workpart from Figure 8.6 redrawn with *x-y* coordinate system and geometric elements labeled.

EXAMPLE 8.2

To illustrate some of these geometry statements we will define the geometry of the workpart shown in Figure 8.6. The drawing of the part is duplicated in Figure 8.9, except that we have added the coordinate axis system and labeled the various geometric elements. We also add the target point P0 to be used in subsequent motion commands.

$$
\begin{aligned}
&\text{P0} &&= \text{POINT}/0,\ -1.0,\ 0\\
&\text{P1} &&= \text{POINT}/6.0,\ 1.125,\ 0\\
&\text{P2} &&= \text{POINT}/0,\ 0,\ 0\\
&\text{P3} &&= \text{POINT}/6.0,\ 0,\ 0\\
&\text{P4} &&= \text{POINT}/1.75,\ 4.5,\ 0\\
&\text{L1} &&= \text{LINE}/\text{P2},\ \text{P3}\\
&\text{C1} &&= \text{CIRCLE}/\text{CENTER},\ \text{P1},\ \text{RADIUS},\ 1.125\\
&\text{L2} &&= \text{LINE}/\text{P4},\ \text{LEFT},\ \text{TANTO},\ \text{C1}\\
&\text{L3} &&= \text{LINE}/\text{P2},\ \text{P4}\\
&\text{PL1} &&= \text{PLANE}/\text{P2},\ \text{P3},\ \text{P4}
\end{aligned}
$$

Motion statements

APT motion statements have a general format, just as the geometry statements do. The general form of a motion statement is

$$\text{motion command/descriptive data} \tag{8.3}$$

An example of a motion statement is

$$\text{GOTO/P1} \tag{8.4}$$

The statement consists of two sections separated by a slash. The first section is the basic motion command, which tells the tool what to do. The second section is comprised of descriptive data, which tell the tool where to go. In the example statement above, the tool is commanded to go to point P1, which has been defined in a preceding geometry statement.

At the beginning of the motion statements, the tool must be given a starting point. This point is likely to be the target point, the location where the operator has positioned the tool at the start of the job. The part programmer keys into this starting position with the following statement:

$$FROM/TARG \qquad\qquad (8.5)$$

The FROM is an APT vocabulary word which indicates that this is the initial point from which others will be referenced. In the statement above, TARG is the symbol given to the starting point. Any other APT symbol could be used to define the target point. Another way to make this statement is

$$FROM/-2.0, \ -2.0, 0.0$$

where the descriptive data in this case are the x, y, and z coordinates of the target point. The FROM statement occurs only at the start of the motion sequence.

It is convenient to distinguish between PTP movements and contouring movements when discussing the APT motion statements.

POINT-TO-POINT MOTIONS. There are only two basic PTP motion commands: GOTO and GODLTA. The GOTO statement instructs the tool to go to a particular point location specified in the descriptive data. Two examples would be:

$$GOTO/P2$$
$$GOTO/2.0, 7.0, 0.0$$

In the first statement, P2 is the destination of the tool point. In the second statement, the tool has been instructed to go to the location whose coordinates are $x = 2.0$, $y = 7.0$, and $z = 0$.

The GODLTA command specifies an incremental move for the tool. For example, the statement

$$GODLTA/2.0, \ 7.0, \ 0.0$$

instructs the tool to move from its present position 2 in. in the x direction and 7 in. in the y direction. The z coordinate remains unchanged.

The GODLTA command is useful in drilling and related operations. The tool can be directed to a particular hole location with the GOTO statement. Then the GODLTA command would be used to drill the hole, as in the following sequence:

> GOTO/P2
> GODLTA/0, 0, −1.5
> GODLTA/0, 0, +1.5

EXAMPLE 8.3

Example 8.1 was a PTP job which was programmed manually. Let us write the APT geometry and motion statements necessary to perform the drilling portion of this job. We will set the plane defined by $z = 0$ about ¼ in. above the part surface. The part will be assumed to be ½ in. thick. The reader should refer back to Figures 8.3 and 8.4.

> P1 = POINT/1.0, 2.0, 0
> P2 = POINT/1.0, 1.0, 0
> P3 = POINT/3.5, 1.5, 0
> P0 = POINT/−1.0, 3.0, 2.0
> FROM/P0
> GOTO/P1
> GODLTA/0, 0, −1.0
> GODLTA/0, 0, +1.0
> GOTO/P2
> GODLTA/0, 0, −1.0
> GODLTA/0, 0, +1.0
> GOTO/P3
> GODLTA/0, 0, −1.0
> GODLTA/0, 0, +1.0
> GOTO/P0

This is not a complete APT program because it does not contain the necessary auxiliary and postprocessor statements. However, the statement sequence demonstrates how geometry and motion statements can be combined to command the tool through a series of machining steps.

CONTOURING MOTIONS. Contouring commands are somewhat more complicated because the tool's position must be continuously controlled throughout the move. To accomplish this control, the tool is directed along two intersecting surfaces as shown in Figure 8.10. These surfaces have very specific names in APT:

1. *Drive surface*. This is the surface (it is pictured as a plane in Figure 8.10) that guides the side of the cutter.
2. *Part surface*. This is the surface (again shown as a plane in the figure) on which the bottom of the cutter rides. The reader should note that the "part surface" may or may not be an actual surface of the workpart. The part programmer must define this plus the drive surface for the purpose of maintaining continuous path control of the tool.

There is one additional surface that must be defined for APT contouring motions:

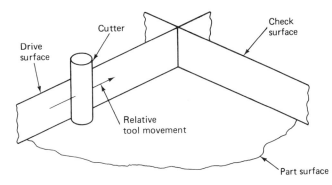

FIGURE 8.10   Three surfaces in APT contouring motions which guide the cutting tool.

3. *Check surface.*   This is the surface that stops the movement of the tool
    in its current direction. In a sense, it checks the forward movement of
    the tool.

There are several ways in which the check surface can be used. This is determined by APT modifier words within the descriptive data of the motion statement.
The three main modifier words are TO, ON, and PAST, and their use with regard
to the check surface is shown in Figure 8.11. A fourth modifier word is TANTO.
This is used when the drive surface is tangent to a circular check surface, as illustrated in Figure 8.12. In this case the cutter can be brought to the point of tangency
with the circle by use of the TANTO modifier word.

The APT contour motion statement commands the cutter to move along the
drive and part surfaces and the movement ends when the tool is at the check surface. There are six motion command words:

GOLFT   GOFWD   GOUP

GORGT   GOBACK   GODOWN

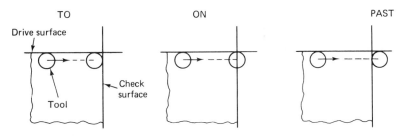

FIGURE 8.11   Use of APT modifier words in a motion statement: TO, ON and PAST. TO
moves tool into initial contact with check surface. ON moves tool until tool center is on check
surface. PAST moves tool just beyond check surface.

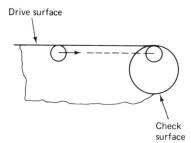

Drive surface

Check surface

FIGURE 8.12 Use of APT modifier word TANTO. TANTO moves tool to point of tangency between two surfaces, at least one of which is circular.

Their interpretation is illustrated in Figure 8.13. In commanding the cutter, the programmer must keep in mind where it is coming from. As the tool reaches the new check surface, does the next movement involve a right turn or an upward turn or what? The tool is directed accordingly by one of the six motion words. In the use of these words, it is helpful for the programmer to assume the viewpoint that the workpiece remains stationary and the tool is instructed to move relative to the piece. To begin the sequence of motion commands, the FROM statement, Eq. (8.5) is used in the same manner as for PTP moves. The statement following the FROM statement defines the initial drive surface, part surface, and check surface. The sequence is of the following form:

$$\text{FROM/TARG}$$
$$\text{GO/TO, PL1, TO, PL2, TO, PL3}$$

The symbol TARG represents the target point where the operator has set up the tool. The GO command instructs the tool to move to the intersection of the drive

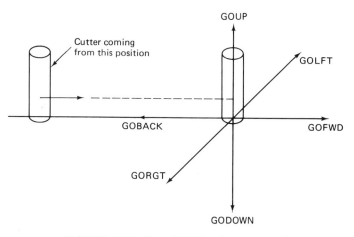

Cutter coming from this position

GOUP

GOLFT

GOBACK

GOFWD

GORGT

GODOWN

FIGURE 8.13 Use of APT motion commands.

surface (PL1), the part surface (PL2), and the check surface (PL3). The periphery of the cutter is tangent to PL1 and PL3, and the bottom of the cutter is touching PL2. This cutter location is defined by use of the modifier word TO. The three surfaces included in the GO statement must be specified in the order: drive surface first, part surface second, and check surface last.

Note that the GO/TO command is different from the GOTO command. GOTO is used only for PTP motions. GO/TO is used to initialize the sequence of contouring motions.

After initialization, the tool is directed along its path by one of the six command words. It is not necessary to repeat the symbol of the part surface after it has been defined. For instance, consider Figure 8.14. The cutter has been directed from TARG to the intersection of surfaces PL1, PL2, and PL3. It is desired to move the tool along plane PL3. The following command would be used:

GORGT/PL3, PAST, PL4

This would direct the tool to move along PL3, using it as the drive surface. The tool would continue until past surface PL4, which is the new check surface. Although the part surface (PL2) may remain the same throughout the motion sequence, the drive surface and check surface are redefined in each new command.

Let us consider an alternative statement to the above which would accomplish the same motion but would lead to easier programming:

GORGT/L3, PAST, L4

We have substituted lines L3 and L4 for planes PL3 and PL4, respectively. When looking at a part drawing, such as Figure 8.6, the sides of the part appear as lines. On the actual part, they are three-dimensional surfaces, of course. However, it would be more convenient for the part programmer to define these surfaces as lines and circles rather than planes and cylinders. Fortunately, the APT computer program allows the geometry of the part to be defined in this way. Hence the lines L3 and L4 in the foregoing motion statement are treated as the drive surface and check surface. This substitution can only be made when the part surfaces are perpendicular to the x-y plane.

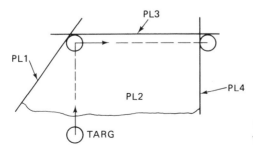

FIGURE 8.14   Initialization of APT contouring motion sequence.

EXAMPLE 8.4

To demonstrate the use of the motion commands in a contouring sequence, we will refer back to the workpart of Example 8.2. The geometry statements were listed in this example for the part shown in Figure 8.9. Using the geometric elements shown in this figure, following is the list of motion statements to machine around the periphery of the part. The sequence begins with tool located at the target point P0.

> FROM/P0
> GO/TO, L1, TO, PL1, TO, L3
> GORGT/L1, TANTO, C1
> GOFWD/C1, PAST, L2
> GOFWD/L2, PAST, L3
> GOLFT/L3, PAST, L1
> GOTO/P0

The reader may have questioned the location of the part surface (PL1) in the APT sequence. For this machining job, the part surface must be defined below the bottom plane of the workpiece in order for the cutter to machine the entire thickness of the piece. Therefore, the part surface is really not a surface of the part at all.

Example 8.4 raises several other questions: How is the cutter size accounted for in the APT program? How are feeds and speeds specified? These and other questions are answered by the postprocessor and auxiliary statements.

Postprocessor statements

To write a complete part program, statements must be written that control the operation of the spindle, the feed, and other features of the machine tool. These are called postprocessor statements. Some of the common postprocessor statements that appear in the appendix at the end of the chapter are:

COOLNT/	RAPID
END	SPINDL/
FEDRAT/	TURRET/
MACHIN/	

The postprocessor statements, and the auxiliary statements in the following section, are of two forms: either with or without the slash (/). The statements without the slash are self-contained. No additional data are needed. The APT words with the slash require descriptive data after the slash. These descriptions are given for each word in the appendix.

The FEDRAT/ statement should be explained. FEDRAT stands for feed rate and the interpretation of feed differs for different machining operations. In a drilling operation the feed is in the direction of the drill bit axis. However, in an end milling operation, typical for NC, the feed would be in a direction perpendicular to the axis of the cutter.

Auxiliary statements

The complete APT program must also contain various other statements, called auxiliary statements. These are used for cutter size definition, part identification, and so on. The following APT words used in auxiliary statements are defined in the appendix to this chapter:

CLPRNT	INTOL/
CUTTER	OUTTOL/
FINI	PARTNO

The offset calculation of the tool path from the part outline is based on the CUTTER/definition. For example, the statement

$$CUTTER/.500$$

would instruct the APT program that the cutter diameter is 0.500 in. Therefore, the tool path must be offset from the part outline by 0.250 in.

EXAMPLE 8.5

We are now in a position to write a complete APT program. The workpart of Example 8.4 will be used to illustrate the format of the APT program.

We will assume that the workpiece is a low-carbon steel plate, which has previously been cut out in the rough shape of the part outline. The tool is a ½-in.-diameter end-milling cutter. Typical cutting conditions might be recommended as follows: cutting speed = 573 rpm and feed = 2.29in./min.

Figure 8.15 presents the program with correct character spacing identified at the top as if it were to be keypunched onto computer cards. Modern NC programming systems utilize a CRT terminal for program entry.

8.7 THE MACRO STATEMENT IN APT

In the preceding section we described the basic ingredients of the APT language. In the present section we describe a very powerful feature of APT, the MACRO statement. The MACRO feature is similar to a subroutine in FORTRAN and other computer programming languages. It would be used where certain motion sequences would be repeated several times within a program. The purpose in using a MACRO subroutine is to reduce the total number of statements required in the APT program, thus making the job of the part programmer easier and less time

```
Column
 1      6 │ 8 │ 10                                                          72
 PARTNO   │   │   EXAMPLE PART
          │   │   MACHIN/MILL, 1
          │   │   CLPRNT
          │   │   INTOL/.001
          │   │   OUTTOL/.001
          │   │   CUTTER/.500
 P0       │ = │   POINT/0, −1.0, 0
 P1       │ = │   POINT/6.0, 1.125, 0
 P2       │ = │   POINT/0, 0, 0
 P3       │ = │   POINT/6.0, 0, 0
 P4       │ = │   POINT/1.75, 4.5, 0
 L1       │ = │   LINE/P2, P3
 C1       │ = │   CIRCLE/CENTER, P1, RADIUS, 1.125
 L2       │ = │   LINE/P4, LEFT' TANTO, C1
 L3       │ = │   LINE/P2, P4
 PL1      │ = │   PLANE/P2, P3, P4
          │   │   SPINDL/573
          │   │   FEDRAT/2.29
          │   │   COOLNT/ON
          │   │   FROM/P0
          │   │   GO/TO, L1, TO, PL1, TO, L3
          │   │   GORGT/L1, TANTO, C1
          │   │   GOFWD/C1, PAST, L2
          │   │   GOFWD/L2, PAST, L3
          │   │   GOLFT/L3, PAST, L1
          │   │   RAPID
          │   │   GOTO/P0
          │   │   COOLNT/OFF
          │   │   FINI
```

FIGURE 8.15   APT program for Example 8.5.

consuming. The MACRO subroutine is defined by a statement of the following format:

$$symbol = MACRO/parameter\ definition(s) \qquad (8.6)$$

The rules for naming the MACRO symbol are the same as for any other APT symbol. It must be six characters or fewer and at least one of the characters must be a letter of the alphabet. The parameter definition(s) following the slash would identify certain variables in the subroutines which might change each time the subroutine was called into use. Equation (8.6) would serve as the title and first line of a MACRO subroutine. It would be followed by the set of APT statements that comprise the subroutine. The very last statement in the set must be the APT word TERMAC. This signifies the termination of the MACRO.

To activate the MACRO subroutine within an APT program, the following call statement would be used:

$$CALL/symbol,\ parameter\ specification \qquad (8.7)$$

The symbol would be the name of the MACRO that is to be called. The parameter specification identifies the particular values of the parameters that are to be used in this execution of the MACRO subroutines.

An example will serve to illustrate the use of the MACRO statement and how the MACRO would be called by the main APT program.

EXAMPLE 8.6

We will refer back to the drilling operations of Example 8.2. In this example the GODLTA sequence was repeated in the program a total of three times, once for each hole. This represents an opportunity to use the MACRO feature in the APT system. The four point locations (P0, P1, P2, and P3) would be defined just as they are in Example 8.2. These points would be used in the MACRO subroutine and main APT program in the following way:

$$DRILL = MACRO/PX$$

$$GOTO/PX$$

$$GODLTA/0, 0, -1.0$$

$$GODLTA/0, 0, +1.0$$

$$TERMAC$$

$$FROM/P0$$

$$CALL/DRILL, PX = P1$$

$$CALL/DRILL, PX = P2$$

$$CALL/DRILL, PX = P3$$

$$GOTO/P0$$

In this example the number of APT motion statements in the main program has been reduced from 11 down to five. (If we include the MACRO subroutine in our line count, the reduction is from 11 statements to 10.) The reader can visualize the power of the MACRO feature for a programming job in which there are a large number of holes to be drilled. The savings in the required number of APT statements would approach 66⅔% in this case, since one call statement replaces three motion statements in the program.

The MACRO feature has many uses in APT. They are limited primarily by the imagination of the part programmer. Some of these uses will be considered in the exercise problems at the end of the chapter. It is even possible to have a CALL/ statement within one MACRO which refers to another MACRO subroutine. This might be used for example in a matrix of holes where both the x and y positions of the holes are changed with each drilling operation.

Another common use of the MACRO feature would be for a roughing and finishing sequence on a given workpart. This is conveniently handled by defining the motion sequence in the MACRO, and calling the MACRO twice with two different cutter diameters. The CALL for the roughing cut would specify a larger cutter diameter than the CALL for the finishing cut, even though the actual cutter is the same for both operations. An example will illustrate this powerful use of the MACRO feature.

EXAMPLE 8.7

In this example we perform a roughing cut and a finishing cut on the workpart of Example 8.5. It is a milling sequence in which the same 0.50-in.-diameter end-milling cutter will be used to rough and finish the part outline. The roughing operation will be performed so that 0.035 in. of material is remaining on the part for the finishing cut. The finishing operation will then cut the part to final size.

The only difference between the roughing sequence and the finishing sequence is that the tool offset is greater for the roughing cut. Therefore, the motion sequence can be defined in a MACRO and the tool offset can be defined by changing the diameter specification each time the MACRO is called. This would be done in the following way.

> MILL = MACRO/DIA
>
> CUTTER/DIA
>
> FROM/PO
>
> GO/TO, L1, TO, PL1, TO, L3
>
> GORGT/L1, TANTO, C1
>
> GOFWD/C1, PAST, L2
>
> GOFWD/L2, PAST, L3
>
> GOLFT/L3, PAST, L1
>
> GOTO/PO

The main part of the program which refers to this MACRO would simply consist of

> CALL/MILL, DIA = .570
>
> CALL/MILL, DIA = .500

The first statement accomplishes the roughing operation with the cutter path offset from the desired part outline by one-half of the stated diameter. This leaves 0.035 in. to be machined by the cutter during the finishing pass. It might be desirable to reduce the feedrate for the finishing pass to achieve a better surface.

The MACRO feature is only one of the many features in APT. We have only scratched the surface in this powerful part programming system. However, enough of the APT language has been included to permit the reader to try some of the programming exercises at the end of the chapter.

8.8 NC PROGRAMMING WITH INTERACTIVE GRAPHICS

There are several innovations in NC part programming which have developed during the middle and late 1970s. Their use is expected to grow substantially during the 1980s. The first involves the use of interactive graphics as a highly productive aid in performing the part programming process. We cover this topic in the present section. The second innovation is voice programming. This involves the input of NC programming statements through oral communication by the human programmer. Voice programming is described in Section 8.9. The third development during the past few years is manual data input (MDI) of the NC part program. In a sense, this involves a step backward in NC programming technology. We explain why in Section 8.10.

The use of interactive graphics in NC part programming is an excellent example of the integration of computer-aided design and computer-aided manufacturing. The programming procedure is carried out on the graphics terminal of a CAD/CAM system. Using the same geometric data which defined the part during the computer-aided design process, the programmer constructs the tool path using high-level commands to the system. In many cases the tool path is automatically generated by the software of the CAD/CAM system. The output resulting from the procedure is a listing of the APT program or the actual CLFILE (cutter location file) which can be postprocessed to generate the NC punched tape.

Let us consider the step-by-step procedure that would be used to generate the NC part program using a CAD/CAM system. We will then illustrate the procedure with an example. All of the major CAD/CAM system vendors offer part programming packages. Although the features of these packages vary between the vendors, they all operate in a similar way. In our description of the procedure we will attempt to portray a composite of the various packages.

Initial steps in the procedure

The CAD/CAM procedure for NC programming begins with the geometric definition of the part. A significant benefit of using a CAD/CAM system is realized when these geometry data have already been created during design. If the geometric model of the part has not been previously created, it must be constructed on the graphics terminal.

With the part displayed on the CRT screen, the programmer would proceed to label the various surfaces and elements of the geometry. The CAD/CAM system would accomplish the labeling in response to a few simple commands by the programmer. After labeling is completed, the APT geometry statements can be generated automatically by the system.

In addition to the ease with which the APT geometry has been defined using the CAD/CAM system, there are several other benefits afforded the user of a graphics system for NC part programming. The part can be displayed at various angles, magnifications, and cross sections to examine potential problem areas in machining. This capability to manipulate the part image on the CRT screen is helpful to the programmer in visualizing the design of the part. Also, with the part defined in the computer, the programmer can overlay the outline of the raw workpart to consider the number of passes required to complete the machining. Alternative methods of fixturing the part can be explored using the graphics terminal.

Tool selection is the next step in the procedure. The CAD/CAM system would typically have a tool library with the various tools used in the shop catalogued according to the type. The programmer could either select one of these tools or create a new tool design by specifying the parameters and dimensions of the new tool (diameter, corner radius, cutter length, etc.).

Generation of the tool path

At this point in the procedure, the programmer has a geometric model of the workpart and the tools needed to machine the part. The next step is to create the cutter path. The method of accomplishing this using interactive graphics depends on the type of operation (e.g., profile milling, turning, sheet metal working) and the complexity of the part. The currently available commercial CAD/CAM systems use an interactive approach, with certain common machining routines being done automatically by the system. These automatic routines might include profile milling around a part outline, end milling a pocket, point-to-point, PTP presswork hole piercing, and surface contouring.

The interactive approach permits the programmer to generate the tool path in a step-by-step manner with visual verification on the graphics display. The procedure begins by defining a starting position for the cutter. The programmer would then command the tool to move along the defined geometric surfaces of the part. As the tool is being moved on the CRT screen, the corresponding APT motion commands are automatically prepared by the CAD/CAM system. The interactive mode provides the user with the opportunity to insert postprocessor statements at appropriate points during program creation. These postprocessor statements would consist of machine tool instructions such as feed rates, speeds, and control of the cutting fluid.

The automatic machining routines are called into operation for frequently encountered part programming situations. These routines are analogous to high-level MACRO subroutines which have been developed as part of the CAD/CAM system software. The part geometry data represent the set of parameter definitions or arguments for the MACRO. Accordingly, these automatic routines can be called with a minimum of user interaction.

Profiling and pocketing are two common examples of automatic machining routines that are available on most CAD/CAM systems. The profiling routine is used to generate the sequence of cutter paths for machining around a series of geometry elements which have been identified by the user. These geometry elements would typically define the outline of the workpart. Let us consider an example to illustrate the graphics approach to NC programming and the automatic profiling feature in particular.

EXAMPLE 8.8

In this example we generate the APT program and CLFILE for the part from our previous Example 8.5. The Applicon CAD/CAM system in the Computer-Aided Manufacturing Lab at Lehigh University was used to accomplish the programming using Applicon's AGS/883 package. Since part programming is performed on a graphics terminal of the CAD/CAM system, the example will be developed by means of figures generated by the system. Figure 8.16 shows the part geometry as created on the Applicon terminal. Labeling of the geometry elements is accomplished automatically by the system under prompting by the user. Figure 8.17 shows the part with labeling added. The machining in this example involves profile milling around the circumference of the part. The Applicon system has an automatic routine for generating the tool path for this situation. The lines representing the cutter path are pictured in Figure 8.18. Figure 8.19 shows the APT program listing as generated on the CAD/CAM system. Note that the list of statements is not identical to those developed in Example 8.5.

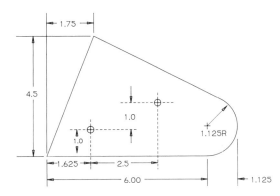

FIGURE 8.16 Part from Figure 8.6 redrawn on the Applicon CAD/CAM system.

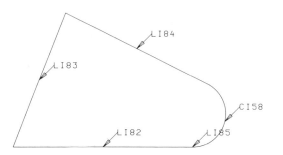

FIGURE 8.17 Workpart with labels added by Applicon CAD/CAM system.

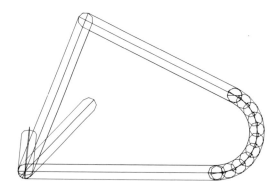

FIGURE 8.18 Cutter path generated automatically by Applicon CAD/CAM system.

```
@VW APT/MPG;

IO 70 0 0 0 0 0 0 2 GD APT/MPG;

INOUT VER. 003.20-I


CLPRNT
LI82    =LINE/6.25,-1.0,2.0,0.25,-1.0,2.0
LI83    =LINE/0.25,-1.0,2.0,2.0,3.5,2.0
LI84    =LINE/2.0,3.5,2.0,6.7525,1.1319,2.0
CI58    =CIRCLE/6.2507,0.125,2.0,1.125
LI85    =LINE/6.2507,-1.0,2.0,6.25,-1.0,2.0
REMARK  START OF CUT SEQUENCE 901
        CUTTER/0.25,0.05,0.075,0.05,0.0,0.0,4.0
        COOLNT/ON
        SPINDL/1200
        FEDRAT/1.0
        OUTTOL/0.005
        TLAXIS/0.0,0.0,1.0
          FROM/0.0,0.0,5.0
          RAPID
          GOTO/-0.1228,-1.255,3.0
        THICK/0.0,0.13
          DNTCUT
          GOTO/-0.1228,-1.255,1.0
          GO/ON,LI82,TO,(PLANE/0.0,0.0,1.0,1.0),TO,LI83
          CUT
          INDIRV/0.3624454,0.932005,0.0
          TLLFT, GOFWD/LI83,PAST,LI84
          GORGT/LI84,TANTO,CI58
          GOFWD/CI58,TANTO,LI85
        THICK/0.0,0.13,0.0
          GOFWD/LI85,ON,(LINE/(POINT/6.25,-1.255,1.0),PERPTO,(LINE/$
          6.2507,-1.255,1.0,6.25,-1.255,1.0))
        THICK/0.0,0.13
          GOFWD/LI82,PAST,LI83
          GORGT/LI83,PAST,LI84
          GORGT/LI84,TANTO,CI58
          GOFWD/CI58,TANTO,LI85
        THICK/0.0,0.13,0.0
          GOFWD/LI85,ON,(LINE/(POINT/6.25,-1.255,1.0),PERPTO,(LINE/$

          6.2507,-1.255,1.0,6.25,-1.255,1.0))
        THICK/0.0,0.13
          GOFWD/LI82,PAST,LI83
          TLON, GORGT/(LINE/-0.1228,-1.255,1.0,2.0,1.0,1.0),ON,(LINE/($
          POINT/2.0,1.0,1.0),PERPTO,(LINE/-0.1228,-1.255,1.0,2.0,1.0$
          ,1.0))
        FINI
```

FIGURE 8.19 APT program listing as generated by Applicon CAD/CAM system.

182

The automatic pocketing routine works in a manner similar to the profiling routine except that the part material is removed from within a closed set of boundary elements. These elements define a pocket or cavity in the workpart which needs to be machined out. The programmer would have to identify the sides and bottom of the pocket. The system would then generate the tool path to machine the pocket. This feature is illustrated in Figure 8.20.

In generating the tool path on a CAD/CAM system the use of color graphics is very helpful to the programmer. The part can be displayed in one color, while the tool path would be shown in a different color. This permits easy visual differentiation of the tool path from the part outline. Another feature which aids visualization of the machining sequence is dynamic tool path simulation on the graphics screen. An example of this feature is Computervision's 3D Dynamic Verification of NC Toolpaths. As illustrated in Figure 8.21, the orientation of the tool and toolholder can be shown on the screen relative to the workpiece. The simulated tool motion can be displayed in any of several modes: (1) high-speed motion, which reduces the time to verify the tool path; (2) actual speed, which shows the tool feed at the commanded rate; (3) freeze mode, which stops the tool motion for close inspection; and (4) stepping mode, which displays the tool path in discrete steps.

Advantages of CAD/CAM in NC programming

The CAD/CAM approach offers several very significant advantages in numerical control part programming. Among these advantages are the following:

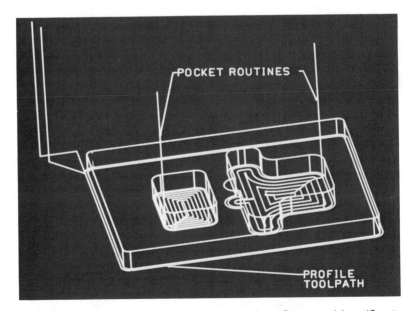

FIGURE 8.20 Automatic pocketing routing generated on Computervision. (Courtesy of Computervision Corp.)

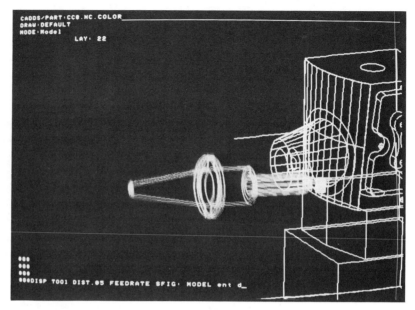

FIGURE 8.21 Dynamic verification of NC tool path in three-dimensional view. (Courtesy of Computervision Corp.)

1. *Savings in geometry definition.* Since the part geometry data have already been created during design using the CAD/CAM graphics system, the part programmer is not required to redefine the geometry of the part. This can be a time-consuming procedure in conventional APT programming.

2. *Immediate visual verification.* The graphics terminal provides a display of the tool path for immediate verification by the part programmer. Most programming errors can be detected by the user and corrected at the time the error is made. With conventional APT or other NC language, there is a delay between writing the program and the verification/correction process.

3. *Use of automatic programming routines.* For common part programming situations such as profiling and pocketing, the use of automatic MACRO-type routines yields a significant reduction in part programming time.

4. *One-of-a-kind jobs.* Because the part programming time is significantly reduced when using a CAD/CAM system, numerical control becomes an economically attractive method for producing one-of-a-kind jobs. Without CAD/CAM, the time required to prepare the part program represents a significant obstacle which often precludes the use of NC for one/off production.

5. *Integration with other related functions.* There is the obvious opportunity to integrate the product design function with part programming. Other opportunities for functional integration within manufacturing also exist. These include tool design, process planning, preparation of operator and setup instructions, grouping of parts into families for programming convenience, and so on. With the tremendous advances made in NC programming over the last three

decades, it is not difficult to imagine that the entire logic of the part programming process will be captured and put on the computer. This would permit NC programming to be accomplished completely and automatically by the computer without human assistance.

8.9   VOICE NC PROGRAMMING

Voice programming of NC machines (abbreviated VNC) involves vocal communication of the machining procedure to a voice-input NC tape-preparation system. VNC allows the programmer to avoid steps such as writing the program by hand, keypunching or typing, and manual verification. One of the principal companies specializing in voice-input systems is Threshold Technology, Inc., of Delran, New Jersey.

To perform the part programming process with VNC, the operator speaks into a headband microphone designed to reduce background acoustical noise. Communication of the programming instructions is in shop language with such terms as "turn," "thread," and "mill line," together with numbers to provide dimensional and coordinate data. Before the voice-input system can be used, it must be "trained" to recognize and accept the individual programmer's voice pattern. This is accomplished by repeating each word of the vocabulary about five times to provide a reference set which can subsequently be compared to voice commands given during actual programming. The entire vocabulary for the Threshold system contains about 100 words. Most NC programming jobs can be completed by using about 30 of these vocabulary words.

In talking to the system, the programmer must isolate each word by pausing before and after the word. The pause must be only one-tenth of a second or longer. This allows the speech recognition system to identify boundaries for the uttered command so that its wave characteristics can be compared with words in the reference set for that programmer. Typical word input rates under this restriction are claimed to be about 70 per minute. As the words are spoken, a CRT terminal in front of the operator verifies each command and prompts the operator for the next command.

EXAMPLE 8.9

A typical dialogue between the VNC system (printing on the CRT screen) and the programmer (speaking) might go as follows for defining a circle:

Programmer: "Define"
System: DEFINITION TYPE
Programmer: "Circle"
System: CIRCLE # =
Programmer: "Three"
System: CENTER PT X =

Programmer: "Five decimal three one, Go"

System: Y =

Programmer: "Two decimal four seven five, Go"

System: RADIUS =

Programmer: "One decimal five, Go"

System: CW/CCW

Programmer: "Counterclockwise"

When all of the programming instructions have been entered and verified, the system prepares the punched tape for the job.

The advantages of VNC lie principally in the savings in programming time and resulting improvements in manufacturing lead time. Savings in programming time up to 50% are claimed. Improvements in accuracy and lower computer-skill requirements for the programmer are also given as benefits of VNC.

8.10   MANUAL DATA INPUT

Manual data input (MDI) involves the entry of part programming data through a CRT display at the machine site; hence the use of punched tape is avoided. The programming process is usually carried out by the machine operator. NC systems equipped with MDI capability possess a computer (microcomputer) as the control unit.

MDI systems are designed to facilitate the part programming process by using an interactive mode to assist the operator through the programming steps. It queries the operator about the details of the machining job so that the operator types in the program responding to the sequence of questions. MDI units use shop language rather than alphanumeric codes. This removes some of the mystery usually surrounding the programming activity. Basically, the operator must be able to read an engineering drawing and be familiar with the machining process. No extensive training is required in NC part programming.

The great advantage of MDI is its simplicity. It represents a relatively easy way for a small shop to make the transition to numerical control. There is a minimum of change in normal shop procedures needed to use NC systems featuring manual data input. Since no punched tape is employed with MDI, the shop is spared the expense of tape punching equipment normally associated with NC.

The limitation on MDI is that the programs should be relatively short and simple. This means that the machining jobs should be uncomplicated. There are several reasons for this limitation. First, since there is no paper copy of the program, there is a limit on the length and complexity of program that the operator is capable of visualizing. Also, the CRT can only display a total of 22 or 25 lines, so this adds to the operator's visualization problem. Finally, one of the biggest disadvantages of manual data input is that the machine tool itself is not productive while programming is being accomplished. In essence, a very expensive piece of equipment is being utilized to prepare the part program. The more complicated the program, the more time is taken when the machine is not cutting metal.

One way of overcoming this last disadvantage, which is beginning to be offered on MDI machines, is for the machine to be operated in a foreground/ background mode. This allows the operator to be entering the next program into the system while the last part is still being produced from the program currently in computer memory. This foreground/background capability reduces the changeover time between jobs.

The particular features of MDI make it appropriate for workparts which are simpler than the usual NC jobs, and small batch sizes which are not likely to be repeated. On this issue of repeated batches, many MDI systems possess the optional feature in which programs can be extracted from computer memory for long-term storage (in anticipation of a repeat order). However, instead of using punched tape as the storage medium so typical in NC, magnetic tape cassettes are the general storage device. The tape cassette can be stored in a part program library and then used to reenter the program into the MDI system's control memory when needed.

REFERENCES

[1]   APPLICON, INC., *AGS/883 NC II User's Manual*, Document A-20805, Burlington, Mass., 1981.

[2]   CHILDS, J. J., *Numerical Control Part Programming*, Industrial Press, Inc., New York, 1973.

[3]   COMPUTERVISION CORP., *CADDS 4, Numerical Control Application Package*, Publication PS-S47-01, Beford, Mass.

[4]   GRINDSTAFF, C. C., "Computer Graphics for CAM Applications," *Commline*, July/August, 1981, pp. 12−14.

[5]   GROOVER, M. P., *Automation, Production Systems, and Computer-Aided Manufacturing*, Prentice-Hall, Inc., Englewood Cliffs N.J., 1980, Chapter 8.

[6]   HATSCHEK, R. L., "NC Programming," Special Report 719, *American Machinist*, February, 1980, pp. 119−134.

[7]   HEGLAND, D. E., "Numerical Control—Your Best Investment in Productivity," *Production Engineering*, March, 1981, pp. 42−47.

[8]   HUBER, R. F., "Tell It to Your Machines," *Production*, June, 1980.

[9]   ILLINOIS INSTITUTE OF TECHNOLOGY RESEARCH INSTITUTE (IITRI), *APT Part Programming*, McGraw-Hill Book Company, New York, 1967.

[10] LERRO, J. P., "CAD/CAM System: More than an Automated Drafting Tool," *Design News*, November 17, 1980.

[11] MACHOVER, C., AND BLAUTH, R. E. (Eds.), *The CAD/CAM Handbook*, Computervision Corp., Bedford, Mass., 1980.

[12] MARTIN, J. M., "Voice Commands Produce NC Tapes," *American Machinist*, June, 1980, pp. 102−104.

[13]   NEIL, R., "CAD/CAM Use in Numerical Control," *Proceedings, Eighteenth Annual Meeting and Technical Conference*, Numerical Control Society, Dallas, Tex., May, 1981, pp. 56–82.

[14]   ROBERTS, A. D., AND PRENTICE, R. C., *Programming for Numerical Control Machines*, 2nd ed., McGraw-Hill Book Company, New York, 1978.

[15]   SCHAFFER, G., "Computer Graphics Goes to Work," Special Report 724, *American Machinist*, July, 1980, pp. 149–164.

[16]   THRESHOLD TECHNOLOGY, INC., "Programming Your N/C Machine Tools with Your Own Voice," Delran, N.J.

[17]   ZWICA, T., "MDI—The Pros and Cons," *Commline*, September/October, 1980, pp. 15, 43.

PROBLEMS

8.1. The workpart of Figure P8.1 is to be completed in an NC drill press. The outline of the part has already been completed and the five holes are to be drilled. The axis system for this sequence is to be located with the origin at the lower left-hand corner of the part. The part is ⅜ in. thick.
(a)   Write the APT geometry statements to define the hole locations.
(b)   Write the sequence of motion statements in APT to perform the drilling sequence. Use a point at $x = -1$ and $y = -3$ as the target point for the FROM statement.

8.2. Write the complete APT program for the problem of Figure P8.1. The postprocessor call statement is MACHIN/DRILL. The drill size is ⅜-in. diameter and the work material is machinable aluminum. Cutting speed and feed rate are 900 revolutions per minute and 3.0 in./ min.

8.3. A profile milling operation is to be performed to generate the outline of the part in Figure P8.3. Disregard the two holes in the part. They have already been drilled and will be used to clamp the part to the machine table. The part is ½ in. thick.

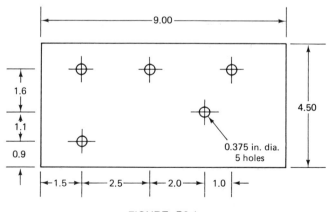

FIGURE P8.1

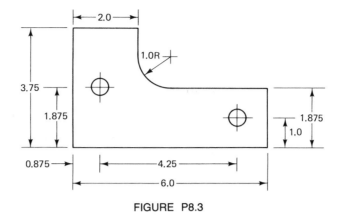

FIGURE P8.3

(a)   Write the APT geometry statements to define the part outline.

(b)   Write the sequence of APT motion statements to perform the profile milling around the periphery of the part. Use a location 3 in. below and 3 in. to the left of the lower left-hand corner of the part as the target point for the FROM statement. Assume that the part has been cut to rough size with a bandsaw. This has left about $\frac{1}{8}$ in. of material to be cut in the final profiling pass.

8.4.   Write the complete APT program for the part of Problem 8.3. The postprocessor call statement is MACHIN/MILL, 05. The inside and outside tolerances on the circular approximation should be 0.001 in. The end mill is 1 in. in diameter. Speed and feed should be 400 rpm and 3.0 in./min. respectively.

8.5.   The outline of the cam shown in Figure P8.5 is to be milled using a two-flute, $\frac{1}{2}$-in.-diameter end mill.

(a)   Write the geometry statements in APT to define the part outline.

(b)   Write the motion statement sequence using the geometry elements defined in part (a).

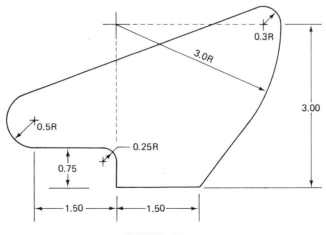

FIGURE P8.5

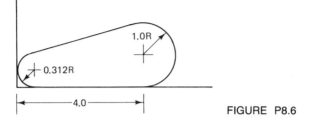

FIGURE P8.6

(c)   Write the complete APT program. Inside and outside tolerance should be 0.0005 in. Feed rate = 3 in./min; speed = 500 rpm. Postprocessor call statement is MACHIN/MILL, 01. Assume that the rough outline for the part has been obtained in a bandsaw operation. Ignore clamping problems with this part.

8.6.   The part outline of Figure P8.6 is to be milled in two passes with the same tool. The tool is a ¾-in.-diameter end mill. The first cut is to leave 0.050 in. of stock on the part outline. The second cut will take the part to size. Write the APT geometry and motion statements to peform the two passes.

8.7.   Write the APT program for Problem 8.2 (Figure P8.1) but use the MACRO feature in a manner similar to that used in Example 8.6.

8.8.   Suppose that the part of Figure P8.1 was to be drilled, reamed, and tapped (three separate tools) at the hole locations indicated on the drawing. Write a complete APT program for this sequence using two MACRO subroutines, one to call the hole locations for the proper tool, the other to call the operation (similar to the MACRO use of Example 8.6). The tool call statement and cutting conditions for each operation in the sequence should be:

For drilling TURRET/04
                       speed = 800 rpm
                       feed = 2.5 in./min
For reaming TURRET/06
                       speed = 500 rpm
                       feed = 4.0 in./min
For tapping TURRET/02
                       speed = 200 rpm
                       feed = 12.5 in./min

8.9.   Write the complete APT program for the part in Figure P8.6 using the information supplied in Problem 8.6. The program should employ the MACRO feature for the two passes. The postprocessor call statement is MACHIN/MILL, 01. Inside and outside tolerances are 0.001 in. Feed rate = 3 in./min and speed = 400 rpm.

8.10. A matrix of holes are to be drilled in the manner illustrated in Figure P8.10. The number of holes in the x direction is 5 and the number of holes in the y direction is 7, thus making 35 holes in all. Write the APT program making use of the MACRO feature to establish the hole coordinates.

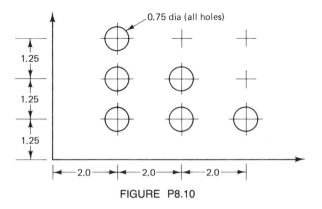

FIGURE P8.10

APPENDIX: APT WORD DEFINITIONS

ATANGL: At angle (descriptive data). Indicates that the data that follow represent a specified angle. Angle is given in degrees. *See* LINE.

CALL: Call. Used to call a MACRO subroutine and to specify parameter values for the MACRO.

$$\text{CALL/DRILL, PX} = \text{P1}$$

CENTER: Center (descriptive data). Used to indicate the center of circle. *See* CIRCLE.

CIRCLE: Circle (geometry type). Used to define a circle in the *xy*-plane. Methods of definition:

1.   By the coordinates of the center and the radius. See Figure A8.1.

$$\text{C1} = \text{CIRCLE/CENTER, 4.0, 3.0, 0.0, RADIUS, 2}$$

$$\text{C1} = \text{CIRCLE/4.0, 3.0, 0.0, 2.0}$$

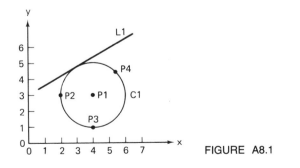

FIGURE A8.1

2. By the center point and the radius. See Figure A8.1.

$$C1=CIRCLE/CENTER, P1, RADIUS, 2.0$$

3. By the center point and tangent to a line. See Figure A8.1.

$$C1=CIRCLE/CENTER, P1, TANTO, L1$$

4. By three points on the circumference. See Figure A8.1.

$$C1=CIRCLE/P1, P3, P4$$

5. By two intersecting lines and the radius. See Figure A8.2.

$$C2=CIRCLE/XSMALL, L2, YSMALL, L3, RADIUS, .375$$
$$C3=CIRCLE/YLARGE, L2, YLARGE, L3, RADIUS, .375$$
$$C4=CIRCLE/XLARGE, L2, YLARGE, L3, RADIUS, .375$$
$$C5=CIRCLE/YSMALL, L2, YSMALL, L3, RADIUS, .375$$

CLPRNT: Cutter location print (auxiliary statement). Can be used to obtain a computer printout of the cutter location sequence on the NC tape.

COOLNT: Coolant (postprocessor statement). Turns coolant on, off, and actuates other coolant options that may be available. Examples:

COOLNT/ON   COOLNT/OFF

COOLNT/FLOOD   COOLNT/MIST

CUTTER: Cutter (auxiliary statement). Defines cutter diameter to be used in tool offset computations. The statement

CUTTER/1.0

defines a 1.0-in.-diameter milling cutter. Cutter path would be offset from part outline by one-half the diameter.

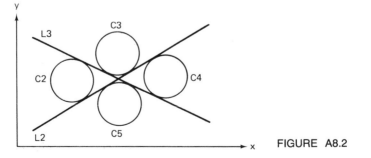

FIGURE A8.2

END: End (postprocessor statement). Used to stop the machine at the end of a section of the program. Can be used to change tools manually. Meaning may vary between machine tools. To continue program, a FROM statement should be used.

FEDRAT: Feed rate (postprocessor statement). Used to specify feet rate in inches per minute.

FEDRAT/6.0

FINI: Finish (auxiliary statement). Must be the last word in the APT program. Used to indicate the end of the complete program.

FROM: From the tool starting location (motion startup command). Used to specify the starting point of the cutter, from which other tool movements will be measured. The starting point is specified by the part programmer and set up by the machine operator. Methods of specification:

1. By a previously defined starting point (TARG).

FROM/TARG

2. By the coordinates of the starting point.

FROM/−1.0, −1.0, 0.0

GO: Go (motion startup command in contouring). Used to bring the tool from the starting point against the drive surface, part surface, and check surface. In the statements

GO/TO, L1, TO, PL1, TO, L2

GO/PAST, L1, TO, PL1, ON, TO, L2

the initial drive surface is the line L1, the part surface is PL1, and the initial check surface is L2.

GODLTA: Go delta (PTP motion command). Instructs the tool to move in increments as specified from the current tool location. In the statement

GODLTA/2.0, 3.0, −4.0

the tool is instructed to move 2.0 in. in the x-direction, 3.0 in. in the y-direction and −4.0 in. in the z-direction from its present position.

GOBACK: Go back (contour motion command). Instructs the tool to move back relative to its previous direction of movement. In the statement

GOBACK/PL5, TO, L1

the tool is instructed to move in the opposite general direction relative to its previous path. It moves on the drive surface PL5 until it reaches L1. The part surface has been specified in a previous GO statement.

In specifying the motion command the part programmer must pretend to be riding on top of the cutter and must give the next move (GOBACK, GOFWD, GOUP, GODOWN, GORGT, GOLFT) according to the tool's preceding motion. Also, the motion command should indicate the largest direction component. For example, if the next tool move was both forward and to the left, the motion command (GOFWD versus GOLFT) would be determined by whichever direction component was larger. See Figure 8.13.

GODOWN: Go down (contour motion command). *See* GOBACK.

GOFWD: Go forward (contour motion command). *See* GOBACK.

GOLFT: Go left (contour motion command). *See* GOBACK.

GORGT: Go right (contour motion command. *See* GOBACK.

GOTO: Go to (PTP motion command). Used to move the tool center to a specified point location. Methods of specification:

1.   By using a previously defined point. GOTO/P1
2.   By defining the coordinates of the point

$$GOTO/2.0, \ 5.0, \ 0.0$$

GOUP: Go up (contour motion command). *See* GOBACK.

INTOF: Intersection of (descriptive data). Indicates that the intersection of two geometry elements is the specified point. *See* POINT.

INTOL: Inside tolerance (auxiliary statement). Indicates the allowable tolerance between the inside of a curved surface and any straight-line segments used to approximate the curve. See Figure A8.3.

INTOL/.005   FIGURE A8.3

LEFT: Left (descriptive data). Used to indicate which of two alternatives, left or right, is desired. *See* LINE.

LINE: Line (geometry type). Used to define a line that is interpreted by APT as a plane perpendicular to the *xy* plane. Methods of definition:

1.   By the coordinates of two points. See Figure A8.4.

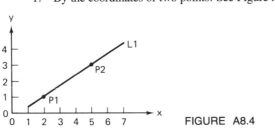

FIGURE A8.4

L1=LINE/2, 1, 0, 5, 3, 0

2. By two points. See Figure A8.4.

L1=LINE/P1,P2

3. By a point and tangent to a circle. See Figure A8.5.

L1=LINE/P1, LEFT, TANTO, C1

L2=LINE/P1, RIGHT, TANTO, C1

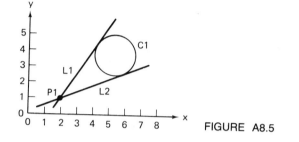

FIGURE A8.5

The descriptive words LEFT AND RIGHT are used by looking from the point toward the circle.

4. By a point and the angle of the line to the *x*-axis or another line. See Figure A8.6.

L3=LINE/P1, LEFT, ATANGL, 20

L4=LINE/P1, LEFT, ATANGL, 30, L3

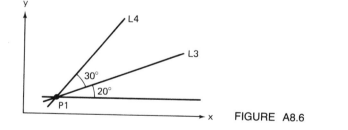

FIGURE A8.6

5. By a point and being parallel to or perpendicular to another line. See Figure A8.7.

L5=LINE/P2, PARLEL, L3

L6=LINE/P2, PERPTO, L3

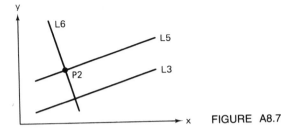

FIGURE A8.7

6. By being tangent to two circles. See Figure A8.8.

$$L7 = LINE/LEFT, TANTO, C3, LEFT, TANTO, C4$$

$$L8 = LINE/LEFT, TANTO, C3, RIGHT, TANTO, C4$$

$$L9 = LINE/RIGHT, TANTO, C3, LEFT, TANTO, C4$$

$$L10 = LINE/RIGHT, TANTO, C3, RIGHT, TANTO, C4$$

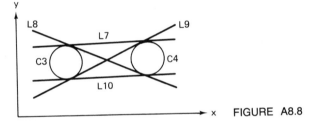

FIGURE A8.8

The descriptive words LEFT and RIGHT are used by looking from the first circle written toward the second circle. For example, another way to specify L7 would be

$$L7 = LINE/RIGHT, TANTO, C4, RIGHT, TANTO, C3$$

MACHIN: Machine (postprocessor statement). Used to specify the machine tool and to call the postprocessor for that machine tool. In the statement

$$MACHIN/MILL, 1$$

the MILL identifies the machine tool type and 1 identifies the particular machine and postprocessor. The APT system then calls the specified postprocessor to prepare the NC tape for that machine.

MACRO: Used to define a suborutine which will be called by the main APT program.

$$DRILL = MACRO/PX$$

where DRILL is the symbol for the subroutine and PX is a parameter in the subroutine whose value will be specified when the subroutine is called from the main program. *See* CALL.

ON: On (motion modifier word) used with three other motion modifier words—TO, PAST, and TANTO—to indicate the point on the check surface where the tool motion is to terminate. See Figure A8.9.

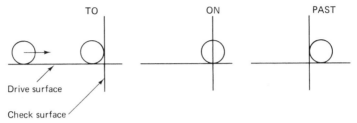

FIGURE A8.9

OUTTOL: Outside Tolerance (auxiliary statement). Indicates the allowable tolerance between the outside of a curved surface and any straight-line segments used to approximate the curve. See Figure A8.10.

OUTTOL/.005   **FIGURE A8.10**

Note: The INTOL and OUTTOL statements can be used together to indicate allowable tolerances on both inside and outside of the curved surface. See Figure A8.11.

INTOL/.0025

OUTTOL/.0025

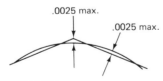

INTOL/.0025
OUTTOL/.0025   **FIGURE A8.11**

PARLEL: Parallel (descriptive data). Used to define a line or plane as being parallel to another line or plane. *See* LINE and PLANE.

PARTNO: Part number (auxiliary statement). Used at start of program to idenfity the part program. PARTNO must be typed in columns 1 through 6 of the first computer card in the deck.

PARTNO MECHANISM PLATE 47320

PAST: Past (motion modifier word). *See* ON.

PERPTO: Perpendicular to (descriptive data). Used to define a line or plane as being perpendicular to some other line or plane. *See* LINE and PLANE.

PLANE: Plane (geometry type). Used to define a plane.

Methods of definition:

1. By three points that do not lie on the same straight line. See Figure A8.12.

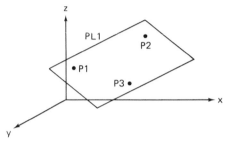

FIGURE A8.12

2. By a point and being parallel to another plane. See Figure A8.13.

PL2 = PLANE/P4, PARLEL, PL1

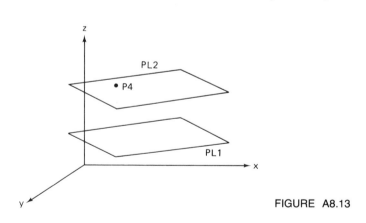

FIGURE A8.13

3. By two points and being perpendicular to another plane. See Figure A8.14.

PL3=PLANE, PERPTO, PL1, P5, P6

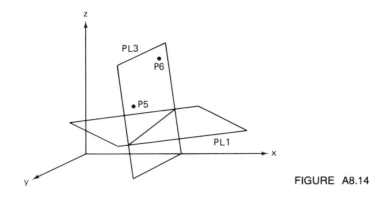

FIGURE A8.14

POINT: Point (geometry type). Used to define a point.

       Methods of definition:

         1.   By the x, y, and z coordinates. See Figure A8.15.

$$P1 = POINT/3.0, \ 1.5, \ 0.0$$

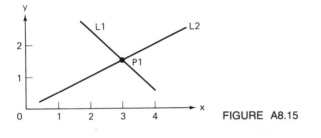

FIGURE A8.15

         2.   By the intersection of two lines. See Figure A8.15.

$$P1 = POINT/L1, \ L2$$

         3.   By the intersection of a line and a circle. See Figure A8.16.

$$P2 = POINT/YLARGE, \ INTOF, \ L3, \ C1$$

$$P3 = POINT/XLARGE, \ INTOF, \ L3, \ C1$$

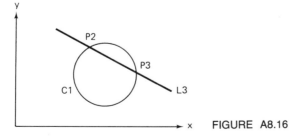

FIGURE A8.16

Any of the descriptive words—XLARGE, XSMALL, YLARGE, YSMALL—can be used to indicate the relative position of the point. For example, for point P2, YLARGE or XSMALL could be used. For point P3, YSMALL or XLARGE could be used.

4. By two intersecting circles. See Figure A8.17.

P4=POINT/YLARGE, INTOF, C1, C2

P5=POINT/YSMALL, INTOF, C1, C2

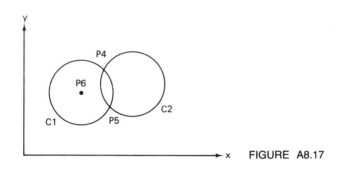

FIGURE A8.17

5. By the center of a circle

P6=POINT/CENTER, C1

RADIUS: Radius (descriptive data). Used to indicate the radius of a circle. *See* CIRCLE.

RIGHT: Right (descriptive data). *See* LEFT and LINE.

TANTO: Tangent to (two uses: descriptive data and motion modifier word).

1. As descriptive data, it is used to indicate tangency of one geometry element to another. *See* CIRCLE and LINE.

2. As a motion modifier word, it is used to indicate that the tool motion is to terminate at the point of tangency between the drive surface and the check surface. See Figure A8.18.

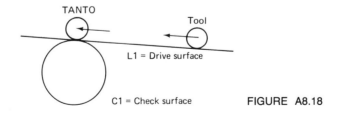

FIGURE A8.18

TERMAC: Termination of MACRO subroutine. Used as the last statement in the MACRO subroutine to indicate a return to the main program at the statement following the CALL. *See* MACRO, CALL.

TO: To (motion modifier word). *See* ON.

TURRET: Turret (postprocessor statement). Used to specify a turret position on a turret lathe or drill or to call a specific tool from an automatic tool changer. Example: TURRET/T30.

XLARGE: In the positive x-direction (descriptive data). Used to indicate the relative position of one geometric element with respect to another when there are two possible alternatives. *See* CIRCLE and POINT.

XSMALL: In the negative x-direction (descriptive data). *See* XLARGE.

YLARGE: In the positive y-direction (descriptive data). *See* XLARGE.

YSMALL: In the negative y-direction (descriptive data). *See* XLARGE.

chapter 9

Computer Controls in NC

9.1 INTRODUCTION

The evolution of numerical control technology has been closely related to and dependent on the development of computer technology. In Chapter 8 we examined the use of computers in NC part programming. As a practical matter, it would not be possible to carry out the part programming function for many part designs without computer-assisted part programming. In addition, the computer is being used to refine and improve the NC part programming procedure through such technologies as interactive graphics and voice programming.

Use of the digital computer has also permitted substantial improvements to be made in the controls for NC. In this chapter we discuss three NC-related control topics:

1. Computer numerical control (CNC)
2. Direct numerical control (DNC)
3. Adaptive control

Computer NC involves the replacement of the conventional hard-wired NC controller unit by a small computer (minicomputer or microcomputer). The small computer is used to perform some or all of the basic NC functions by programs stored in its read/write memory. One of the distinguishing features of CNC is that one computer is used to control one machine tool. This contrasts with the second type of computer control, direct numerical control. DNC involves the use of a larger computer to control a number of separate NC machine tools.

The third control topic, adaptive control, does not require a digital computer for implementation. Adaptive control machining denotes a control system that measures one or more process variables (such as cutting force, temperature, horsepower, etc.) and manipulates feed and/or speed in order to compensate for undesirable changes in the process variables. Its objective is to optimize the machining process, something that NC alone is unable to accomplish. Many of the initial adaptive control projects relied on analog controls rather than digital computers. Today, these systems employ microprocessor technology to implement the adaptive control strategy.

Before describing the three types of control systems, it is appropriate to examine some of the problems related to the use of conventional numerical control which have influenced the changeover to computer control.

9.2 PROBLEMS WITH CONVENTIONAL NC

There are a number of problems inherent in conventional NC which have motivated machine tool builders to seek improvements in the basic NC system. Among the difficulties encountered in using conventional numerical control are the following:

1. *Part programming mistakes.* In preparing the punched tape, part programming mistakes are common. The mistakes can be either syntax or numerical errors, and it is not uncommon for three or more passes to be required before the NC tape is correct. Another related problem in part programming is to achieve the best sequence of processing steps. This is mainly a problem in manual part programming. Some of the computer-assisted part programming languages provide aids to achieve the best operation sequences.

2. *Nonoptimal speeds and feeds.* In conventional numerical control, the control system does not provide the opportunity to make changes in speeds and feeds during the cutting process. As a consequence, the programmer must set the speeds and feeds for worst-case conditions. The result is lower than optimum productivity.

3. *Punched tape.* Another problem related to programming is the tape itself. Paper tape is especially fragile, and its susceptibility to wear and tear causes

it to be an unreliable NC component for repeated use in the shop. More durable tape materials, such as Mylar, are utilized to help overcome this difficulty. However, these materials are relatively expensive.

4. *Tape reader*. The tape reader that interprets the punched tape is generally acknowledged among NC users to be the least reliable hardware component of the machine. When a breakdown is encountered on an NC machine, the maintenance personnel usually begin their search for the problem with the tape reader.

5. *Controller*. The conventional NC controller unit is hard-wired. This means that its control features cannot be easily altered to incorporate improvements into the unit. Use of a computer as the control device would provide the flexibility to make improvements in such features as circular interpolation when better software becomes available.

6. *Management information*. The conventional NC system is not equipped to provide timely information on operational performance to management. Such information might include piece counts, machine breakdowns, and tool changes.

Machine tool builders and control engineers have been continually improving NC technology by designing systems which help to solve these problems. Much of this improvement has been provided by advances in electronics. In the following section we explore the developments in electronics and solid-state technology which have lead the way in NC controller evolution.

9.3 NC CONTROLLER TECHNOLOGY

The hardware technology in NC controls has changed dramatically over the years. At least seven generations of controller hardware can be identified.

1. Vacuum tubes (circa 1952)
2. Electromechanical relays (circa 1955)
3. Discrete semiconductors (circa 1960)
4. Integrated circuits (circa 1965)
5. Direct numerical control (circa 1968)
6. Computer numerical control (circa 1970)
7. Microprocessors and microcomputers (circa 1975)

The initial NC prototype machine built in the MIT Servomechanism Laboratories used vacuum tubes for the controller hardware. These components were so large that the control unit consumed more space than the machine tool. But that was the state of the technology in controls at that time. By the time the first NC machines were sold to the commercial market serveral years later, electromechanical relays were substituted for the vacuum tubes. The problem with these relay-based controls was their large size and poor reliability. Even the relatively simple point-to-point logic required several large cabinets filled with relays. The relays were susceptible to wear, and controls requiring a large number of these components were inherently unreliable.

The use of transistors based on discrete semiconductor technology formed the next generation of NC controllers. The use of transistors helped to reduce the number of electromechanical relays required. Accordingly, this increased the reliability because the use of transistors avoided the wear problem. It also contributed to a downsizing of the controller cabinet and allowed systems designers to build more complex circuitry into the NC controller. Features such as circular interpolation became practical with these controls.

Size and reliability still remained as problems with NC controls which used discrete semiconductors. Also, the electronics were sensitive to heat, and fans or air conditioners were required in the cabinets to operate under factory conditions.

Around 1965, integrated circuits were introduced for use in NC controls. This type of electronic hardware brought significant improvements in size and reliability. The number of separate components could be reduced by 90%. There were corresponding savings in cost to the user. The trend toward LSI (large-scale integrated) circuits has allowed more control features to be packaged into smaller control cabinets. Among these features are circular and hyperbolic interpolation routines, inch-to-metric conversions, and vector feedrate computations.

The next development in NC control marked the introduction of digital computers in NC controller technology. This constituted a fundamental change in NC evolution. All of the previous controls were made up of hard-wired components. The functions that were performed by these control systems could not be easily changed, due to the fixed nature of the hard-wired design. Digital computers, on the other hand, are based on a different approach. In this new approach, the control functions were programmed into the computer memory and could be changed by altering the program.

DNC was the first of the computer control systems to be introduced, around 1968. In the evolution of computer technology, the computers of that era were quite large and expensive, and the only feasible approach seemed to be to use one large computer to control a number of machine tools on a time-shared basis. The advantage of DNC was that it established a direct control link between the computer and the machine tool, hence eliminating the necessity for using punched tape input. The tape and tape reader were turning out to be the least reliable components in the conventional NC systems.

With the recognized trend toward smaller, less expensive computers, it soon became practical to apply a single small computer to one machine tool. This concept came to be called computer numerical control (CNC). The CNC systems were first commercially introduced around 1970, and they applied the soft-wired controller approach to good advantage. One standard computer control unit could be adapted to various types of machine tools by programming the control functions into the computer memory for that particular machine. Today, because of the advantages of CNC, very few conventional hard-wired NC systems are sold in the United States.

Advances in computer technology have continued to provide smaller and smaller digital control devices which have greater speed and capacity at lower cost. This has permitted the machine tool builders to design the CNC control panel as an integral part of the machine tool rather than as a separate stand-alone cabinet. This reduces floor space requirements for the machine. The VLSI (very large scale integrated) circuits used in these controllers are advantageous to the machine tool designer and to the machine user. Fewer components in the controller means it is easier and less expensive for the machine tool builder to fabricate. Fewer circuit boards, which are readily replaced, reduce the burden on the user for maintenance and repair.

Now that we have traced the evolution of NC controls from the original vacuum-tube controller at MIT to the modern microcomputer-based controls, we will next examine the technology of CNC and DNC in more detail.

9.4   COMPUTER NUMERICAL CONTROL

Computer numerical control is an NC system that utilizes a dedicated, stored program computer to perform some or all of the basic numerical control functions. Because of the trend toward downsizing in computers, most of the CNC systems sold today use a microcomputer-based controller unit. Over the years, minicomputers have also been used in CNC controls.

The external appearance of a CNC machine is very similar to that of a conventional NC machine. Part programs are initially entered in a similar manner. Punched tape readers are still the common device to input the part program into the system. However, with conventional numerical control, the punched tape is cycled through the reader for every workpiece in the batch. With CNC, the program is entered once and then stored in the computer memory. Thus the tape reader is used only for the original loading of the part program and data. Compared to regular NC, CNC offers additional flexibility and computational capability. New system options can be incorporated into the CNC controller simply by reprogramming the unit. Because of this reprogramming capacity, both in terms of part programs and system control options, CNC is often referred to by the term ''soft-wired'' NC. Figure 9.1 illustrates the general configuration of a CNC system.

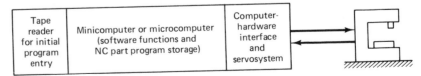

FIGURE 9.1   General configuration of computer numerical control (CNC) system.

Functions of CNC

There are a number of functions which CNC is designed to perform. Several of these functions would be either impossible or very difficult to accomplish with conventional NC. The principal functions of CNC are:

1. Machine tool control
2. In-process compensation
3. Improved programming and operating features
4. Diagnostics

MACHINE TOOL CONTROL. The primary function of the CNC system is control of the machine tool. This involves conversion of the part program instructions into machine tool motions through the computer interface and servosystem. The capability to conveniently incorporate a variety of control features into the soft-wired controller unit is the main advantage of CNC. Some of the control functions, such as circular interpolation, can be accomplished more efficiently with hard-wired circuits than with the computer. This fact has lead to the development of two alternative controller designs in CNC:

1. Hybrid CNC
2. Straight CNC

In the hybrid CNC system, illustrated in the diagram of Figure 9.2, the controller consists of the soft-wired computer plus hard-wired logic circuits. The hard-wired components perform those functions which they do best, such as feed rate generation and circular interpolation. The computer performs the remaining control functions plus other duties not normally associated with a conventional hard-wired controller. There are several reasons for the popularity of the hybrid CNC configuration. As mentioned previously, certain NC functions can be performed more efficiently with the hard-wired circuits. These are functions which are common to

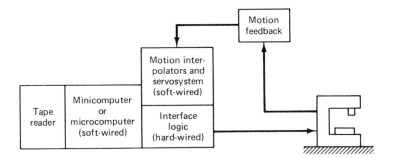

FIGURE 9.2 Hybrid CNC.

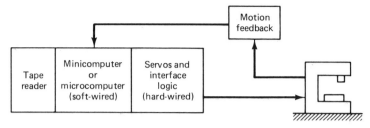

FIGURE 9.3   Straight CNC.

most NC systems. Therefore, the circuits that perform these functions can be produced in large quantities at relatively low cost. Use of these hard-wired circuits saves the computer from performing these calculation chores. Hence a less expensive computer is required in the hybrid CNC controller.

The straight CNC system uses a computer to perform all the NC functions. The only hard-wired elements are those required to interface the computer with the machine tool and the operator's console. Interpolation, tool position feedback, and all other functions are performed by computer software. Accordingly, the computer required in a straight CNC system must be more powerful than that needed for a hybrid system. The advantage gained in the straight CNC configuration is additional flexibility. It is possible to make changes in the interpolation programs, whereas the logic contained in the hard-wired circuits of hybrid CNC cannot be altered. A diagram of the straight CNC design is shown in Figure 9.3.

IN-PROCESS COMPENSATION. A function closely related to machine tool control is in-process compensation. This involves the dynamic correction of the machine tool motions for changes or errors which occur during processing. Some of the options included within the category of CNC in-process compensation are:

Adjustments for errors sensed by in-process inspection probes and gauges. (These are discussed in Section 9.8)

Recomputation of axis positions when an inspection probe is used to locate a datum reference on a workpart.

Offset adjustments for tool radius and length.

Adaptive control adjustments to speed and/or feed. (We consider adaptive control machining in Section 9.7.)

Computation of predicted tool life and selection of alternative tooling when indicated.

IMPROVED PROGRAMMING AND OPERATING FEATURES. The flexibility of soft-wired control has permitted the introduction of many convenient programming and operating features. Included among these features are the following:

Editing of part programs at the machine. This permits correction or optimization of the program.

Graphic display of the tool path to verify the tape.

Various types of interpolation: circular, parabolic, and cubic interpolation.

Support of both U.S. customary units and metric units (International System).

Use of specially written subroutines.

Manual data input (MDI).

Local storage of more than one part program.

DIAGNOSTICS. NC machine tools are complex and expensive systems. The complexity increases the risk of component failures which lead to system downtime. It also requires that the maintenance personnel be trained to a higher level of proficiency in order to make repairs. The higher cost of NC provides a motivation to avoid downtime as much as possible. CNC machines are often equipped with a diagnostics capability to assist in maintaining and repairing the system. These diagnostics features are still undergoing development and future systems will be much more powerful in their capabilities than current CNC units. Ideally, the diagnostics subsystem would accomplish several functions. First, the subsystem would be able to identify the reason for a downtime occurrence so that the maintenance personnel could make repairs more quickly. Second, the diagnostics subsystem would be alert to signs that indicate the imminent failure of a certain component. Hence maintenance personnel could replace the faulty component during a scheduled downtime, thus avoiding an unplanned interruption of production. A third possible function which goes beyond the normal diagnostics capability is for the CNC system to contain a certain amount of redundancy of components which are considered unreliable. When one of these components fails, the diagnostics subsystem would automatically disconnect the faulty component and activate the redundant component. Repairs could thus be accomplished without any breaks in normal operations.

ADVANTAGES OF CNC

Computer numerical control possesses a number of inherent advantages over conventional NC. The following list of benefits will serve also as a summary of our preceding discussion:

1. *The part program tape and tape reader are used only once* to enter the program into computer memory. This results in improved reliability, since the tape reader is commonly considered the least reliable component of a conventional NC system.

2. *Tape editing at the machine site.* The NC tape can be corrected and even optimized (e.g., tool path, speeds, and feeds) during tape tryout at the site of the machine tool.

3. *Metric conversion.* CNC can accommodate conversion of tapes prepared in units of inches into the International System of units.

4. *Greater flexibility.* One of the more significant advantages of CNC over conventional NC is its flexibility. This flexibility provides the opportunity to introduce new control options (e.g., new interpolation schemes) with relative ease at low cost. The risk of obsolescence of the CNC system is thereby reduced.

5. *User-written programs.* One of the possibilities not originally anticipated for CNC was the generation of specialized programs by the user. These programs generally take the form of MACRO subroutines stored in CNC memory which can be called by the part program to execute frequently used cutting sequences.

6. *Total manufacturing system.* CNC is more compatible with the use of a computerized factory-wide manufacturing system. One of the stepping stones toward such a system is the concept of direct numerical control.

9.5 DIRECT NUMERICAL CONTROL

Direct numerical control can be defined as a manufacturing system in which a number of machines are controlled by a computer through direct connection and in real time. The tape reader is omitted in DNC, thus relieving the system of its least reliable component. Instead of using the tape reader, the part program is transmitted to the machine tool directly from the computer memory. In principle, one large computer can be used to control more than 100 separate machines. The DNC computer is designed to provide instructions to each machine tool on demand. When the machine needs control commands, they are communicated to it immediately. DNC also involves data collection and processing from the machine tool back to the computer.

Components of a DNC system

Figure 9.4 illustrates the configuration of the basic DNC system. A direct numerical control system consists of four basic components:

1. Central computer
2. Bulk memory, which stores the NC part programs
3. Telecommunication lines
4. Machine tools

The computer calls the part program instructions from bulk storage and sends them to the individual machines as the need arises. It also receives data back from the

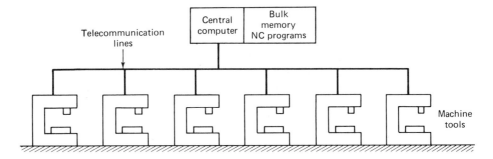

FIGURE 9.4   General configuration of direct numerical control (DNC) system.

machines. This two-way information flow occurs in real time, which means that each machine's requests for instructions must be satisfied almost instantaneously. Similarly, the computer must always be ready to receive information from the machines and to respond accordingly. The remarkable feature of the DNC system is that the computer is servicing a large number of separate machine tools, all in real time.

Depending on the number of machines and the computational requirements that are imposed on the computer, it is sometimes necessary to make use of satellite computers, as shown in Figure 9.5. These satellites are minicomputers, and they serve to take some of the burden off the central computer. Each satellite controls several machines. Groups of part program instructions are received from the central computer and stored in buffers. They are then dispensed to the individual machines as required. Feedback data from the machines are also stored in the satellite's buffer before being collected at the central computer.

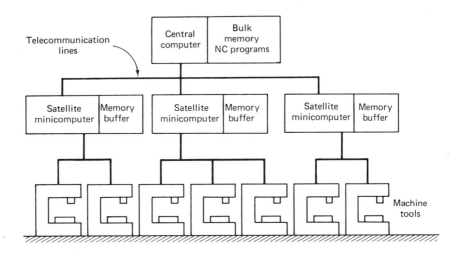

FIGURE 9.5   DNC with satellite minicomputers.

Two types of DNC

There are two alternative system configurations by which the communication link is established between the control computer and the machine tool. One is called a behind-the-tape reader system; the other configuration makes use of a specialized machine control unit.

BEHIND-THE-TAPE-READER (BTR) SYSTEM. In this arrangement, pictured in Figure 9.6, the computer is linked directly to the regular NC controller unit. The replacement of the tape reader by the telecommunication lines to the DNC computer is what gives the BTR configuration its name. The connection with the computer is made between the tape reader and the controller unit—behind the tape reader.

Except for the source of the command instructions, the operation of the system is very similar to conventional NC. The controller unit uses two temporary storage buffers to receive blocks of instructions from the DNC computer and convert them into machine actions. While one buffer is receiving a block of data, the other is providing control instructions to the machine tool.

SPECIAL MACHINE CONTROL UNIT. The other strategy in DNC is to eliminate the regular NC controller altogether and replace it with a special machine control unit. The configuration is illustrated in Figure 9.7. This special MCU is a device that is specifically designed to facilitate communication between the machine tool and the computer. One area where this communication link is important is in circular interpolation of the cutter path. The special MCU configuration

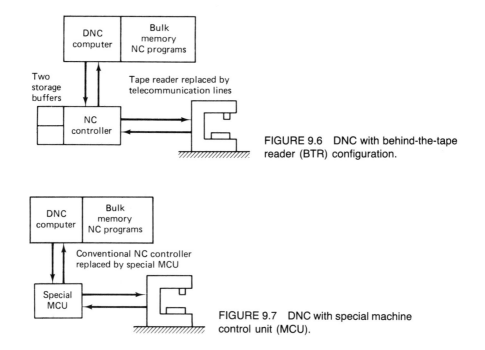

FIGURE 9.6 DNC with behind-the-tape reader (BTR) configuration.

FIGURE 9.7 DNC with special machine control unit (MCU).

achieves a superior balance between accuracy of the interpolation and fast metal removal rates than is generally possible with the BTR system.

The special MCU is soft-wired, while the conventional NC controller is hard-wired. The advantage of soft-wiring is its flexibility. Its control functions can be altered with relative ease to make improvements. It is much more difficult to make changes in the regular NC controller because rewiring is required.

At present, the advantage of the BTR configuration is that its cost is less, since only minor changes are needed in the conventional NC system to bring DNC into the shop. BTR systems do not require the replacement of the conventional control unit by a special MCU. However, this BTR advantage is a temporary one, since most NC machines are sold with computer numerical control. The CNC controller serves much the same purpose as a special MCU when incorporated into a DNC system.

Functions of DNC

There are several functions which a DNC system is designed to perform. These functions are unique to DNC and could not be accomplished with either conventional NC or CNC. The principal functions of DNC are:

1. NC without punched tape
2. NC part program storage
3. Data collection, processing, and reporting
4. Communications

NC WITHOUT PUNCHED TAPE. One of the original objectives in direct numerical control was to eliminate the use of punched tape. Several of the problems with conventional NC discussed in Section 9.2 are related to the use of punched tape (the relatively unreliable tape reader, the fragile nature of paper tape, the difficulties in making corrections and changes in the program contained on punched tape, etc.). There is also the expense associated with the equipment that produces the punched tape. All of these costs and inconveniences can be eliminated with the DNC approach.

NC PART PROGRAM STORAGE. A second important function of the DNC system is concerned with storing the part programs. The program storage subsystem must be structured to satisfy several purposes. First, the programs must be made available for downloading to the NC machine tools. Second, the subsystem must allow for new programs to be entered, old programs to be deleted, and existing programs to be edited as the need arises. Third, the DNC software must accomplish the postprocessing function. The part programs in a DNC system would typically be stored as the CLFILE. The CLFILE must be converted into instructions for a particular machine tool. This conversion is performed by the postprocessor. Fourth, the storage subsystem must be structured to perform certain data processing and management functions, such as file security, display of programs, manipulation of data, and so on.

The DNC program storage subsystem usually consists of an active storage and a secondary storage. The active storage would be used to store NC programs which are frequently used. A typical mass storage device for this purpose would be a disk. The active storage can be readily accessed by the DNC computer to drive an NC machine in production. The secondary storage would be used for NC programs which are not frequently used. Sometimes, even though it is anticipated that a particular program will probably never be used again, it may be decided to save that program if the storage costs are not excessive. Examples of secondary storage media used in DNC include magnetic tape, tape cassettes, floppy disks, disk packs, and even punched tape. (Unfortunately, the last alternative resurrects the several disadvantages mentioned earlier.)

DATA COLLECTION, PROCESSING, AND REPORTING. The two previous functions for DNC both concerned the direct link from the central computer to the machine tools in the factory. Another important function of DNC involves the opposite link, the transfer of data from the machine tools back to the central computer. DNC involves a two-way transfer of data.

The basic purpose behind the data collection, processing, and reporting function of DNC is to monitor production in the factory. Data are collected on production piece counts, tool usage, machine utilization, and other factors that measure performance in the shop. These data must be processed by the DNC computer, and reports are prepared to provide management with information necessary for running the plant. The scope of this DNC function has been broadened over the years to include data collection not only from the NC machines, but from all other work centers in the factory. The term used to describe this broader scope activity is shop floor control. We shall be discussing shop floor control and computer process monitoring in subsequent chapters of this book.

COMMUNICATIONS. A communications network is required to accomplish the previous three functions of DNC. Communication among the various subsystems is a function that is central to the operation of any DNC system. The essential communication links in direct numerical control are between the following components of the system:

Central computer and machine tools
Central computer and NC part programmer terminals
Central computer and bulk memory, which stores the NC programs

Optional communication links may also be established between the DNC system and any of the following additional systems:

Computer-aided design (CAD) system

Shop floor control system

Corporate data processing computer

Remote maintenance diagnostics system

Other computer-automated systems in the plant

These types of communications are becoming more common as DNC technology moves toward the computer-integrated factory of the future.

Advantages of DNC

Just as CNC had certain advantages over a conventional NC system, there are also advantages associated with the use of direct numerical control. The following list will recapitulate much of our previous discussion of DNC:

1. *Elimination of punched tapes and tape readers.* Direct numerical control eliminates the least reliable element in the conventional NC system. In some DNC systems, the hard-wired control unit is also eliminated, and replaced by a special machine control unit designed to be more compatible with DNC operation.

2. *Greater computational capability and flexibility.* The large DNC computer provides the opportunity to perform the computational and data processing functions more effectively than traditional NC. Because these functions are implemented with software rather than with hard-wired devices, there exists the flexibility to alter and improve the method by which these functions are carried out. Examples of these functions include circular interpolation and part programming packages with convenient editing and diagnostics features.

3. *Convenient storage of NC part programs in computer files.* This compares with the more manually oriented storage of punched tapes in conventional NC.

4. *Programs stored as CLFILE.* Storage of part programs in DNC is generally in the form of cutter path data rather than postprocessed programs for specific machine tools. Storing of the programs in this more general format affords the flexibility in production scheduling to process a job on any of several different machine tools.

5. *Reporting of shop performance.* One of the important features in DNC involves the collection, processing, and reporting of production performance data from the NC machines.

6. *Establishes the framework for the evolution of the future computer-automated factory.* The direct numerical control concept represents a first step in the development of production plants which will be managed by computer systems.

9.6   COMBINED DNC/CNC SYSTEMS

The direct numerical control systems that were marketed in the late 1960s and early 1970s were extremely expensive. Their high cost, combined with an unfavorable economic climate at that time, caused business managers to resist the temptation to plunge into the new DNC technology. Also, the DNC systems available then were somewhat rigid in terms of management reporting formats and hardware requirements. The advent of CNC systems, together with lower-cost computers and improvements in software, have resulted in the development of hierarchical computer systems in manufacturing. In these hierarchical systems, CNC computers have direct control over the production machines and report to satellite minicomputers, which in turn report to other computers, and so on. There are advantages to this hierarchical approach over the DNC packages that were offered around 1970. The common theme in these advantages is flexibility. The information system can be tailored to the specific needs and desires of the firm. This contrasts with many of the early DNC systems, in which the reporting formats were fixed, in some cases providing more data than management wanted and in other cases omitting details that management needed. Another advantage of the hierarchy approach is the ability to gradually build the system instead of implementing the entire DNC configuration all at once. This piece-by-piece installation of the computer-integrated manufacturing system is a more versatile and economic approach. It permits changes and corrections to be made more easily as the system is being built. It also allows the company to spread the cost of the system over a longer time period and to obtain benefits from each subsystem as it is installed. The hierarchical computer arrangement embraces the DNC philosophy, which is to provide useful reports on production operations to management in real time. One might say that DNC has not really been replaced by this new approach; it has simply altered its physical form. This evolution in the configuration of DNC and its inclusion of computer numerical control have resulted in the introduction of the term "distributed numerical control" for the initials DNC.

The combination of DNC and CNC provides the opportunity to add new capabilities and refine existing capabilities in these computerized manufacturing systems. Certainly, an obvious advantage derived from the combination of CNC and DNC is the elimination of the use of punched tape as the input media for CNC machines. The DNC computer downloads the program directly to the CNC computer memory. Unlike the machine control unit in a conventional DNC system, the CNC controller has sufficient storage capacity to accept the entire part program. The part program is downloaded once rather than in a block-by-block procedure. This reduces the amount of communication required between the central computer and each machine tool. A likely future trend in these combined DNC/CNC systems will be that the postprocessor is built into the software of the CNC controller. This would allow the part program to be downloaded from the DNC computer in CLFILE form without postprocessing.

A second advantage created by combining CNC with DNC is redundancy. If the central DNC computer fails, this will not necessarily cause the individual machines in the system to be down. It is possible to provide the necessary backup to permit the CNC machines to operate on a stand-alone basis. This backup capability consists of two elements. The first is a file of punched tapes which duplicate the programs contained in the DNC computer files. The second is that each CNC machine must be equipped with a tape reader or be capable of connecting to a portable tape reader for the purpose of entering the program from the punched tape. There are, of course, costs associated with providing this backup feature.

A third improvement that develops from combined DNC/CNC systems is improved communication between the central computer and the shop floor. With digital computers located at both ends of the communication links, many of the constraints in the design of powerful factory management information systems are removed. It is easier for computers to communicate with other computers than with hard-wired devices.

9.7 ADAPTIVE CONTROL MACHINING SYSTEMS

Adaptive control (abbreviated AC) machining originated out of research in the early 1960s sponsored by the U.S. Air Force at the Bendix Research Laboratories. The initial adaptive control systems were based on analog control devices, representing the state of technology at that time. Today, AC uses microprocessor-based controls and it is typically integrated with an existing CNC system. Accordingly, the topic of adaptive control is appropriate to include in this chapter on computer controls in NC.

For a machining operation, the term *adaptive control* denotes a control system that measures certain output process variables and uses these to control speed and/or feed. Some of the process variables that have been used in adaptive control machining systems include spindle deflection or force, torque, cutting temperature, vibration amplitude, and horsepower. In other words, nearly all the metal-cutting variables that can be measured have been tried in experimental adaptive control systems. The motivation for developing an adaptive machining system lies in trying to operate the process more efficiently. The typical measures of performance in machining have been metal removal rate and cost per volume of metal removed.

Where to use adaptive control

One of the principal reasons for using numerical control (including DNC and CNC) is that NC reduces the nonproductive time in a machining operation. This time savings is achieved by reducing such elements as workpiece handling time, setup of the job, tool changes, and other sources of operator and machine delay. Because these nonproductive elements are reduced relative to total production time, a larger proportion of the time is spent in actually machining the workpart. Although NC has a significant effect on downtime, it can do relatively little to reduce the in-process time compared to a conventional machine tool. The most promising answer for reducing the in-process time lies in the use of adaptive control. Whereas numerical control guides the sequence of tool positions or the path of the tool during machining, adaptive control determines the proper speeds and/or feeds during machining as a function of variations in such factors as work-material hardness, width or depth of cut, air gaps in the part geometry, and so on. Adaptive control has the capability to respond to and compensate for these variations during the process. Numerical control does not have this capability.

Adaptive control (AC) is not appropriate for every machining situation. In general, the following characteristics can be used to identify situations where adaptive control can be beneficially applied:

1. The in-process time consumes a significant portion of the machining cycle time. Mathias [12] uses the rule of thumb that adaptive control can best be justified when the cutter is engaged in the workpiece more than 40% of the time it is on the machine.
2. There are significant sources of variability in the job for which adaptive control can compensate. In effect, AC adapts feed and/or speed to these variable conditions. We examine these sources of variability in the following subsection.
3. The cost of operating the machine tool is high. The high operational cost results mainly from the high investment in equipment.
4. The typical jobs are ones involving steel, titanium, and high-strength alloys. Cast iron and aluminum are also attractive candidates for AC, but these materials are generally easier to machine.

Sources of variability in machining

The following are the typical sources of variability in machining where adaptive control can be most advantageously applied. Not all of these sources of variability need be present to justify the use of AC. However, it follows that the greater the variability, the more suitable the process will be for using adaptive control.

1. *Variable geometry of cut in the form of changing depth or width of cut.* In these cases, feed rate is usually adjusted to compensate for the variability. This type of variability is often encountered in profile milling or contouring operations.
2. *Variable workpiece hardness and variable machinability.* When hard spots or other areas of difficulty are encountered in the workpiece, either speed or feed is reduced to avoid premature failure of the tool.

3. *Variable workpiece rigidity.* If the workpiece deflects as a result of insufficient rigidity in the setup, the feed rate must be reduced to maintain accuracy in the process.

4. *Toolwear.* It has been observed in research that as the tool begins to dull, the cutting forces increase. The adaptive controller will typically respond to tool dulling by reducing the feed rate.

5. *Air gaps during cutting.* The workpiece geometry may contain shaped sections where no machining needs to be performed. If the tool were to continue feeding through these so-called air gaps at the same rate, time would be lost. Accordingly, the typical procedure is to increase the feed rate by a factor or 2 or 3, when air gaps are encountered.

These sources of variability present themselves as time varying and, for the most part, unpredictable changes in the machining process. We shall now examine how adaptive control can be used to compensate for these changes.

Two types of adaptive control

In the development of adaptive control machining systems, two distinct approaches to the problem can be distinguished. These are:

1. Adaptive control optimization (ACO)
2. Adaptive control constraint (ACC)

ADAPTIVE CONTROL OPTIMIZATION. This is represented by the early Bendix research on adaptive control machining. In this form of adaptive control, an index of performance is specified for the system. This performance index is a measure of overall process performance, such as production rate or cost per volume of metal removed. The objective of the adaptive controller is to optimize the index of performance by manipulating speed and/or feed in the operation.

Most adaptive control optimization systems attempt to maximize the ratio of work material removal rate to tool wear rate. In other words, the index of performance is

$$IP = \text{a function of } \frac{MRR}{TWR}$$

where MRR = material removal rate

TWR = tool wear rate

The trouble with this performance index is that TWR cannot be measured on-line with today's measurement technology. Hence the IP above cannot really be monitored during the process. Eventually, sensors will be developed to a level at which the true process performance can be measured on-line. When this occurs, adaptive control optimization systems will become more prominent. However, because of the sensor problems encountered in the design of ACO systems, nearly all adaptive control machining is of the second type, adaptive control constraint systems.

ADAPTIVE CONTROL CONSTRAINT. The systems developed for actual production were somewhat less sophisticated (and less expensive) than the research ACO systems. The production AC systems utilize constraint limits imposed on certain measured process variables. Accordingly, these are called adaptive control constraint (ACC) systems. The objective in these systems is to manipulate feed and/or speed so that these measured process variables are maintained at or below their constraint limit values. The following subsection describes the operation of the most common commercially available ACC system.

Operation of an ACC System

Typical applications of adaptive control machining are in profile or contour milling jobs on an NC machine tool. Feed is used as the controlled variable, and cutter force and horsepower are used as the measured variables. It is common to attach an adaptive controller to an NC machine tool. Numerical control machines are a natural starting point for AC for two reasons. First, NC machine tools often possess the required servomotors on the table axes to accept automatic control. Second, the usual kinds of machining jobs for which NC is used possess the sources of variability that make AC feasible. Several large companies have retrofitted their NC machines to include adaptive control. One company, Macotech Corporation in Seattle, Washington, specializes in retrofitting NC machine tools for other manufacturing firms. The adaptive control retrofit package consists of a combination of hardware and software components. The typical hardware components are:

1. Sensors mounted on the spindle to measure cutter deflection (force).
2. Sensors to measure spindle motor current. This is used to provide an indication of power consumption.
3. Control unit and display panel to operate the system.
4. Interface hardware to connect the AC system to the existing NC or CNC control unit.

The software in the AC package consists of a machinability program which can be called as an APT MACRO statement. The relationship of the machinability program in the part programming process is shown in Figure 9.8. The inputs to the program include cutting parameters such as cutter size and geometry, work material hardness, size of cut, and machine tool characteristics. From calculations

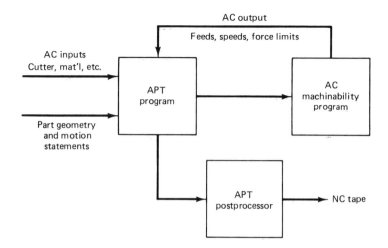

FIGURE 9.8   Relationship of adaptive control (AC) software to APT program.

based on these parameters, the outputs from the program are feed rates, spindle speeds, and cutter force limits for each section of the cut. The objective in these computations is to determine cutting conditions which will maximize metal removal rates. The NC part programmer would ordinarily have to specify feeds and speeds for the machining job. With adaptive control, these conditions are computed by the machinability program based on the input data supplied by the part programmer.

In machining, the AC system operates at the force value calculated for the particular cutter and machine tool spindle. Maximum production rates are obtained by running the machine at the highest feed rate consistent with this force level. Since force is dependent on such factors as depth and width of cut, the end result of the control action is to maximize metal removal rates within the limitations imposed by existing cutting conditions.

Figure 9.9 shows a schematic diagram illustrating the operation of the AC system during the machining process. When the force increases due to increased workpiece hardness or depth or width of cut, the feed rate is reduced to compensate. When the force decreases, owing to decreases in the foregoing variables or air gaps in the part, feed rate is increased to maximize the rate of metal removal.

Figure 9.9 shows an air-gap override feature which monitors the cutter force and determines if the cutter is moving through air or through metal. This is usually sensed by means of a low threshold value of cutter force. If the actual cutter force is below this threshold level, the controller assumes that the cutter is passing through an air gap. When an air gap is sensed, the feed rate is doubled or tripled to minimize the time wasted traveling across the air gap. When the cutter reengages metal on the other side of the gap, the feed reverts back to the cutter force mode of control.

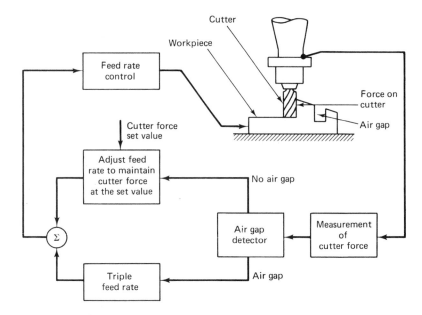

FIGURE 9.9   Configuration of typical adaptive control machining system that uses cutter forces as the measured process variable.

More than one process variable may be measured in an adaptive control machining system. Originally, attempts were made to employ three measured signals in the Bendix system: temperature, torque, and vibration. Currently, the Macotech system has used both cutter load and horsepower generated at the machine motor. The purpose of the power sensor is to protect the motor from overload when the metal removal rate is constrained by spindle hoursepower rather than spindle force.

Benefits of adaptive control machining

A number of potential benefits accrue to the user of an adaptive control machine tool. The advantage gained will depend on the particular job under consideration. There are obviously many machining situations for which it cannot be justified. Adaptive control has been successfully applied in such machining processes as milling, drilling, tapping, grinding, and boring. It has also been applied in turning, but with only limited success. Following are some of the benefits gained from adaptive control in the successful applications.

1. *Increased production rates.* Productivity improvement was the motivating force behind the development of adpative control machining. On-line adjustments to allow for variations in work geometry, material, and tool wear provide the machine with the capability to achieve the highest metal removal rates that

are consistent with existing cutting conditions. This capability translates into more parts per hour. Given the right application, adaptive control will yield significant gains in production rate compared to conventional machining or numerical control.

The production rate advantage of adaptive control over NC machining is illustrated in Table 9.1 for milling and drilling operations on a variety of work materials. Savings in cycle time reported in this table range from 20% up to nearly 60% for milling and 33 to 38% for drilling.

2. *Increased tool life.* In addition to higher production rates, adaptive control will generally provide a more efficient and uniform use of the cutter throughout its tool life. Because adjustments are made in the feed rate to prevent severe loading of the tool, fewer cutters will be broken.

3. *Greater part protection.* Instead of setting the cutter force constraint limit on the basis of maximum allowable cutter and spindle deflection, the force limit can be established on the basis of work size tolerance. In this way, the part is protected against an out-of-tolerance condition and possible damage.

4. *Less operator intervention.* The advent of adaptive control machining has transferred control over the process even further out of the hands of the machine operator and into the hands of management via the part programmer.

TABLE 9.1   Comparison of Machining Times—NC versus Adaptive Control

Operation	Description	Work material	NC time		AC time		Percent saving
Profile milling	Aircraft flap ribs	Aluminum	152	min	81	min	46
Profile milling	Aircraft flap ribs	Aluminum	641	min	319	min	50
Profile milling	Aerospace component	Stainless steel	9.6 h		7.5 h		22
Profile milling	Aerospace component	Stainless steel	11.8 h		9.4 h		20
Profile milling	Space shuttle engine ring	Inconel 718					35
Profile milling	Engine Mounting ring	Inconel 718					45
Profile milling	Aircraft component	Titanium	64	min	35	min	48
End milling	Aerospace component	4330 Steel	61	min	25	min	59
Drilling	0.433'' diameter × 1.0'' deep	1019 Steel	8	s	5	s	38
Drilling	0.433'' diameter × 1.75'' deep	Cast iron	10.5 s		7	s	33

Source: Data courtesy of Macotech Corp.

5. *Easier part programming.*   A benefit of adaptive control which is not so obvious concerns the task of part programming. With ordinary numerical control, the programmer must plan the speed and feed for the worst conditions that the cutter will encounter. The program may have to be tried out several times before the programmer is satisfied with the choice of conditions. In adaptive control part programming, the selection of feed is left to the controller unit rather than to the part programmer. The programmer can afford to take a less conservative approach than with conventional NC programming. Less time is needed to generate the program for the job, and fewer tryouts are required.

9.8   TRENDS AND NEW DEVELOPMENTS IN NC

We will conclude these three chapters on numerical control by discussing some of the important trends and new developments in NC technology. Without question, the most important general trend in NC involves the expanding use of computer technology. The use of computers has already provided significant improvements in part programming procedures (e.g., computer-assisted programming, interactive graphics, and voice programming). The control of NC machinery has also been dramatically enhanced through computer technology (e.g., CNC, DNC, and adaptive control). We have covered these topics in Chapter 8 on programming and in the current chapter on computerized NC. In the following sections, we discuss some additional topics likely to influence the future evolution of numerical control.

Replacement of punched tape

The standard medium used to contain the part program in NC has been 1-in.-wide punched tape. As already discussed there are problems with the use of this medium. With the greater use of computers in today's numerically controlled processes, there is a movement among machine builders and controls people to replace the punched tape and tape reader with a medium that is more compatible with modern computer systems. The most likely substitute is a magnetic tape cartridge. Because of the large inventory of punched tapes and associated equipment in industry presently, the transition to a new part program medium will take a long time.

Inspection probes

The use of in-process inspection probes is becoming more common in modern NC machine tool systems. These inspection probes are sophisticated dial indicators which can be mounted in the machine tool spindle. In machines with automatic tool changers, the probe would be placed in the tool storage drum and loaded into the spindle when needed, just like any of the regular cutting tools. Sensors in the probe detect when contact has been made with a surface of the workpart (or other

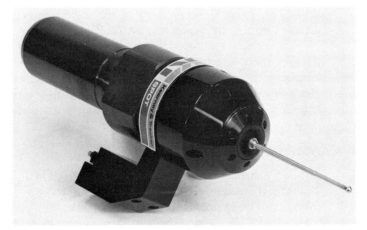

FIGURE 9.10   Inspection probe used in NC machine tools. (Courtesy of Kearney & Trecker Corp.)

object) being checked. The sensor signals are transmitted inductively to the machine tool, thus avoiding the need for supplementary direct wiring between the inspection probe and the NC computer control unit. The software in the controller performs the necessary computations to interpret the signals from the probe. One of the commercially available inspection probes is illustrated in Figure 9.10.

Some typical applications of the inspection probes include the following:

In-process inspection of parts while still fixtured on the machine table

Self-correction of tool locations to compensate for machine errors

Location of a datum reference on the part after initial machining to achieve greater accuracy in subsequent machining operations

Inspection of cutting tools to determine the condition of the cutter (e.g., broken teeth on cutter)

The prinicpal benefits that derive from the use of inspection probes are time savings and improved accuracy. Time can be saved in several activities during part production. The most obvious of these is reducing the need for the time-consuming manual inspection procedures which normally follow the machining operations. As in-process inspection techniques are perfected, this tedious human activitiy will be substantially decreased. Other time savings result from a reduction in the setup and alignment of parts on the machine table (the probe is used to determine compensations for part location errors), and savings in remachining time (in-process inspection is performed with the probe while the job is still set up on the machine). Improved accuracy in the measurement process results from the inherent accuracy in the rigid machine tool structure. Measurements taken with the spindle probe are generally more accurate than traditional techniques used to measure part dimensions. Furthermore, the accuracy of the inspection probe system is substantially greater than the machining process itself.

When used for in-process measurement with resulting compensations in cutter position, these inspection probes represent a form of adaptive control. The process variable is the deviation from specified part dimension caused by cutter wear, fixture problems, or other errors. To correct for these errors, the system makes the necessary adjustments in the calculated tool path.

Advanced NC systems

The APT language for NC part programming is more than 20 years old. It was originally developed for the milling process and has been modified and enhanced over the years to be compatible with other manufacturing operations. Many of the concepts of APT geometry definition were utilized to develop the current geometric modeling technology of CAD/CAM. In many respects, developments in computerized geometric modeling have outpaced and outdated the APT geometry concepts. Advanced NC systems would attempt to make use of the latest CAD/CAM geometry concepts and not be constrained by the limitations inherent in APT. One of the important research and development efforts in this area is CAM-I's[1] Advanced NC Project. The objectives are to develop an advanced NC system that would use the latest concepts and technologies in CAD/CAM and data base management. It is beyond the scope and purpose of this book to provide a lengthy discussion of this R&D project. The interested reader should consult the latest project reports published by CAM-I[3]. However, it seems appropriate to give a summary of this important project.

In APT programming, the workpart is defined with geometric elements such as lines, planes, and circles. These lines, planes, and circles are "unbounded" in the sense that the lines and planes are infinite and the circles are complete circles. The workpart, of course, is bounded, so the APT geometry elements do not really provide an accurate and comprehensive definition of the part geometry. It is by means of a sequence of APT motion statements that the tool is directed around the actual surface of the part, ignoring the portions of the circles and lines that do not relate to the part. In the CAD/CAM approach to geometric modeling, the part is defined by surfaces and edges that construct a solid geometric description of the part. The surfaces and edges do not extend infinitely in their respective directions. The term given to this method of part definiton is "bounded geometry," which contrasts with the unbounded approach used in APT. One of the important goals of CAM-I's advanced NC project is to utilize the concepts embodied in the bounded geometry approach to part definition. Some of the important objectives of the project can be outlined as follows:

---

[1]CAM-I stands for Computer-Aided Manufacturing—International, a nonprofit firm based in Arlington, Texas.

1. *New language set.* The objective is to develop a new language set which would use the concepts of bounded geometry. An attempt will be made to make the new language set compatible with the APT language. The new language would be a higher level than APT.

2. *Multiple applications.* The advanced NC system would not be restricted to machining operations but would be suitable for many non-machining applications, such as inspection and pressworking (shearing and forming operations).

3. *Modular design.* To facilitate the multiple applications, the advanced NC system would possess a modular design. Separate processors or subroutines would be designed to accomplish the basic functions, such as profiling, pocketing, point-to-point operations, turning, inspection, and so forth. This approach is similar to the one used in modern CAD/CAM part programming packages, where automatic routines have been designed to accomplish special functions.

4. *Automation of NC programming function.* NC programming includes, in addition to generation of the tool path, tool selection, feed, and speed selection, sequencing of operations, and other details. Advanced NC systems would automate the determination of these parameters.

5. *Interference checking.* Subroutines would be built into the advanced NC programming packages to check for possible collision between the tool and fixtures holding the part, as well as other potential interference problems.

6. *Interface with CAD/CAM data base.* The advanced numerical control sytems would be interfaced with a common design and manufacturing data base. The data base would contain data relating to the part geometry (raw stock and finished part dimensions), fixtures, machine tools, available tooling, and cost data. The NC programming function would be interfaced with the more general computer-automated process planning packages. We will be covering the various automated process planning procedures in a subsequent chapter.

Flexible manufacturing systems

One of the important developments in DNC was the introduction of the flexible manufacturing sytem (FMS). An FMS is a group of NC machines (or other automated workstations) which are interconnected by a materials handling system. All of the machines and the work handling system are controlled by computer. Flexible manufacturing systems were first introduced around 1970. Owing to the very high cost of these systems (several million dollars per FMS), there were only about a dozen of these systems installed by the end of 1980. However, the FMS offers such a high potential for productivity improvement in batch manufacturing that the number of installations is expected to grow substantially during the 1980s.

The FMS represents an important step in the evolution of the computer-automated factory of the future, and we will examine these systems in Chapter 20.

Robotics

In terms of control technology and programming, industrial robots share much in common with numerical control machines. Robots are used for moving workparts and tools in the performance of industrial tasks. An important number of these tasks are concerned with the loading and unloading of production machines, including NC machines. The robot and the machine form an automatic work cell, with raw workparts being fed into the cell by conveyor and completed parts leaving the cell by conveyor. All this is accomplished with little or no human attention. Because robotics constitutes such an important CAD/CAM topic, we devote the next major part of this book (two chapters) to the subject of industrial robots.

REFERENCES

[1] BEERCHECK, R.C., "Machine Tools: Cutting Edge of Technology," *Machine Design*, January 25, 1979, pp. 18−47.

[2] CAMPBELL, F. S., "Adaptive Control at McAir," *CAM Directions* (McDonnell Douglas), October, 1977.

[3] COMPUTER-AIDED MANUFACTURING-INTERNATIONAL, INC., *CAM-I Advanced N/C Project,* PR-81-ASPP-D1.4, Arlington, Tex., 1982.

[4] DIRKSON, G. F., "CNC and DNC—The Marriage," *Proceedings, Eighteenth Annual Meeting and Technical Conference*, Numerical Control Society, Dallas, Tex., May, 1981, pp. 132−138.

[5] FOLKMAN, J., "DNC, What Is It?" *Tooling and Production*, January, 1982, pp. 70−73.

[6] GROOVER, M. P., "Adaptive Control and Adaptive Control Machining," *Educational Module*, Manufacturing Productivity Educational Committee, Purdue Research Foundation, West Lafayette, Ind., 1977.

[7] GROOVER, M. P., *Automation, Production Systems, and Computer-Aided Manufacturing*, Prentice-Hall, Inc., Englewood Cliffs, N. J., 1980, Chapter 9.

[8] HATSCHEK, R. L., "NC Diagnostics," Special Report 744, *American Machinist*, April, 1982, pp. 161−168.

[9] HEGLAND, D. E., "Numerical Control—Your Best Investment in Productivity," *Production Engineering*, March, 1981, pp. 42−47.

[10] "Manufacturing Turns to Technology," *American Machinist*, January, 1981, pp. 101−108.

[11] MATHIAS, R. A., "Adaptive Control for the Eighties," *Paper MS80-242*, Society of Manufacturing Engineers, Dearborn, Mich., 1980.

[12] MATHIAS, R. A., "Determining Where Adaptive Control Can Most Benefit Your Machining Operations," *Paper MS81-272*, Society of Manufacturing Engineers, Dearborn, Mich., 1981.

[13] MATHIAS, R. A., "Adaptive Control, Key to Productivity," paper presented at the Conference on Computer Aided Manufacturing and Productivity, London, October, 1981.

[14] MURRAY, D., "CAM-I's Advanced Numerical Control Project Update," *Commline*, March/April, 1980, p. 40.

[15] PRESSMAN, R. S., AND WILLIAMS, J. E., *Numerical Control and Computer-Aided Manufacturing*, John Wiley & Sons, Inc., New York, 1977, Chapter 10.

[16] SMITH, D. N., AND EVANS, L., *Management Standards for Computer and Numerical Controls*, University of Michigan, Ann Arbor, 1977.

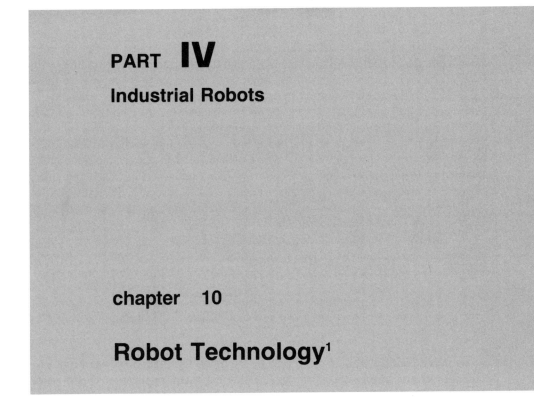

PART IV

Industrial Robots

chapter 10

Robot Technology[1]

10.1 INTRODUCTION

An industrial robot is a general-purpose, programmable machine possessing certain anthropomorphic characteristics. The most typical anthropomorphic, or human-like, characteristic of a robot is its arm. This arm, together with the robot's capacity to be programmed, makes it ideally suited to a variety of production tasks, including machine loading, spot welding, spray painting, and assembly. The robot can be programmed to perform a sequence of mechanical motions, and it can repeat that motion sequence over and over until reprogrammed to perform some other job.

An industrial robot shares many attributes in common with a numerical control machine tool. The same type of NC technology used to operate machine tools is used to actuate the robot's mechanical arm. The robot is a lighter, more portable piece of equipment than an NC machine tool. The uses of the robot are more general, typically involving the handling of workparts. Also, the programming of the robot is different from NC part programming. Traditionally, NC programming has

[1]Portions of this chapter and the following chapter are based on Groover [6, 7].

been performed off-line with the machine commands being contained on a punched tape. Robot programming has usually been accomplished on-line, with the instructions being retained in the robot's electronic memory. In spite of these differences, there are definite similarities between robots and NC machines in terms of power drive technologies, feedback systems, the trend toward computer control, and even some of the industrial applications.

The popular concept of a robot has been fashioned by science fiction novels and movies such as "Star Wars." These images tend to exaggerate the robot's similarity to human anatomy and behavior. The human analogy has sometimes been a troublesome issue in industry. People tend to associate the future use of advanced robots in factories with high unemployment and the subjugation of human beings by these machines.

Largely in response to this humanoid conception associated with robots, there have been attempts to develop definitions which reduce the anthropomorphic impact. The Robot Institute of America has developed the following definition:

> A robot is a programmable, multi-function manipulator designed to move material, parts, tools, or special devices through variable programmed motions for the performance of a variety of tasks.

Attempts have even been made to rename the robot. George Devol, one of the original inventors in robotics technology, called his patent application by the name "programmed article transfer." For many years, the Ford Motor Company used the term "universal transfer device" instead of "robot." Today the term "robot" seems to have become entrenched in the language, together with whatever human-like characteristics people have attached to the device.

10.2   ROBOT PHYSICAL CONFIGURATIONS

Industrial robots come in a variety of shapes and sizes. They are capable of various arm manipulations and they possess different motion systems. This section discusses the various physical configurations of robots. The following section deals with robot motion systems. Section 10.4 is concerned with most of the remaining technical features by which industrial robots are distinguished. Other topics, such as robot programming and gripper devices, are covered in the remaining sections.

Almost all present-day commercially available industrial robots have one of the following four configurations:

1. Polar coordinate configuration
2. Cylindrical coordinate configuration
3. Jointed arm configuration
4. Cartesian coordinate configuration

The four types are schematically illustrated in Figure 10.1 and described below.

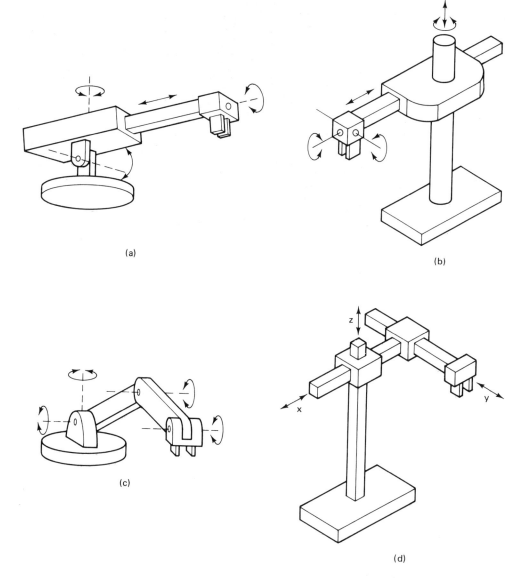

FIGURE 10.1 The four most common robot configurations: (a) polar coordinate; (b) cylindrical coordinate; (c) jointed arm configuration; (d) cartesian coordinate. (Reprinted from Toepperwein et al. [15].)

233

FIGURE 10.2 Unimate 2000 series robot—polar coordinate configuration. (Courtesy of Unimation Inc.)

Polar coordinate configuration

This configuration also goes by the name "spherical coordinate," because the workspace within which it can move its arm is a partial sphere. As shown in Figure 10.1(a), the robot has a rotary base and a pivot that can be used to raise and lower a telescoping arm. One of the most familiar robots, the Unimate Model 2000 series, was designed around this configuration, and is pictured in Figure 10.2.

Cylindrical coordinate configuration

In this configuration, the robot body is a vertical column that swivels about a vertical axis. The arm consists of several orthogonal slides which allow the arm to be moved up or down and in and out with respect to the body. This is illustrated schematically in Figure 10.1(b). The Prab Versatran Model FC, pictured in Figure 10.3, is an example of the cylindrical coordinate design.

Jointed arm configuration

The jointed arm configuration is similar in appearance to the human arm, as shown in Figure 10.1(c). The arm consists of several straight members connected by joints which are analogous to the human shoulder, elbow, and wrist. The robot arm

FIGURE 10.3   Prab FC model robot—cylindrical configuration. (Courtesy of Prab Conveyors Inc.)

is mounted to a base which can be rotated to provide the robot with the capacity to work within a quasi-spherical space. The Cincinnati Milacron T^3 model (Figure 10.4) and the Unimate PUMA model (Figure 10.5) are examples of this general configuration.

Cartesian coordinate configuration

A robot which is constructed around this configuration consists of three orthogonal slides, as pictured in Figure 10.1(d). The three slides are parallel to the x, y, and z axes of the cartesian coordinate system. By appropriate movements of these slides, the robot is capable of moving its arm to any point within its three-dimensional rectangularly shaped workspace.

10.3   BASIC ROBOT MOTIONS

Whatever the configuration, the purpose of the robot is to perform a useful task. To accomplish the task, an end effector, or hand, is attached to the end of the robot's arm. It is this end effector which adapts the general-purpose robot to a particular

FIGURE 10.4 Cincinnati Milacron T^3 robot—jointed arm configuration. (Courtesy of Cincinnati Milacron, Inc.)

FIGURE 10.5 Unimation PUMA 500 robot—jointed arm configuration. (Courtesy of Unimation, Inc.)

task. To do the task, the robot arm must be capable of moving the end effector through a sequence of motions and/or positions.

Six degrees of freedom

There are six basic motions, or degrees of freedom, which provide the robot with the capability to move the end effector through the required sequence of motions. These six degrees of freedom are intended to emulate the versatility of movement possessed by the human arm. Not all robots are equipped with the ability to move in all six degrees. The six basic motions consist of three arm and body motions and three wrist motions, as illustrated in Figure 10.6 for the polar-type robot. These motions are described below.

Arm and body motions:

1. *Vertical traverse:*   up-and-down motions of the arm, caused by pivoting the entire arm about a horizontal axis or moving the arm along a vertical slide
2. *Radial traverse*:   extension and retraction of the arm (in-and-out movement)
3. *Rotational traverse*:   rotation about the vertical axis (right or left swivel of the robot arm)

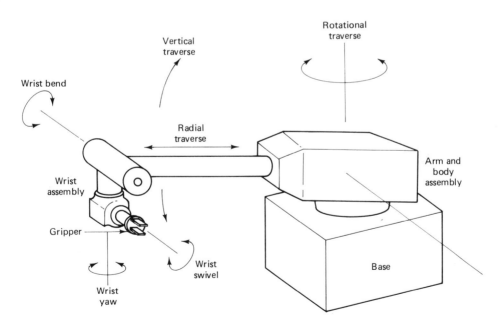

FIGURE 10.6 Typical six degrees of freedom in robot motion.

Wrist motions:

4. *Wrist swivel*:   rotation of the wrist
5. *Wrist bend*:   up-or-down movement of the wrist, which also involves a rotational movement
6. *Wrist yaw*:   right-or-left swivel of the wrist

Additional axes of motion are possible, for example, by putting the robot on a track or slide. The slide would be mounted in the floor or in an overhead track system, thus providing a conventional six-axis robot with a seventh degree of freedom. The gripper device is not normally considered to be an additional axis of motion.

Motion systems

Similar to NC machine tool systems, the motion systems of industrial robots can be classified as either point-to-point (PTP) or contouring (also called continuous path).

In PTP, the robot's movement is controlled from one point location in space to another. Each point is programmed into the robot's control memory and then played back during the work cycle. No particular attention is given to the path followed by the robot in its move from one point to the next. Point-to-point robots would be quite capable of performing certain kinds of productive operations, such as machine loading and unloading, pick-and-place activities, and spot welding.

Contouring robots have the capability to follow a closely spaced locus of points which describe a smooth compound curve. The memory and control requirements are greater for contouring robots than for PTP because the complete path taken by the robot must be remembered rather than merely the end points of the motion sequence. However, in certain industrial operations, continuous control of the work cycle path is essential to the use of the robot in the operation. Examples of these operations are paint spraying, continuous welding processes, and grasping objects moving along a conveyor.

10.4   OTHER TECHNICAL FEATURES

In addition to the robot's physical configuration and basic motion capabilities, there are numerous other technical features of an industrial robot which determine its efficiency and effectiveness at performing a given task. The following are some of the most important among these technical features:

1. Work volume
2. Precision of movement

3. Speed of movement
4. Weight-carrying capacity
5. Type of drive system

These features are described in this section.

Work volume

The term "work volume" refers to the space within which the robot can operate. To be technically precise, the work volume is the spatial region within which the end of the robot's wrist can be manipulated. Robot manufacturers have adopted the policy of defining the work volume in terms of the wrist end, with no hand or tool attached.

The work volume of an industrial robot is determined by its physical configuration, size, and the limits of its arm and joint manipulations. The work volume of a cartesian coordinate robot will be rectangular. The work volume of a cylindrical coordinate robot will be cylindrical. A polar coordinate configuration will generate a work volume which is a partial sphere. The work volume of a jointed arm robot will be somewhat irregular, the outer reaches generally resembling a partial sphere. Robot manufacturers usually show a diagram of the particular model's work volume in their marketing literature, providing a top view and side view with dimensions of the robot's motion envelope.

Precision of movement

The precision with which the robot can move the end of its wrist is a critical consideration in most applications. In robotics, precision of movement is a complex issue, and we will describe it as consisting of three attributes:

1. Spatial resolution
2. Accuracy
3. Repeatability

These attributes are generally interpreted in terms of the wrist end with no end effector attached and with the arm fully extended.

SPATIAL RESOLUTION. The term "spatial resolution" refers to the smallest increment of motion at the wrist end that can be controlled by the robot. This is determined largely by the robot's control resolution, which depends on its position control system and/or its feedback measurement system. In addition, mechanical inaccuracies in the robot's joints would tend to degrade its ability to position its arm. The spatial resolution is the sum of the control resolution plus these mechanical inaccuracies. The factors determining control resolution are the range of movement of the arm and the bit storage capacity in the control memory for that movement. The arm movement must be divided into its basic motions or degrees of freedom, and the resolution of each degree of freedom is figured separately. Then the total control resolution is the vector sum of each component. An example will serve to illustrate this.

EXAMPLE 10.1

Assume that we want to find the spatial resolution for a cartesian coordinate robot that has two degrees of freedom. The two degrees of freedom are manifested by two orthogonal slides. Each slide has a range of 0.4 m (about 15.75 in.), hence giving the robot a work volume which is a plane square, with 0.4 m on a side. Suppose that the robot's control memory has a 10-bit storage capacity for each axis.

To determine the control resolution, we must first determine the number of control increments of which the control memory is capable. For the 10-bit storage, there are $2^{10} = 1024$ control increments (the number of distinct zones into which the slide range of 0.4 m can be divided).

Then the control resolution would be found by dividing the slide range by the number of control increments:

$$\text{control resolution} = \frac{0.4\text{m}}{1024} = 0.3906 \text{ mm}$$

Since there are two orthogonal slides, the control resolution of this robot would be a square with 0.39 mm per side. Any mechanical inaccuracies would be added to this figure to get the spatial resolution.

This example shows that the spatial resolution can be improved by increasing the bit capacity of the robot's control memory. Also, for a given memory capacity, a larger robot would have a poorer (larger) spatial resolution than a small robot. In reality, the spatial resolution would be worse (larger) than the control resolution computed in Example 10.1 because of mechanical inaccuracies in the slides.

ACCURACY. The accuracy of the robot refers to its capability to position its wrist end (or a tool attached to the wrist) at a given target point within its work volume. Accuracy is closely related to spatial resolution, since the robot's ability to reach a particular point in space depends on its ability to divide its joint movements into small increments. According to this relation, the accuracy of the robot would be one-half the distance between two adjacent resolution points. This definition is illustrated in Figure 10.7. The robot's accuracy is also affected by mechanical inaccuracies, such as deflection of its components, gear inaccuracies, and so forth.

REPEATABILITY. This refers to the robot's ability to position its wrist end (or tool) back to a point in space that was previously taught. Repeatability is different from accuracy. The difference is illustrated in Figure 10.8. The robot was initially

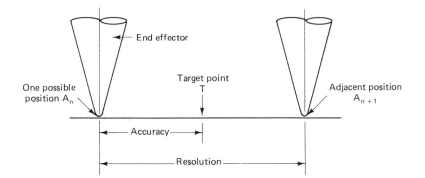

FIGURE 10.7   Illustration of accuracy versus resolution.

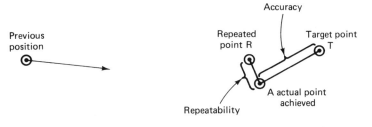

FIGURE 10.8   Illustration of repeatability versus accuracy.

programmed to move the wrist end to the target point T. Because it is limited by its accuracy, the robot was only capable of achieving point A. The distance between points A and T is the accuracy. Later, the robot is instructed to return to this previously programmed point A. However, because it is limited by it repeatability, it is only capable of moving to point R. The distance between points R and A is a measure of the robot's repeatability. As the robot is instructed to return to the same position in subsequent work cycles, it will not always return to point R, but instead will form a cluster of positions about point A. Repeatability errors form a random variable. In general, repeatability will be better (less) than accuracy. Mechanical inaccuracies in the robot's arm and wrist components are principal sources of repeatability errors.

Speed of movement

The speed with which the robot can manipulate the end effector ranges up to a maximum of about 1.5 m/s. Almost all robots have an adjustment to set the speed to the desirable level for the task performed. This speed should be determined by such factors as the weight of the object being moved, the distance moved, and the precision with which the object must be positioned during the work cycle. Heavy objects cannot be moved as fast as light objects because of inertia problems. Also, objects must be moved more slowly when high positional accuracy is required.

Weight-carrying capacity

The weight-carrying capacity of commercially available robots covers a wide range. At the upper end of the range, there are robots capable of lifting over 1000 lb. The Versatran FC model has a maximum load-carrying capacity rated at 2000 lb. At the lower end of the range, the Unimate PUMA Model 250 has a load capacity of only 2.5 lb. What complicates the issue for the low-weight-capacity robots is that the rated capacity includes the weight of the end effector. For example, if the gripper for the PUMA 250 weighs 1 lb, the net capacity of the robot is only 1.5 lb.

Type of drive system

There are three basic drive systems used in commercially available robots:

1. Hydraulic
2. Electric motor
3. Pneumatic

Hydraulically driven robots are typified by the Unimate 2000 series robots (Figure 10.2) and the Cincinnati Milacron T^3 (Figure 10.4). These drive systems are usually associated with large robots, and the hydraulic drive system adds to the floor space required by the robot. Advantages which this type of system gives to the robot are mechanical simplicity (hydraulic systems are familiar to maintenance personnel), high strength, and high speed.

Robots driven by electric motors (dc stepping motors or servomotors) do not possess the physical strength or speed of hydraulic units, but their accuracy and repeatability is generally better. Less floor space is required due to the absence of the hydraulic power unit.

Pneumatically driven robots are typically smaller and technologically less sophisticated than the other two types. Pick-and-place tasks and other simple, high-cycle-rate operations are examples of the kinds of applications usually reserved for these robots.

10.5 PROGRAMMING THE ROBOT

There are various methods by which robots can be programmed to perform a given work cycle. We divide these programming methods into four categories:

1. Manual method
2. Walkthrough method

3. Leadthrough method
4. Off-line programming

Manual method

This method is not really programming in the conventional sense of the word. It is more like setting up a machine rather than programming. It is the procedure used for the simpler robots and involves setting mechanical stops, cams, switches, or relays in the robot's control unit. For these low-technology robots used for short work cycles (e.g., pick-and-place operations), the manual programming method is adequate.

Walkthrough method

In this method the programmer manually moves the robot's arm and hand through the motion sequence of the work cycle. Each movement is recorded into memory for subsequent playback during production. The speed with which the movements are performed can usually be controlled independently so that the programmer does not have to worry about the cycle time during the walkthrough. The main concern is getting the position sequence correct. The walkthrough method would be appropriate for spray painting and arc welding robots.

Leadthrough method

The leadthrough method makes use of a teach pendant to power drive the robot through its motion sequence. The teach pendant is usually a small hand-held device with switches and dials to control the robot's physical movements. Each motion is recorded into memory for future playback during the work cycle. The leadthrough method is very popular among robot programming methods because of its ease and convenience.

Off-line programming

This method involves the preparation of the robot program off-line, in a manner similar to NC part programming. Off-line robot programming is typically accomplished on a computer terminal. After the program has been prepared, it is entered into the robot memory for use during the work cycle. The advantage of off-line robot programming is that production time of the robot is not lost to delays in teaching the robot a new task. Programming off-line can be done while the robot is still in production on the preceding job. This means higher utilization of the robot and the equipment with which it operates.

Another benefit associated with off-line programming is the prospect of integrating the robot into the factory CAD/CAM data base and information system. In future manufacturing systems, robot programming will be performed by advanced CAD/CAM systems, just as NC part programs can be generated by today's CAD/CAM technology.

10.6   ROBOT PROGRAMMING LANGUAGES

Non-computer-controlled robots do not require a programming language. They are programmed by the walkthrough or leadthrough methods while the simpler robots are programmed by manual methods. With the introduction of computer control for robots came the opportunity and the need to develop a computer-oriented robot programming language. In this section we discuss two of these languages: VAL, developed for the Unimation PUMA™ robot; and MCL, and APT-based language developed by McDonnell-Douglas Corporation.

The VAL™ language

The VAL language was developed by Victor Scheinman for the PUMA robot, an assembly robot produced by Unimation Inc. Hence, VAL stands for Victor's Assembly Language. It is basically an off-line language in which the program defining the motion sequence can be developed off-line, but the various point locations used in the work cycle are most conveniently defined by leadthrough.

   VAL statements are divided into two categories, Monitor Commands and Programming Instructions.

   The Monitor Commands are a set of administrative instructions that direct the operation of the robot system. The Monitor Commands would be used for such functions as:

   Preparing the system for the user to write programs for the PUMA
   Defining points in space
   Commanding the PUMA to execute a program
   Listing programs on the CRT

Some of the important Monitor Commands are given in Table 10.1, together with a description of the command and one or more examples showing how it would be entered.

   The Program Instructions are a set of statements used to write robot programs. Programs in VAL direct the sequence of motions of the PUMA. One statement usually corresponds to one movement of the robot's arm or wrist. Examples of Program Instructions include:

   Move to a point.
   Move to a point in a straight-line motion.
   Open gripper.
   Close gripper.

TABLE 10.1   Common Monitor Commands in VAL

EDIT	This statement, abbreviated ED, is used to create or modify the program named in the statement.
	    Examples: EDIT TEST
	                 ED TOM

EXECUTE	This is used to command the robot to execute the program names. The statement can be abbreviated EX. The user can specify how many times the program is to be executed by giving the number of executions following the program name.
	    Examples: EX TEST
	                 EX TEST, 5

SPEED	This command, abbreviated SP, specifies the speed of all subsequent robot motions under program control. This command would precede the EXECUTE command.
	    Examples: SPEED 50
	                 SP 30

HERE	This statement would be used to define the current position of the robot by a symbol specified by the user.
	    Examples: HERE A
	                 HERE PT1
	The user would move the robot to the desired position under manual control by means of the teach pendant. When the robot is in the desired location, the user defines that position with the HERE statement. The CRT displays the PUMA's six axis positions. The user then depresses the carriage return button to proceed.

Source: Adapted from Ref. [16].

The Program Instructions are entered into memory to form programs by first using the Monitor Command EDIT. This prepares the system to receive the Program Instruction statements in the proper order. Some of the important Program Instructions are given in Table 10.2. Each instruction is described and examples are given to illustrate typical usage.

The statements listed in Tables 10.1 and 10.2 are only a few of the many VAL commands. Reference [16] provides a comprehensive presentation of VAL and its many statements. The following example will illustrate the programming language.

EXAMPLE 10.2

This example deals with a pick-and-place operation, in which the robot is instructed to pick up a workpart from one conveyor (point A) and place it on another conveyor (point B). The VAL program for this work cycle is presented in Figure 10.9. Figure 10.10 shows the PUMA locations for points A and B in the program. The listing gives the PUMA's *x, y,* and *z* coordinate positions (joints 1, 2, and 3), as well as the three wrist joint positions (joints 4, 5, and 6) for the two points, A and B.

The MCL language

The MCL stands for Machine Control Language and was developed by McDonnell-Douglas Corporation under contract to the U.S. Air Force ICAM

TABLE 10.2 Common Programming Instructions in VAL

MOVE	This moves the robot to the location and orientation specified by the symbol. Example: MOVE A
MOVES	This moves the robot along a straight-line trajectory to the location and orientation specified by the symbol. Example: MOVES A
APPRO	This statement moves the end effector or tool to the position defined by the symbol, but offsets it along the tool z axis by the distance given in millimeters. Example: APPRO A, 50 The tool would be moved to a distance of 50 mm away from position A in the tool z direction.
APPROS	This is similar to the APPRO instruction except that the move to the vicinity of the desired point is made along a straight-line trajectory. Example: APPROS A, 50
DEPART	This moves the tool the distance given along the current z axis of the tool. Example: DEPART 50 This moves the tool back from its current position by 50 mm.
OPENI and CLOSEI	These instructions cause the pneumatic control valves to receive a signal to open or close the gripper immediately. Examples: OPENI CLOSEI
EXIT	This causes an exit from the program and control is returned to the monitor mode. The EXIT must be abbreviated E. Example: E

Source: Adapted from Ref. [16].

```
.
.LISTP PICK
.PROGRAM PICK
    1.        APPRO A, 50.00
    2.        MOVES A
    3.        CLOSEI 0.00
    4.        DEPART 50.00
    5.        APPRO B, 50.00
    6.        MOVES B
    7.        OPENI 0.00
    8.        DEPART 50.00
.END
```

FIGURE 10.9 Simple robot program listing for Example 10.2 using Unimation's VAL language.

```
.LISTL A
           X/JT1    Y/JT2    Z/JT3    O/JT4    A/JT5    T/JT6
    A     -543.94   292.31  -374.00 -104.249  85.638 -179.956
.
.LISTL B
           X/JT1    Y/JT2    Z/JT3    O/JT4    A/JT5    T/JT6
    B     -411.56  -355.41  -269.13 -104.480  85.622  179.819
```

FIGURE 10.10 Listing of joint positions for Unimate PUMA 600 robot in Example 10.2.

(Integrated Computer-Aided Manufacturing) Program [10]. The language is based on the APT NC language, but is designed to control a complete manufacturing cell, including a cell with robots. MCL is an enhancement of APT which possesses additional options and features needed to do off-line programming of a robotic work cell.

Additional vocabulary words were developed to provide the supplementary capabilities intended to be covered by the MCL language. These capabilities include vision, inspection, and the control of signals to and from the various devices that constitute the robotic workstation. A list of some of these vocabulary words is presented in Table 10.3. MCL also permits the user to define MACRO-like statements that would be convenient to use for specialized applications.

After the MCL program has been written, it is compiled to produce the CLFILE as output. The definition of the CLFILE has been extended to accommodate the new MCL features that go beyond the conventional cutter location data in APT. The extensions include such capabilities as:

The definition of the various devices within the work cell and the tasks which are performed by these devices

Predefined frames of reference which are associated with the different machines or devices in the cell

User-defined frames of reference which could be used for defining the geometry of the workpart

The part identification and acquisition within the work cell

MCL represents a significant enhancement of APT which can be used to perform off-line programming of complex robotic work cells.

TABLE 10.3   Some Representative Programming Words in MCL

DEVICE	This command is used to activate devices in the work cell.
SEND	This word causes an output signal to be sent to a defined destination.
RECEIV	Similar to SEND, but this word causes an input signal to be accepted from some defined source.
WORKPT	This statement is used to define a sophisticated tool tip (work point).
ABORT	This word is used to halt all task activity.
TASK	This word allows the program to define certain portions of the program as tasks which can then be called into execution either singly or in combination with other tasks.
REGION	This statement is used to define a closed two-dimensional path for vision processing.
LOCATE	The LOCATE command is used with a vision system to seek out and identify a specified part within the field of view, and to determine the position and orientation of the part.

Source: Adapted from Ref. [11].

10.7   END EFFECTORS

In the terminology of robotics, an end effector can be defined as a device which is attached to the robot's wrist to perform a specific task. The task might be workpart handling, spot welding, spray painting, or any of a great variety of other functions. The possibilities are limited only by the imagination and ingenuity of the applications engineers who design robot systems. (Economic considerations might also impose a few limitations.) The end effector is the special-purpose tooling which enables the robot to perform a particular job. It is usually custom engineered for that job, either by the company that owns the robot or by the company that sold the robot. Most robot manufacturers have engineering groups which design and fabricate end effectors or provide advice to their customers on end effector design.

For purposes of organization, we will divide the various types of end effectors into two categories: grippers and tools. The following two sections discuss these two categories.

Grippers

Grippers are used to hold either workparts (in pick-and-place operations, machine loading, or assembly work) or tools. There are numerous alternative ways in which the gripper can be designed. The most appropriate design depends on the workpart or substance being handled. The following is a list of the most common grasping methods used in robot grippers:

> Mechanical grippers, where friction or the physical configuration of the gripper retains the object
> Suction cups (also called vacuum cups), used for flat objects
> Magnetized gripper devices, used for ferrous objects
> Hooks, used to lift parts off conveyors
> Scoops or ladles, used for fluids, powders, pellets, or granular substances

Several alternative gripper designs are illustrated in Figure 10.11.

Tools as end effectors

There are a limited number of applications in which a gripper is used to grasp a tool and use it during the work cycle. In most applications where the robot manipulates a tool during the cycle, the tool is fastened directly to the robot wrist and becomes the end effector. A few examples of tools used with robots are the following:

> Spot welding gun
> Arc welding tools (and wire-feed mechanisms)

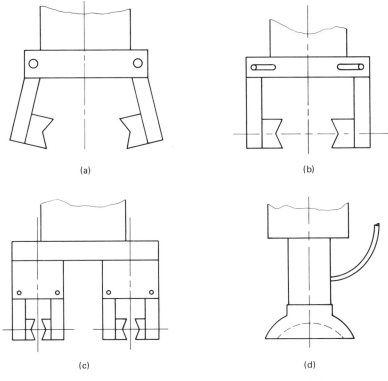

FIGURE 10.11   Sample gripper designs: (a) pivot action gripper; (b) slide action gripper; (c) double gripper—pivot action mechanism; (d) vacuum-operated hand.

Spray painting gun
Drilling spindle
Routers, grinders, wire brushes
Heating torches

In some of these examples, quick-change mechanisms can be incorporated into the tool design to allow for fast changeover from one size or type to the next.

10.8   WORK CELL CONTROL AND INTERLOCKS

Work cell control

Industrial robots usually work with other things: processing equipment, workparts, conveyors, tools, and perhaps, human operators. A means must be provided for coordinating all of the activities which are going on within the robot workstation.

Some of the activities occur sequentially, while others take place simultaneously. To make certain that the various activities are coordinated and occur in the proper sequence, a device called the work cell controller is used (another name for this is workstation controller). The work cell controller usually resides within the robot and has overall responsibility for regulating the activities of the work cell components. To illustrate the kinds of problems and issues that would have to be managed by the work cell controller, consider the following example of a relatively simple machine loading application.

EXAMPLE 10.3

The workstation consists of the robot; the machine tool, which operates on a semiautomatic cycle; and two conveyors, one for incoming raw workparts and the other for outgoing finished pieces. The setup is shown in Figure 10.12. The work cycle consists of the following activities:

1. Incoming conveyor delivers raw workpart to fixed position.
2. Robot picks up part from conveyor and loads it into machine.
3. Machine processes workpart.
4. Robot unloads finished part from machine and places it on outgoing conveyor.
5. Outgoing conveyor delivers part out of work cell, and robot returns to ready position near incoming conveyor.

As this work cycle is described, most of the activities occur sequentially. The work cell controller would have to make sure that certain steps are completed before subsequent steps are initiated. For example, the machine tool must finish processing the workpart before the robot attempts to reach in and grasp the part for unloading. Similarly, the machine cycle must not begin until the robot has loaded the raw workpiece and removed its arm. The robot cannot pick up the raw part from the incoming conveyor unless and until the part has been

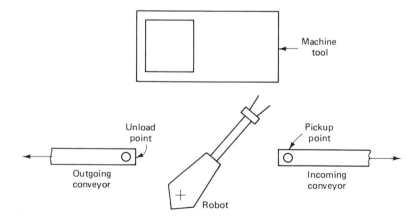

FIGURE 10.12 Workplace layout for robot cell of Example 10.3.

positioned for gripping. The purpose of the work cell controller in this example is to ensure that the work elements are sequenced correctly and that each step is finished before the next one begins.

Although most of the activities occur sequentially, some of the devices perform their functions simultaneously. For example, the conveyors continue to deliver workparts into and out of the work cell while the machine tool is operating. In two of the exercise problems at the end of the chapter, we require the reader to consider improvements that might be made in this work cycle to increase production rate.

This example illustrates several of the functions that must be performed by the work cell controller. These functions include:

1. Controlling the sequence of activities in the work cycle.
2. Controlling simultaneous activities.
3. Making decisions to proceed based on incoming signals.

Additional functions that might be required of the work cell controller (in more sophisticated work cells) are:

4. Making logical decisions.
5. Performing computations.
6. Dealing with exceptional events, such as broken tools or equipment breakdowns.
7. Performing irregular cycles, such as periodically changing tools.

To enable it to carry out these functions, the work cell controller must receive signals from other devices in the work cell, and it must communicate signals to the various components of the work cell. These signals are accomplished by means of interlocks and sensors.

Interlocks

An interlock is the feature of work cell control which prevents the work cycle sequence from continuing until a certain condition or set of conditions has been satisfied. In a robotic work cell, there are two types: outgoing and incoming. The outgoing interlock is a signal sent from the workstation controller to some external machine or device that will cause it to operate or not operate. For example, this would be used to prevent a machine from initiating its process until it was commanded to proceed by the work cell controller. An incoming interlock is a signal from some external machine or device to the work controller which determines whether or not the programmed work cycle sequence will proceed. For example, this would be used to prevent the work cycle program from continuing until the machine signaled that it had completed its processing of the workpiece.

The use of interlocks provides an important benefit in the control of the work cycle because it prevents actions from happening when they shouldn't, and it causes actions to occur when they should. Interlocks are needed to help coordinate the activities of the various independent components in the work cell and to help avert damage of one component by another.

In the planning of interlocks in the robotic work cell, the applications engineer must consider both the normal sequence of activities that will occur during the work cycle, and the potential malfunctions that might occur. Then these normal activities are linked together by means of limit switches, pressure switches, photoelectric devices, and other system components. Malfunctions that can be anticipated are prevented by means of similar devices.

10.9 ROBOTIC SENSORS

For certain robot applications, the type of workstation control using interlocks is not adequate. The robot must take on more humanlike senses and capabilities in order to perform the task in a satisfactory way. These senses and capabilities include vision and hand–eye coordination, touch, and hearing. Accordingly, we will divide the types of sensors used in robotics into the following three categories:

1. Vision sensors
2. Tactile and proximity sensors
3. Voice sensors

It is beyond the scope of this chapter to provide more than a brief survey of this fascinating topic. For additional study, Refs. [1], [3], [5], [9], [10], [15], and [17] are recommended.

Vision sensors

This is one of the areas that is receiving a lot of attention in robotics research. Computerized visions systems will be an important technology in future automated factories. Robot vision is made possible by means of a video camera, a sufficient light source, and a computer programmed to process image data. The camera is mounted either on the robot or in a fixed position above the robot so that its field of vision includes the robot's work volume. The computer software enables the vision system to sense the presence of an object and its position and orientation. Vision capability would enable the robot to carry out the following kinds of operations:

Retrieve parts which are randomly oriented on a conveyor.
Recognize particular parts which are intermixed with other objects.
Perform visual inspection tasks.
Perform assembly operations which require alignment.

All of these operations have been accomplished in research laboratories. It is merely a matter of time and economics before vision sensors become a common feature in robot applications.

Tactile and proximity sensors

Tactile sensors provide the robot with the capability to respond to contact forces between itself and other objects within its work volume. Tactile sensors can be divided into two types:

1. Touch sensors
2. Stress sensors (also called force sensors)

Touch sensors are used simply to indicate whether contact has been made with an object. A simple microswitch can serve the purpose of a touch sensor. Stress sensors are used to measure the magnitude of the contact force. Strain gage devices are typically employed in force-measuring sensors.

Potential uses of robots with tactile sensing capabilities would be in assembly and inspection operations. In assembly, the robot could perform delicate part alignment and joining operations. In inspection, touch sensing would be useful in gauging operations and dimensional-measuring activities. Proximity sensors are used to sense when one object is close to another object. On a robot, the proximity sensor would be located on or near the end effector. This sensing capability can be engineered by means of optical-proximity devices, eddy-current proximity detectors, magnetic-field sensors, or other devices.

In robotics, proximity sensors might be used to indicate the presence or absence of a workpart or other object. They could also be helpful in preventing injury to the robot's human coworkers in the factory.

Voice sensors

Another area of robotics research is voice sensing or voice programming. Voice programming can be defined as the oral communication of commands to the robot or other machine. (Voice programming is also used in NC part programming and the reader is invited to refer back to Section 8.9 for a description of this interesting application.) The robot controller is equipped with a speech recognition system which analyzes the voice input and compares it with a set of stored word patterns. When a match is found between the input and the stored vocabulary word, the robot performs some action which corresponds to that word.

Voice sensors would be useful in robot programming to speed up the programming procedure, just as it does in NC programming. It would also be beneficial in especially hazardous working environments for performing unique operations such as maintenance and repair work. The robot could be placed in the hazardous environment and remotely commanded to perform the repair chores by means of step-by-step instructions.

REFERENCES

[1] AGIN, G. J., "Real-Time Robot Control with a Mobile Camera," *Robotics Today,* Fall, 1979, pp. 35−39.

[2] ALLAN, J. J. (Ed.), *A Survey of Industrial Robots*, Productivity International, Inc., Dallas, Tex., 1980.

[3] CARLISLE, B., ROTH, S., GLEASON, J., AND MCGHIE, D., "The PUMA/VS-100 Robot Vision System," paper presented at the First International Conference on Robot Vision and Sensory Controls, Stratford-upon-Avon, England, April, 1981.

[4] ENGELBERGER, J. F., *Robotics in Practice*, AMACOM (American Management Association), New York, 1980.

[5] GLEASON, J., AND AGIN, G. J., "The SRI Vision Module," *Robotics Today*, Winter, 1980−1981, pp. 36−40.

[6] GROOVER, M. P., "Industrial Robots: A Primer on the Present Technology," *Industrial Engineering*, November, 1980, pp. 54−61.

[7] GROOVER, M. P., *Automation, Production Systems, and Computer-Aided Manufacturing*, Prentice-Hall, Inc., Englewood Cliffs, N.J., 1980, Chapter 9.

[8] HEER, E., "Robots and Manipulators," *Mechanical Engineering*, November, 1981, pp. 42−49.

[9] MOVICH, R. C., "Robot Drilling and Riveting Using Computer Vision," *Robotics Today*, Winter, 1980−1981, pp. 20−29.

[10] NAGEL, R., VANDERBURG, G., ALBUS, J., AND LOWENFELD, E., "Experiments in Part Acquisition Using Robot Vision," *Robotics Today*, Winter, 1980−1981, pp. 30−35.

[11] OLDROYD, L. A., "MCL: An APT Approach to Robotic Manufacturing," paper presented at SHARE 56, Houston, Tex., March, 1981.

[12] OTTINGER, L. V., "Robotics for the IE: Terminology, Types of Robots," *Industrial Engineering*, November, 1981, pp. 28−35.

[13] TANNER, W. R. (Ed.), *Industrial Robots,* Vol I: *Fundamentals*, Society of Manufacturing Engineers, Dearborn, Mich., 1979.

[14] TARVIN, R. L., "An Off-Line Programming Approach," *Robotics Today*, Summer, 1981, pp. 32−35.

[15] TOEPPERWEIN, L. L., BLACKMAN, M. T., et al., "ICAM Robotics Application Guide," *Technical Report AFWAL-TR-80-4042*, Vol. II, Materials Laboratory, Air Force Wright Aeronautical Laboratories, Ohio, April, 1980.

[16] UNIMATION, INC., *User's Guide to VAL* (398H2A), Version 12, Danbury, Conn., June, 1980.

[17] VANDERBURG, G., ALBUS, J., AND BERKMEYER, E., "A Vision System for Real Robot Control," *Robotics Today,* Winter, 1979 −1980, pp. 20-22.

[18] WINSHIP, J. T., "Update on Industrial Robots," *American Machinist*, January, 1979, pp. 121–124.

PROBLEMS

10.1. One of the axes of a robot is a telescoping arm with a total range of 0.7 m (about 27.5 in.). The robot's control memory has a 12-bit storage capacity for this axis of motion. Determine the robot's control resolution for this axis.

10.2. The telescoping arm of a certain industrial robot obtains its vertical motion by pivoting about a horizontal axis. The total range of rotation is 120°. The robot possesses a 10-bit storage capacity for this axis of motion. Determine the robot's control resolution for this axis in degrees of rotation.

10.3. In Problem 10.2, suppose that the robot's telescoping arm, when fully extended, measured 1.1 m in length from the pivot point. Determine the robot's control resolution on a linear scale in this fully extended position.

10.4. Solve Problem 10.1 but use (a) an 8-bit storage capacity, (b) a 16-bit storage capacity.

10.5. A large, hydraulically operated, cartesian coordinate robot has one orthogonal slide with a total range of 1.2 m. One of the specifications on the robot's precision of movement is that it have a control resolution of 0.5 mm on this slide. Determine the number of bits of storage capacity which the robot's control memory must possess to provide at least this precision.

10.6. Using the VAL programming words from Table 10.2, write a program to pick up parts from a fixed position on a conveyor (using a mechanical stop to locate the parts in a known position) and insert them into a cardboard carton which is 127 mm (5.0 in.) tall. Each carton holds two parts. Open cartons are presented to the robot on a start-and-stop conveyor. The conveyor subsequently goes through a carton closing and sealing machine. Before doing any programming, make a rough top-view sketch of the work cell layout. Assume that a small robot such as the PUMA (Figure 10.5) will be used. (*Note*: You will, or course, be unable to define the various positions used in your program. Nevertheless, identify and name these points in your sketch as if you were documenting your program for a technician to subsequently set up the job.)

10.7. Write the initials for your first and last name in large rectangular letters on a sheet of paper (about 6 in. high). Using the VAL commands from Table 10.2, write a program to move a pen mounted as the robot's end effector from a neutral position to the paper and print the initials. To do this, you must first define the various corners and end points in the letters as named positions so that the robot can be directed from one position to the next.

10.8. For the situation described in Problem 10.6, make a list of the various interlocks and sensors that would be required for the work cycle to operate smoothly in the proper sequence. For each item in the list, identify the potential malfunction that could occur if the interlock or sensor were not employed.

10.9. Consider the workstation layout of Example 10.3. It is desired to redesign the layout so as to shorten the distances the robot must move. The strategy is that shorter distances moved will mean a shorter cycle time. Sketch a layout of the work cell showing the improvements you would make to the arrangement of equipment in Figure 10.12. You may add new equipment (holding tables, etc.) and rearrange existing pieces, but you cannot delete any current piece of equipment. For your cell layout, describe the work cycle in the step-by-step format used in Example 10.3.

10.10. Consider Example 10.3. The use of a double gripper (capable of holding a raw workpart and a finished piece simultaneously) would permit a significant reduction in cycle time. Describe the sequence of work cycle activities in the station with the robot using a double gripper. Use a step-by-step format to describe the work cycle, similar to the format in Example 10.3.

chapter 11

Robot Applications

11.1 GENERAL CONSIDERATIONS IN ROBOT APPLICATIONS

In this chapter we consider how robots are applied in industry. The applications can be divided into seven major categories. Before examining these seven categories, it is appropriate to discuss some of the general considerations in robotic applications.

General application characteristics

There are certain general characteristics of an industrial situation which tend to make the installation of a robot economical and practical. These general characteristics include the following.

1. *Hazardous or uncomfortable working conditions.* In job situations where there are potential dangers or health hazards due to heat, radiation, or toxicity, or where the workplace is uncomfortable and unpleasant, a robot should be considered as a substitute for the human worker. This sort of application has a high

probability for worker acceptance of the robot. Examples of these job situations include hot forging, die casting, spray painting, and foundry operations.

2. *Repetitive tasks.* If the work cycle consists of a sequence of elements which do not vary from cycle to cycle, it is possible that a robot could be programmed to perform the task. This is especially likely if the task is accomplished within a limited workspace. Pick-and-place operations and machine loading are obvious examples of repetitive tasks.

3. *Difficult handling.* If the workpart or tool involved in the operation is awkward or heavy, it might be possible for a robot to perform the task. Operations involving the handling of heavy workparts are a good example of this case. A human worker would need some form of mechanical assistance to lift the part, which would add to the production cycle time. Some industrial robots are capable of lifting payloads weighing several hundred (or even more than a thousand) pounds.

4. *Multishift operation.* If the initial investment cost of the robot can be spread over two or three shifts, the labor savings will result in a quicker payback. This could mean the difference between whether or not the investment can be justified. Plastic injection molding and other processes which must be operated continuously are examples of multishift robot applications.

Selecting the right applications

Not all robot installations have been successful. There are enough case histories of misapplications, poorly selected equipment, and nonacceptance by factory personnel to make a prospective user very careful, especially in an initial application. The ideal application would be one which possessed the four general characteristics described in the preceding section. The candidate for robot installations should be subjected to the same kinds of economic criteria as any other investment proposal concerned with productivity improvement or cost reduction. If at all possible, the application should be a simple one, especially on a first robot installation, where success is important for general acceptance of the technology.

The General Electric Company uses the following criteria in performing a plant survey in search of practical and economical robot applications[1]:

Simple repetitive operations are needed.

Cycle times are greater than 5s.

Parts can be delivered in proper location and orientation.

Part weight is suitable (1100 lb is typically used as the upper limit).

No inspection is required.

One or two persons can be replaced in a 24-hour period.

Setups and changeovers are not frequent.

If the potential application meets all of these criteria, they know they have a prime candidate.

Application areas for industrial robots

Industrial robots have been applied to a great variety of production situations. For purposes of organization, we will divide the applications into the following seven categories:

1. Material transfer
2. Machine loading
3. Welding
4. Spray coating
5. Processing operations
6. Assembly
7. Inspection

In Sections 11.2 through 11.8 we explore the various applications within these seven categories.

11.2 MATERIAL TRANSFER

Material transfer applications are those in which the robot is used to move workparts from one location to another. In some cases a reorientation of the part may be required in this material handling function. Examples of material transfer robot operations include the following:

Simple pick-and-place operations

Transfer of workparts from one conveyor to another conveyor (basically a pick-and-place task)

Palletizing operations, in which the robot takes parts from a conveyor and loads them onto a pallet in a required pattern and sequence

Stacking operations, similar to palletizing

Loading parts from a conveyor into cartons or boxes (similar to palletizing)

Depalletizing operations, in which the robot takes parts which are arranged on a pallet and loads them onto a conveyor

Material transfer operations are often among the easiest and most straightforward of robot applications (e.g., pick-and-place, transfer from conveyor to conveyor).

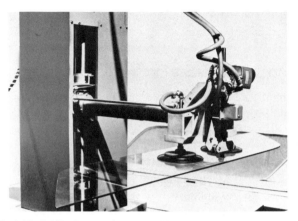

FIGURE 11.1   Prab Model E robot stacks automobile windshield glass in material transfer application. (Courtesy of Prab Conveyors, Inc.)

Robots used for these tasks usually possess a relatively low level of technological sophistication. However, in other cases (e.g., palletizing, depalletizing, etc.), the motion pattern required of the robot can become somewhat complicated. For example, in palletizing, each part must be positioned in its own location on the pallet for each layer, and often multiple layers must be stacked on the pallet. The programming necessary to execute such a motion sequence can become quite involved unless a computer controlled robot is used.

Figure 11.1 shows a Prab Model E performing a typical material transfer application. The robot is used to stack automobile windshield glass with vacuum-actuated grippers.

11.3   MACHINE LOADING

Machine loading applications are material handling operations in which the robot is required to supply a production machine with raw workparts and/or to unload finished parts from the machine. Machine loading is distinguished from a material transfer operation by the fact that the robot works directly with the processing equipment. In material transfer functions, it does not.

In the typical application, the robot would grasp a raw workpart from a conveyor and load it into the machine. In some cases, the robot holds the part in position during processing. When processing is completed, the robot unloads the part from the machine and places it onto another conveyor.

Production operations in which robots have been successfully applied to perform the machine loading and unloading function include the following:

Die casting
Injection (plastic) molding
Transfer (plastic) molding
Hot forging
Upsetting or upset forging
Stamping press operations
Machining operations such as turning and milling

In some of these operations (die casting and plastic molding), the robot only unloads the finished parts. For machining processes, the robot both loads and unloads the machine tool. In upsetting and stamping operations, the robot holds the workpart while it is being processed by the machine.

Some machine loading applications consist of several processing machines in a manufacturing cell, with the robot tending two, three, or even four separate machines. One of the more recent innovations in machine loading applications is to form a flexible manufacturing system using several robots to augment the conveyor system normally used in these production cells.

EXAMPLE 11.1

In this example, a Cincinnati Milacron T³ robot is used to service two turning centers and an automatic inspection station. A portion of the work cell is shown in Figure 11.2. Parts enter the cell as raw castings and leave completely machined and inspected. The sequence of work cycle activities is as follows:

1. The robot picks up a raw workpart from the pallet on the conveyor and takes it to the first turning center (not shown in Figure 11.2). The robot enters from the rear of the machine, removes the part just finished, and loads the raw part.
2. The robot transfers the part just completed at the first machine to an automatic gaging station. If the part is determined to be within tolerance, the robot transfers it to the second turning center, ready for loading.
3. The finished part is unloaded from the second machine and the part just gaged from the first machine is loaded. The unloading and loading take place from the rear as illustrated in Figure 11.2.
4. The robot takes the part just finished from the second machine and carries it to the automatic gaging station. If the part is within tolerance, it is placed on the pallet.

The robot is then ready to pick up another raw casting and repeat the work cycle. An interesting feature of this application is that the robot performs the loading and loading from the rear of each machine in the cell. This leaves the front of the machines clear for tool replacement, access to controls, and observation.

FIGURE 11.2 Cincinnati Milacron T^3 robot in machine loading application. (Courtesy of Cincinnati Milacron, Inc.)

11.4 WELDING

The welding processes are a very important application area for industrial robots. The applications logically divide into two basic categories, spot welding and arc welding.

Spot welding

Spot welding is a process in which metal parts (sheets or plates) are fused together at localized points by passing a large electric current through the two parts at the points of contact. The process is implemented by means of electrodes which squeeze the parts together and conduct the current to the contact point. The typical pair of electrodes have the form of tongs, which can conveniently be mounted on a large robot's wrist as the end effector. Using the welding "gun," as the electrode assembly is sometimes called, the robot accomplishes a spot weld by means of the following sequence:

1. Position the welding gun in the desired location against the two pieces (prior fixturing or matching of the pieces is required).
2. Squeezing the two electrodes against the mating pieces.
3. Weld and hold, when the current is applied to cause heating and fusion of the two surfaces in contact.
4. Release and cool. The electrodes open and sufficient time is allowed to cool the electrodes in anticipation of the next spot weld. (Usually, water is circulated within the electrodes to speed the cooling.)

This is a sequence which has become an ideal task for a point-to-point robot.

Spot welding has become one of the largest application areas for industrial robots, especially in the automotive industry. During the late 1960s, the first spot welding robots had been installed for producing the Vega model automobile. Today, nearly all automobile manufacturers are using robots for spot welding car bodies. The robot population for this purpose numbers over 1200 units. Other applications for which spot welding robots are used include motorcycle and bicycle frames, truck cabs, and appliances.

Figure 11.3 shows an automobile body welding line in which spot welding is performed automatically by Unimate Model 4000s. Robots have made a fundamental change in the way automobiles are assembled, as the following case study illustrates.

FIGURE 11.3   Unimate 4000 series robots in spot welding application on automobile assembly line. (Courtesy of Unimation, Inc.)

EXAMPLE 11.2

A relatively recent example of a spot welding robot automotive line is Chrysler's Robogate System [13]. This system was installed in 1979 in the company's Belvidere, Illinois, assembly plant to produce the Plymouth Horizon and Dodge Omni subcompact models. The Robogate is really a giant clamping fixture used to hold the car's underbody panel and two sides to ensure dimensional stability during tack welding. The tack welding operations are performed by eight Unimate robots. Three of these robots are mounted on each side (a total of six Unimate 4000 models) and two more robots (Unimate 2000 models) are mounted vertically above.

Following the Robogate, four more Unimate 4000s perform spot welding around the door sills and openings. After this section, the car bodies flow through the "respot" line, where over 700 spot welds are accomplished by 24 more Unimate Model 4000 robots.

The plant, using a total of 36 spot welding robots in its car body assembly operations, is capable of a production rate of 75 vehicles per hour. Two different body styles, a four-door sedan and a two-door coupe, are produced on the line. Control over the system is accomplished by a programmable controller which keeps track of which body styles are at the various stations so that each robot will perform the appropriate weld sequence for that style.

Arc welding

Several types of continuous arc welding processes can be accomplished by industrial robots capable of continuous-path operation. These processes include gas metal arc welding (GMAW, also called metal inert gas or MIG welding) and gas tungsten arc welding (GTAW, also called tungsten inert gas or TIG welding). These kinds of operations are traditionally performed by welders, who must often work under conditions which are hot, uncomfortable, and sometimes dangerous. Such conditions make this a logical candidate for the application of industrial robots. However, there are several problems associated with arc welding which have hindered the widespread use of robots in this process. First, arc welding is a fabricating process often used on low-volume products. Hence the economics involved in these cases make the use of any automation difficult, robots included. Second, dimensional variations in the components being arc welded are common. Human welders can compensate for these variations. Robots cannot, at least with current technology. Third, human welders are often required to perform their trade in areas which are difficult to access (inside vessels, tanks, ship hulls, etc.). Fourth and finally, sensor technologies capable of monitoring the variations in the arc welding process have not yet been fully developed.

As a result of these problems, robot arc welding applications have been fairly limited to operations involving high or medium volumes where the components can be conveniently handled and the dimensional variations can be reasonable managed. A typical robotic arc welding station would consist of the following components:

1. A robot, capable of continuous path control
2. A welding unit, consisting of the welding tool, power source, and the wire feed system
3. A workpart manipulator, which fixtures the components and positions them for welding

The workstation controller is equipped to coordinate the wire feed and arc voltage with the robot's arm movement. The activities of the workpart manipulator must also be coordinated by the controller. A human worker would be used to load and unload the workparts from the manipulator. This type of work cell for arc welding is illustrated in Figure 11.4. Some robot welding stations have two part manipulators, so that the human worker can be unloading and loading one manipulator while the robot is welding at the other. This increases the utilization of the equipment.

There are several advantages attributed to a robot welding station compared with its manually operated counterpart. Among these are the following:

1. Higher productivity
2. Improved safety
3. More consistent welds

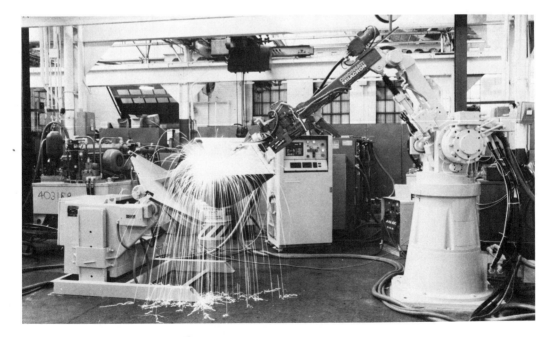

FIGURE 11.4 Cincinnati Milacron T³ robot in an arc welding application. (Courtesy of Cincinnati Milacron, Inc.)

The higher productivity results from several factors. First, a human welder may weld with an average arc-on time of 20 to 30%, while a robotic workstation can operate with an average of 60% or 70% arc-on time. Hence more welding is taking place at the workstation with a robot. Second, the use of a workpart manipulator speeds up the loading and unloading time. Two fixture stations save even more time. Third, the fatigue factor of the welder is reduced. Manual welding is a rather tiring operation for the welder because of the hand—eye coordination needed and the uncomfortable working conditions. Consequently, frequent rest periods are necessary. These are not required by a robot. Finally, many manual welding stations use two workers, a welder and a fitter. A robot welding station eliminates the need for one of those workers. Note, however, that the robotic workstation still requires a human operator to perform the welding process.

11.5 SPRAY COATING

Many large consumer products (e.g., automobiles and appliances) and most industrial products require the application of some form of paint. When human workers apply this paint, the most common method is spray painting. However, the spray painting process poses certain health hazards to the human operator. Among these are:

1. *Fumes and mist from the spraying operation.*   These create an uncomfortable and sometimes toxic atmosphere.
2. *Noise from the spray nozzle.*   This noise is loud, and prolonged exposure can impair hearing.
3. *Fire hazard.*   The mist of paint in the air within the factory can result in flash fires.
4. *Possible cancer dangers.*   Certain of the ingredients used in the paint are suspected of being carcinogenic.

Because of these health hazards, human workers are unenthusiastic about being exposed to the spray painting environment, and companies have been forced by federal law to construct elaborate ventilating systems to protect their workers.

For these and other reasons, specialized industrial robots are being used more and more frequently to perform spray painting and related processes. Spray painting requires a robot capable of executing a smooth motion pattern which will apply the paint or other fluid evenly and avoid runs. To accomplish this, the robot is equipped with continuous-path control. The paint spray nozzle becomes the end effector. To teach the robot, the walkthrough method is commonly employed. An

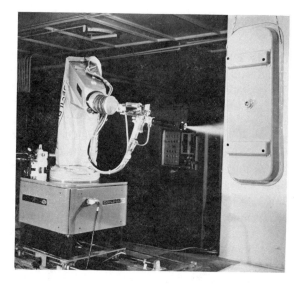

FIGURE 11.5   DeVilbiss/Trallfa robot in spray coating application. (Courtesy of DeVilbiss Company.)

operator-programmer manually leads the robot's end effector through the desired paint spray path. This defines the motion sequence and relative speed for the work cycle. During playback, the robot repeats the cycle to accomplish the paint spray operation. A spray painting robot is illustrated in Figure 11.5.

Among the many advantages of using robots for spray coating applications are the following:

1. *Safety.*   The many safety hazards encountered when human operators perform the spray painting process are reduced.

2. *Coating consistency.*   Once the program is established, the robot will deposit the paint or other coating with the same speed, pattern, and spray rate on every cycle.

3. *Lower material usage.*   The robot's repeatability and consistency reduce wasted paint. Savings in this category seem to range between 10 and 50%.

4. *Less energy used.*   This results from reduced ventilation requirements since the human operator is removed from the actual process.

5. *Greater productivity.*   The paint spraying robot can perform the operation faster than its human counterpart. It can also be used at this faster pace for three shifts per day.

11.6 PROCESSING OPERATIONS

This is a miscellaneous category in which the robot is used to perform some manufacturing process other than welding or spray painting. Assembly and inspection operations are also excluded, and they are covered in the following sections.

Just as in welding and spray painting, the processing operation is performed by a specialized tool attached to the robot's wrist as its end effector. The end effector is typically a powered spindle which holds and rotates a tool such as a drill. The robot would be used to bring the tool into contact with a stationary workpart during processing. In some applications which we will include within this category, the robot's hand is used for gripping the workpart and bringing it into contact with a tool held in a fixed position. In the latter case, we begin to overlap with the types of machine loading applications covered earlier in this chapter.

Some of the processing operations which have been performed by industrial robots include drilling, riveting, grinding, polishing, deburring, wire brushing, and waterjet cutting. The following example will serve to illustrate the applications in this category.

EXAMPLE 11.3

This example involves drilling and routing operations performed on aircraft components [5]. The components were sheet metal fuselage panels of various sizes and shapes for the F-16 fighter plane, manufactured by General Dynamics. A Cincinnati Milacron T^3 robot was used in the workstation. A portion of the robotic work cell is pictured in Figure 11.6.

FIGURE 11.6 Cincinnati Milacron T^3 robot performing a processing operation. (Courtesy of Cincinnati Milacron, Inc.)

This application was developed under the U.S. Air Force ICAM (Integrated Computer-Aided Manufacturing) Program. The work cell has been producing fuselage panels since October, 1979, in a batch production mode. The workstation possesses the following operating features:

Integrated computer control of the workstation components.

Automatic part identification. A commercially available optical character reader is used to identify which workpart is to be processed next.

Mass data storage for part programs. After the workpart has been identified, the correct program for that part is selected from mass storage.

Automatic parts positioning. A weld positioner was used as the work cell parts positioner. It rotates the workpart into the proper registration for the robot to perform its drilling or routing process. While the robot is working on one side of the positioner, a human operator is unloading a finished part and loading a new part on the opposite side.

Automatic tool changing. A tool rack is used to store drills and routing tools for the operation. The appropriate tool for the task is selected automatically for the robot. It should be noted that this work cell was the object of considerable sponsored research and represents a more ambitious robot project than most of the applications included within this category on processing operations.

11.7 ASSEMBLY

Assembly operations are seen as an area with big potential for robot applications [8]. Batch-type assembly operations seem to offer the most promise for using robots. The reason for this is based on economics and the technological capabilities of the robot. For mass production assembly, the most economical method involves fixed automation, where the equipment is designed specifically to produce the particular product. A robot would probably be too slow for mass production, and one of the robot's most important attributes, its programmability, would hardly be used. In batch assembly, there are variations in products and the demand for each product is significantly lower than in mass production. Consequently, the assembly line in batch manufacturing must be capable of dealing with these product variations and the line changeovers that are necessitated. What is basically required for batch production is a flexible assembly system. The term that some companies use for such a system is adaptable-programmable assembly system (APAS), and robot-type arms constitute an important component of these systems.

The APAS would be composed of both conventional material handling devices (conveyors, parts feeders, etc.) and robot arms, probably in an in-line arrangement as shown in Figure 11.7. The robot arms would be used for some parts handling duties, but its main function would be assembly. Robot assembly operations would typically require an extension of the robot's material transfer capability. Many subassemblies consist of a stack of components on top of a base part. To put together the subassembly requires the placement of one part on top of

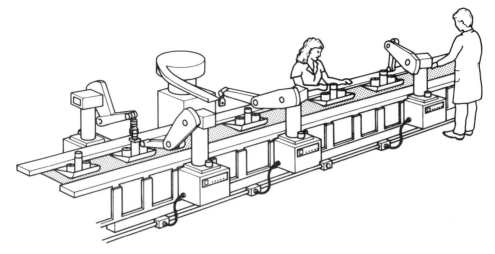

FIGURE 11.7 Possible arrangement of an adaptable-programmable assembly system using robots. (Reprinted with permission from Engelberger [4].)

the base, then another part on top of that, and so forth. The robot is certainly capable of this sort of work cycle. Assembly tasks requiring a special skill or judgment, of which the robot is not capable, would be performed by human workers. Human assembly operators are shown on the APAS line in Figure 11.7.

The features of an industrial robot that make it suitable as a component of an APAS line are its programmability and its adaptability. Programmability is required so that a relatively complex motion cycle can be carried out during the assembly operation. Also, the APAS must be capable of storing multiple program sets to facilitate the differences in products assembled on the line. In this sense the system would be adaptable to changes in product style. Adaptability is also required in the sense that the assembly system would have to compensate for changes in the environment. These environmental variations include:

Variations in the position and orientation of assembly components

Out-of-tolerance and defective parts

The current state of completion of the subassembly

Detection of human beings or objects intruding on the robot's work volume

The PUMA (Programmable Universal Machine for Assembly) is a robot produced by Unimation, Inc. It was designed with assembly operations intended as one of its main functions. The PUMA is a relatively small robot (it requires a space about equal to that which a human operator would need) with a relatively low load-carrying capacity. One of the available PUMA models is pictured in Figure 11.8 performing a part insertion task.

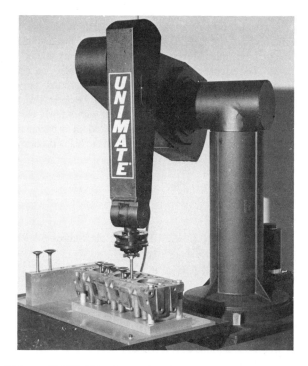

FIGURE 11.8 Unimate PUMA 500 robot performing part insertion in an assembly operation. (Courtesy of Unimation, Inc.)

11.8 INSPECTION

Like assembly, inspection is a relatively new area for the application of industrial robots. Traditionally, the inspection function has been a very labor intensive activity. The activity is slow, tedious, and boring, and is usually performed by human beings on a sampling basis rather than by 100% inspection. With ever-increasing emphasis on quality in manufacturing, there is a trend toward automating the inspection process and toward the use of 100% inspection by machines instead of sampling inspection by human beings. An important role in this area of inspection automation will be played by industrial robots.

Robots equipped with mechanical probes, optical sensing capabilities, or other measuring devices can be programmed to perform dimensional checking and other forms of inspection operations. The following example will serve to illustrate a robotic inspection application.

EXAMPLE 11.4

In this application, four Cincinnati Milacron T³ robots are used in an Automatic Body Checking (ABC) system at Ford Motor Company's Wixom, Michigan, assembly plant [11].

The system performs dimensional inspections on Lincoln Continental and Mark VI automobile bodies. The ABC system is capable of inspecting five or six bodies per hour, whereas the previous manual methods could inspect only one or two bodies per shift.

The robots use electronic probes as their end effectors. With these probes, approximately 150 dimensional checks are made around the windshield, door, and back window openings. These checks are important because the openings receive mating parts (the windshield, doors, etc.) which must fit within fairly close tolerances.

Although the application resulted in a tremendous productivity improvement, the principal justification of a system such as this is product quality improvement.

REFERENCES

[1] BEHUNIAK, J. A., "Planning the Successful Robot Installation," *Robotics Today*, Summer, 1981, pp. 36–37.

[2] CHRISTIAN, J. E., "Putting the Robot to Work," *Robotics Today*, Spring, 1981, pp. 36–37.

[3] COUSINEAU, D. T., "Robots Are Easy, It's Everything Else That's Hard," *Robotics Today*, Spring, 1981, pp. 28–35.

[4] ENGELBERGER, J. F., *Robotics in Practice*, AMACOM (American Management Associations), New York, 1980.

[5] GOLDEN, H. D., et al., "ICAM Robotics System for Aerospace Batch Manufacturing—Task A," *Technical Report AFWAL-TR-80-4042*, Vol. I, Materials Laboratory, Air Force Wright Aeronautical Laboratories, Ohio, April, 1980.

[6] GROOVER, M. P., "Industrial Robots: A Primer on the Present Technology," *Industrial Engineering*, November, 1980, pp. 54–61.

[7] HOLMES, J. G., AND RESNICK, B. J., "A Flexible Robot Arc Welding System," *Proceedings, Fourth National Industrial Robots Conference*, Society of Manufacturing Engineers, November, 1979.

[8] JABLONOWSKI, J., "Robots That Assemble," Special Report 739, *American Machinist*, November, 1981, pp. 175–190.

[9] KONDOLEON, A. S., "Assessing Cycle Times for Robot Assembly Systems," *Robotics Today,* Summer, 1981, pp. 38–42.

[10] LOCKETT, J. H., "The Robotic Work Station in Small Batch Production," *Robotics Today*, Winter, 1979–1980, pp. 17–19.

[11] MACRI, G. C., AND CALENGOR, C. S., "Robots Combine Speed and Accuracy in Dimensional Checks of Automotive Bodies," *Robotics Today*, Summer, 1980, pp. 16–19.

[12] "Robot Teamwork Boosts Production and Lowers Cost at General Electric," *Robotics Today*, Winter, 1981–82, pp. 33–34.

[13] STAUFFER, R. N., "Robogate and Unimates Team Up to Improve Quality and Efficiency," *Robotics Today*, Summer, 1980, pp. 24–30.

[14] STAUFFER, R. N., "Industrial Robots, What They're Doing, Where They're Going," *Robotics Today*, Fall, 1980, pp. 39–42.

[15] TANNER, W. R. (Ed.), *Industrial Robots*, Vol. I: *Fundamentals*1, Vol. II: *Applications*, Society of Manufacturing Engineers, Dearborn, Mich., 1979.

[16] TOEPPERWEIN, L. L., BLACKMAN, M. T., "ICAM Robotics Application Guide," *Technical Report AFWAL-TR-80-4042*, Vol. II, Materials Laboratory, Air Force Wright Aeronautical Laboratories, Ohio April, 1980.

[17] UNIMATION INC., *Unimate Industrial Robot System Planbook*, Danbury, Conn.

[18] UNIMATION INC., *Unimate Industrial Robot Welding Casebook*, Danbury, Conn.

[19] WINSHIP, J. T., "Update on Industrial Robots," *American Machinist*, January, 1979, pp. 121−124.

PART V

Group Technology and Process Planning

chapter 12

Group Technology

12.1 INTRODUCTION

Group technology (abbreviated GT) is a manufacturing philosophy in which similar parts are identified and grouped together to take advantage of their similarities in manufacturing and design. Similar parts are arranged into part families. For example, a plant producing 10,000 different part numbers may be able to group the vast majority of these parts into 50 or 60 distinct families. Each family would possess similar design and manufacturing characteristics. Hence, the processing of each member of a given family would be similar, and this results in manufacturing efficiencies. These efficiencies are achieved in the form of reduced setup times, lower in-process inventories, better scheduling, improved tool control, and the use of standardized process plans. In some plants where GT has been implemented, the production equipment is arranged into machine groups, or cells, in order to facilitate work flow and parts handling.

In product design, there are also advantages obtained by grouping parts into families. For example, a design engineer faced with the task of developing a new part design must either start from scratch or pull an existing drawing from the files and make the necessary changes to conform to the requirements of the new part.

The problem is that finding a similar design may be quite difficult and time consuming. For a large engineering department, there may be thousands of drawings in the files with no systematic way to locate the desired drawing. As a consequence, the designer may decide that it is easier to start from scratch in developing the new part. This decision is replicated many times over in the company, thus consuming valuable time creating duplicate or near-duplicate part designs. If an effective design-retrieval system were available, this waste could be avoided by permitting the engineer to determine quickly if a similar part already exists. A simple change in an existing design would be much less time consuming than starting from scratch. This design-retrieval system is a manifestation of the group technology principle applied to the design function. To implement such a system, some form of parts classification and coding is required.

Parts classification and coding is concerned with identifying the similarities among parts and relating these similarities to a coding system. Part similarities are of two types: *design attributes* (such as geometric shape and size), and *manufacturing attributes* (the sequence of processing steps required to make the part). While the processing steps required to manufacture a part are usually correlated with the part's design attributes, this is not always the case. Accordingly, classification and coding systems are often devised to allow for differences between a part's design and its manufacture.

Whereas a parts classification and coding system is required in a design-retrieval system, it can also be used in computer-aided process planning (CAPP). Computer-aided process planning involves the automatic generation of a process plan (or route sheet) to manufacture the part. The process routing is developed by recognizing the specific attributes of the part in question and relating these attributes to the corresponding manufacturing operations.

In the present chapter we develop the topics of group technology and parts classification and coding. In the following chapter we present a discussion of computer-aided process planning and several related issues. Group technology and parts classification and coding are based on the concept of a part family.

12.2 PART FAMILIES

A *part family* is a collection of parts which are similar either because of geometric shape and size or because similar processing steps are required in their manufacture. The parts within a family are different, but their similarities are close enough to merit their identification as members of the part family. Figures 12.1 and 12.2 show two part families. The two parts shown in Figure 12.1 are similar from a design viewpoint but quite different in terms of manufacturing. The parts shown in Figure 12.2 might constitute a part family in manufacturing, but their geometry characteristics do not permit them to be grouped as a design part family.

The part family concept is central to design-retrieval systems and most current computer-aided process planning schemes. Another important manufactur-

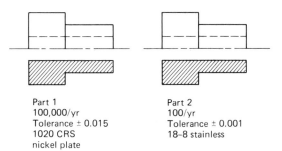

Part 1
100,000/yr
Tolerance ± 0.015
1020 CRS
nickel plate

Part 2
100/yr
Tolerance ± 0.001
18–8 stainless

FIGURE 12.1   Two parts of identical shape and size but different manufacturing requirements.

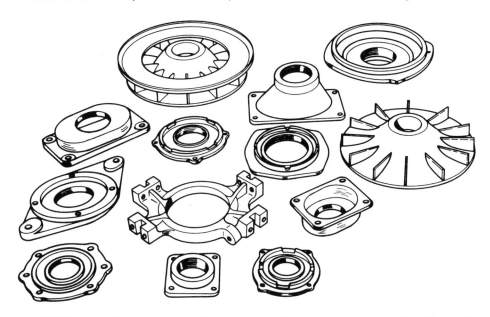

FIGURE 12.2   Thirteen parts with similar manufacturing process requirements but different design attributes.

ing advantage derived from grouping workparts into families can be explained with reference to Figures 12.3 and 12.4. Figure 12.3 shows a process-type layout for batch production in a machine shop. The various machine tools are arranged by function. There is a lathe section, milling machine section, drill press section, and so on. During the machining of a given part, the workpiece must be moved between sections, with perhaps the same section being visited several times. This results in a significant amount of material handling, a large in-process inventory, usually more setups than necessary, long manufacturing lead times, and high cost. Figure 12.4 shows a production shop of supposedly equivalent capacity, but with the machines arranged into cells. Each cell is organized to specialize in the manufacture of a particular part family. Advantages are gained in the form of

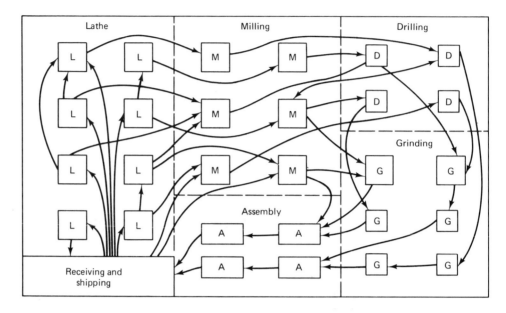

FIGURE 12.3 Process-type layout.

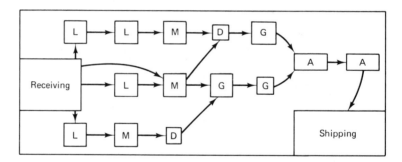

FIGURE 12.4 Group technology layout.

reduced workpiece handling, lower setup times, less in-process inventory, less floor space, and shorter lead times. Some of the manufacturing cells can be designed to form production flow lines, with conveyors used to transport workparts between machines in the cell.

The biggest single obstacle in changing over to group technology from a traditional production shop is the problem of grouping parts into families. There are three general methods for solving this problem. All three methods are time consuming and involve the analysis of much data by properly trained personnel. The three methods are:

1. Visual inspection
2. Production flow analysis (PFA)
3. Parts classification and coding system

The visual inspection method is the least sophisticated and least expensive method. It involves the classification of parts into families by looking at either the physical parts or photographs and arranging them into similar groupings. This method is generally considered to be the least accurate of the three.

The second method, production flow analysis, was developed by J. L. Burbidge [3,6]. PFA is a method of identifying part families and associated machine tool groupings by analyzing the route sheets for parts produced in a given shop. It groups together the parts that have similar operation sequences and machine routings. The disadvantage of PFA is that it accepts the validity of existing route sheets, with no consideration given to whether these process plans are logical or consistent. The production flow analysis approach does not seem to be used much at all in the United States.

The third method, parts classification and coding, is the most time consuming and complicated of the three methods. However, it is the most frequently applied method and is generally recognized to be the most powerful of the three.

12.3 PARTS CLASSIFICATION AND CODING

This method of grouping parts into families involves an examination of the individual design and/or manufacturing attributes of each part. The attributes of the part are uniquely identified by means of a code number. This classification and coding may be carried out on the entire list of active parts of the firm, or a sampling process may be used to establish the part families. For example, parts produced in the shop during a certain given time period could be examined to identify part family categories. The trouble with any sampling procedure is the risk that the sample may be unrepresentative of the entire population. However, this risk may be worth taking, when compared to the relatively enormous task of coding all the company's parts.

Many parts classification and coding systems have been developed throughout the world, and there are several commercially available packages being sold to industrial concerns. It should be noted that none of them has been universally adopted. One of the reasons for this is that a classification and coding system should be custom-engineered for a given company or industry. One system may be best for one company while a different system is more suited to another company. In Section 12.4 we review several of the most familiar parts classification and coding systems. It is the purpose of this section to explain the general structure of these systems.

TABLE 12.1   Design and Manufacturing Part Attributes Typically Included in a Group Technology Classification System

---

Part design attributes

Basic external shape	Major dimensions
Basic internal shape	Minor dimensions
Length/diameter ratio	Tolerances
Material type	Surface finish
Part function	

Part manufacturing attributes

Major process	Operation sequence
Minor operations	Production time
Major dimensions	Batch size
Length/diameter ratio	Annual production
Surface finish	Fixtures needed
Machine tool	Cutting tools

Design systems versus manufacturing systems

Parts classification and coding systems divide themselves into one of three general categories:

1. Systems based on part design attributes
2. Systems based on part manufacturing attributes
3. Systems based on both design and manufacturing attributes

Systems in the first category are useful for design retrieval and to promote design standardization. Systems in the second category are used for computer-aided process planning, tool design, and other production-related functions. The third category represents an attempt to combine the functions and advantages of the other two systems into a single classification scheme. The types of design and manufacturing parts attributes typically included in classification schemes are listed in Table 12.1. It is clear that there is a certain amount of overlap between the design and manufacturing attributes of a part.

Coding system structure

A parts coding scheme consists of a sequence of symbols that identify the part's design and/or manufacturing attributes. The symbols in the code can be all numeric, all alphabetic, or a combination of both types. However, most of the common classification and coding systems use number digits only. There are three basic code structures used in group technology applications:

1. Hierarchical structure
2. Chain-type structure
3. Hybrid structure, a combination of hierarchical and chain-type structures

With the hierarchical structure, the interpretation of each succeeding symbol depends on the value of the preceding symbols. Other names commonly used for this structure are monocode and tree structure. The hierarchical code provides a relatively compact structure which conveys much information about the part in a limited number of digits.

In the chain-type structure, the interpretation of each symbol in the sequence is fixed and does not depend on the value of preceding digits. Another name commonly given to this structure is polycode. The problem associated with polycodes is that they tend to be relatively long. On the other hand, the use of a polycode allows for convenient identification of specific part attributes. This can be helpful in recognizing parts with similar processing requirements.

To illustrate the difference between the hierarchical structure and the chain-type structure, consider a two-digit code, such as 15 or 25. Suppose that the first digit stands for the general part shape. The symbol 1 means round workpart and 2 means flat rectangular geometry. In a hierarchical code structure, the interpretation of the second digit would depend on the value of the first digit. If preceded by 1, the 5 might indicate some length/diameter ratio, and if preceded by 2, the 5 might be interpreted to specify some overall length. In the chain-type code structure, the symbol 5 would be interpreted the same way regardless of the value of the first digit. For example, it might indicate overall part length, or whether the part is rotational or rectangular.

Most of the commercial parts coding systems used in industry are a combination of the two pure structures. The hybrid structure is an attempt to achieve the best features of monocodes and polycodes. Hybrid codes are typically constructed as a series of short polycodes. Within each of these shorter chains, the digits are independent, but one or more symbols in the complete code number are used to classify the part population into groups, as in the hierarchical structure. This hybrid coding seems to best serve the needs of both design and production.

12.4   THREE PARTS CLASSIFICATION AND CODING SYSTEMS

When implementing a parts classification and coding system, most companies elect to purchase a commercially available package rather than develop their own. Inyong Ham [8] recommends that the following factors be considered in selecting a parts coding and classification system:

Objective. The prospective user should first define the objective for the system. Will it be used for design retrieval or part-family manufacturing or both?

Scope and application. What departments in the company will use the system? What specific requirements do these departments have? What kinds of information must be coded? How wide a range of products must be coded? How complex are the parts, shapes, processes, tooling, and so forth?

Costs and time. The company must consider the costs of installation, training, and maintenance for their parts classification and coding system. Will there be consulting fees, and how much? How much time will be required to install the system and train the staff to operate and maintain it? How long will it be before the benefits of the system are realized?

Adapability to other systems. Can the classification and coding system be readily adapted to the existing company computer systems and data bases? Can it be readily integrated with other existing company procedures, such as process planning, NC programming, and production scheduling?

Management problems. It is important that all involved management personnel be informed and supportive of the system. Also, will there be any problems with the union? Will cooperation and support for the system be obtained from the various departments involved?

In the sections below, we review three parts classification and coding systems which are widely recognized among people familiar with GT:

1. Opitz system
2. MICLASS system
3. CODE system

The Opitz classification system

This parts classification and coding system was developed by H. Opitz of the University of Aachen in West Germany. It represents one of the pioneering efforts in the group technology area and is perhaps the best known of the classification and coding schemes.

The Opitz coding system uses the following digit sequence:

$$12345 \quad 6789 \quad ABCD$$

The basic code consists of nine digits, which can be extended by adding four more digits. The first nine digits are intended to convey both design and manufacturing data. The general interpretation of the nine digits is indicated in Figure 12.5. The first five digits, 12345, are called the "form code" and describe the primary design attributes of the part. The next four digits, 6789, constitute the "supple-

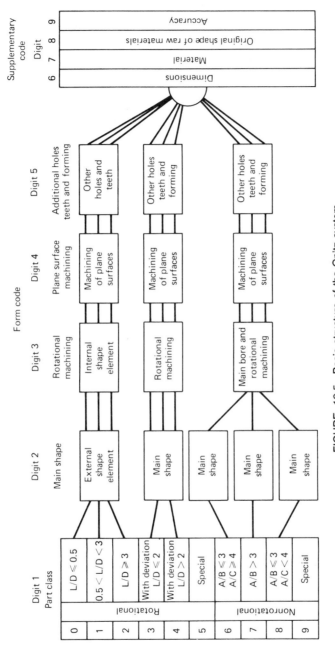

FIGURE 12.5 Basic structure of the Opitz system.

FIGURE 12.6 Form code (digits 1 through 5) for rotational parts in the Opitz system.

Code	Digit 1: Part class	Digit 2: External shape, external shape elements	Digit 3: Internal shape, internal shape elements	Digit 4: Plane surface machining	Digit 5: Auxiliary holes and gear teeth
0	L/D ≤ 0.5 (Rotational parts)	Smooth, no shape elements	No hole, no breakthrough	No surface machining	No auxiliary hole (No gear teeth)
1	0.5 < L/D < 3 (Rotational parts)	No shape elements (Stepped to one end or smooth)	No shape elements (Smooth or stepped to one end)	Surface plane and/or curved in one direction, external	Axial, not on pitch circle diameter (No gear teeth)
2	L/D ≥ 3 (Rotational parts)	Thread (Stepped to one end or smooth)	Thread (Smooth or stepped to one end)	External plane surface related by graduation around a circle	Axial on pitch circle diameter (No gear teeth)
3	(Rotational parts)	Functional groove (Stepped to one end or smooth)	Functional groove (Smooth or stepped to one end)	External groove and/or slot	Radial, not on pitch circle diameter (No gear teeth)
4	(Rotational parts)	No shape elements (Stepped to both ends)	No shape elements (Stepped to both ends)	External spline (polygon)	Axial and/or radial and/or other direction (No gear teeth)
5	(Rotational parts)	Thread (Stepped to both ends)	Thread (Stepped to both ends)	External plane surface and/or slot, external spline	Axial and/or radial on PCD and/or other directions (No gear teeth)
6	(Nonrotational parts)	Functional groove (Stepped to both ends)	Functional groove (Stepped to both ends)	Internal plane surface and/or slot	Spur gear teeth (With gear teeth)
7	(Nonrotational parts)	Functional cone	Functional cone	Internal spline (polygon)	Bevel gear teeth (With gear teeth)
8	(Nonrotational parts)	Operating thread	Operating thread	Internal and external polygon, groove and/or slot	Other gear teeth (With gear teeth)
9	(Nonrotational parts)	All others	All others	All others	All others (With gear teeth)

284

mentary code.'' It indicates some of the attributes that would be of use to manufacturing (dimensions, work material, starting raw workpiece shape and accuracy). The extra four digits, ABCD, are referred to as the ''secondary code'' and are intended to identify the production operation type and sequence. The secondary code can be designed by the firm to serve its own particular needs.

The complete coding system is too complex to provide a comprehensive description here. Opitz wrote an entire book on his system [12]. However, to obtain a general idea of how the Opitz system works, let us examine the first five digits of the code, the form code. The first digit identifies whether the part is a rotational or a nonrotational part. It also describes the general shape and proportions of the part. We will limit our survey to rotational parts possessing no unusual features, those with code values 0, 1, or 2. See Figure 12.5 for definitions. For this general class of workparts, the coding of the first five digits is given in Figure 12.6. An example will demonstrate the coding of a given part.

EXAMPLE 12.1

Given the part design of Figure 12.7, define the ''form code'' (the first five digits) using the Opitz system.

The overall length/diameter ratio, $L/D = 1.5$, so the first digit code = 1. The part is stepped on both ends with a screw thread on one end, so the second digit code would be 5. The third digit code is 1 because of the through-hole. The fourth and fifth digits are both 0, since no surface machining is required and there are no auxiliary holes or gear teeth on the part. The complete form code in the Opitz system is 15100. To add the supplementary code, we would have to properly code the sixth through ninth digits with data on dimensions, material, starting workpiece shape, and accuracy.

The MICLASS System

MICLASS stands for Metal Institute Classification System and was developed by TNO, the Netherlands Organization for Applied Scientific Research [18]. It was started in Europe about five years before being introduced in the United States in

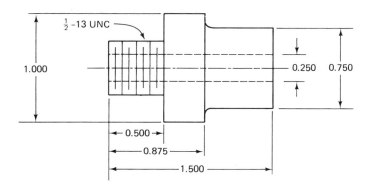

FIGURE 12.7   Workpart for Example 12.1

1974. Today, it is marketed in the United States by the Organization for Industrial Research in Waltham, Massachussets. The MICLASS system was developed to help automate and standardize a number of design, production, and management functions. These include:

Standardization of engineering drawings
Retrieval of drawings according to classification number
Standardization of process routing
Automated process planning
Selection of parts for processing on particular groups of machine tools
Machine tool investment analysis

The MICLASS classification number can range from 12 to 30 digits. The first 12 digits are a universal code that can be applied to any part. Up to 18 additional digits can be used to code data that are specific to the particular company or industry. For example, lot size, piece time, cost data, and operation sequence might be included in the 18 supplementary digits.

The workpart attributes coded in the first 12 digits of the MICLASS number are as follows:

1st digit	Main shape
2nd and 3rd digits	Shape elements
4th digit	Position of shape elements
5th and 6th digits	Main dimensions
7th digit	Dimension ratio
8th digit	Auxiliary dimension
9th and 10th digits	Tolerance codes
11th and 12th digits	Material codes

One of the unique features of the MICLASS system is that parts can be coded using a computer interactively. To classify a given part design, the user responds to a series of questions asked by the computer. The number of questions depends on the complexity of the part. For a simple part, as few as seven questions are needed to classify the part. For an average part, the number of questions ranges between 10 and 20. On the basis of the responses to its questions, the computer assigns a code number to the part. Because the system developer, TNO, is an international organization, the program was written to converse in any of four languages: English, French, German, or Dutch. Also, it can operate in either inches or metric, or both. An example will illustrate the use of the MICLASS computer system for parts classification.

EXAMPLE 12.2

Figure 12.8 shows a rotational part to be classified and coded using MICLASS. Dimensions are given in inches. Figure 12.9 shows a copy of the user interrogation by the computer.

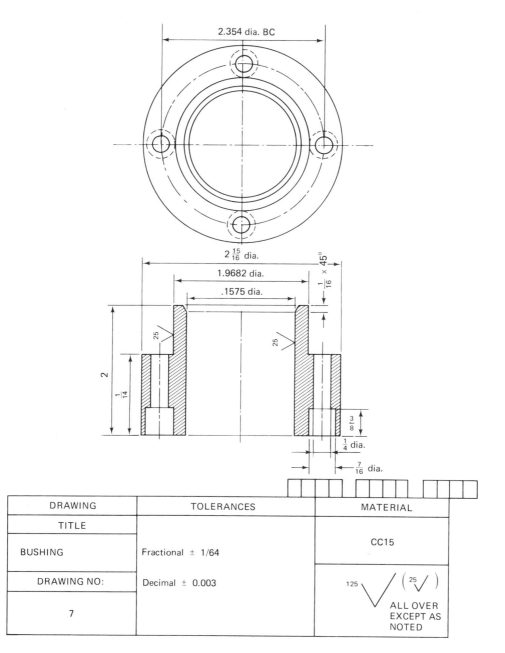

FIGURE 12.8 Workpart for Example 12.2. (Reprinted from Ref. [14].)

The labels on the drawing:

2.354 dia. BC

$2\frac{15}{16}$ dia.

$45°$

$\frac{1}{16} \times$

1.9682 dia.

.1575 dia.

25

25

2

$\frac{1}{14}$

$\frac{3}{8}$

$\frac{1}{4}$ dia.

$\frac{7}{16}$ dia.

DRAWING	TOLERANCES	MATERIAL
TITLE		
BUSHING	Fractional ± 1/64	CC15
DRAWING NO:	Decimal ± 0.003	125 $\sqrt{}$ $\left(\overset{25}{\sqrt{}}\right)$
7		ALL OVER EXCEPT AS NOTED

VERSION —A—

3 MAIN DIMENSIONS (WHEN ROT. PART D.L AND 0) 2.9375 2 0
    DEVIATION OF ROTATIONAL FORM?   NO
    CONCENTRIC SPIRAL GROOVES?   NO

TURNING ON OUTERCONTOUR (EXCEPT ENDFACES)?   YES
    SPECIAL GROOVES OR CONE(S) IN OUTERCONTOUR?   NO
    ALL MACH. DIAM. AND FACES VISIBLE FROM ONE END (EXC. ENDFACE + GROOVES)

TYPING ERROR, ANSWER AGAIN?   YES

INTERNAL TURNING?   YES
    INTERNAL SPECIAL GROOVES OR CONE(S)?   NO
    ALL INT. DIAM. + FACES VISIBLE FROM 1 END (EXC. GROOVES)?   YES

ALL DIAM. + FACES (EXC. ENDFACE) VISIBLE FROM ONE SIDE?   YES

ECC. HOLING AND/OR FACING AND/OR SLOTTING?   YES
    IN INNERFORM AND/OR FACES (INC. ENDFACES)?   YES
    IN OUTERFORM?   NO

ONLY KEYWAYING ETC.?   NO

MACHINED ONLY ONE SENSE?   YES
    ONLY HOLES ON A BOLTCIRCLE AT LEAST 3 HOLES?   YES

FORM–OR THREADING TOLERANCE?   NO

DIAM. ROUGHNESS LESS THAN 33 RU (MICRO–INCHES)?   YES
    SMALLEST POSITIONING TOL. FIELD?.   .016
    SMALLEST LENGTH TOL. FIELD?   .0313

MATERIAL NAME?   CC15

CLASS. NR. = 1271   3231   3144
* *

DRAWING NUMBER MAX 10 CHAR?   7
NOMENCLATURE MAX 15 CHAR?   BUSHING
CONTINUE (1), STOP (2), SECOND PART AGAIN (3)?   2

PROGRAM STOP AT 4690

USED _____ UNITS

FIGURE 12.9 Computerized MICLASS system determination of code number for workpart of Example 12.2. (Reprinted from Ref. [14].)

288

Most of the questions require "Yes" and "No" answers determined from a user analysis of the part drawing. When the interrogation is completed, the computer prints the proper code number. For this part, the universal code (the first 12 digits) is 1271 3231 3144.

The CODE system

The CODE system is a parts classification and coding system developed and marketed by Manufacturing Data Systems, Inc. (MDSI), of Ann Arbor, Michigan. Its most universal application is in design engineering for retrieval of part design data, but it also has applications in manufacturing process planning, purchasing, tool design, and inventory control.

The CODE number has eight digits. For each digit there are 16 possible values (zero through 9 and A through F) which are used to describe the part's design and manufacturing characteristics. The initial digit position indicates the basic geometry of the part and is called the Major Division of the CODE system. This digit would be used to specify whether the shape was a cylinder, flat piece, block, or other. The interpretation of the remaining seven digits depends on the value of the first digit, but these remaining digits form a chain-type structure. Hence the CODE system possesses a hybrid structure.

The second and third digits provide additional information concerning the basic geometry and principal manufacturing process for the part. Digits 4, 5, and 6 specify secondary manufacturing processes such as threads, grooves, slots, and so forth. Digits 7 and 8 are used to indicate the overall size of the part (e.g., diameter and length for a turned part) by classifying it into one of 16 size ranges for each of two dimensions. Figure 12.10 shows a portion of the definitions for digits 2 through 8, given that the part has initially been classified as a cylindrical geometry (Major Division 1 for concentric parts other than profiled). The following example will illustrate the CODE system for a part of this general type.

EXAMPLE 12.3

Figure 12.11 shows a rotational part which is to be classified using the CODE system. The coding analyst would first establish the Major Division into which the part should be classified. In this case, the part is obviously a rotational part, which places it in Major Division 1. Then, by referring to the illustrations in the coding chart, the analyst would develop the values for the remaining seven digits. For this part the CODE is 13188D75.

12.5   GROUP TECHNOLOGY MACHINE CELLS

The traditional view of group technology includes the concept of GT machine cells—groups of machines arranged to produce similar part families. This cellular arrangement of production equipment is designed to achieve an efficient work flow within the cell. It also results in labor and machine specialization for the particular part families produced by the cell. This presumably raises the productivity of the cell.

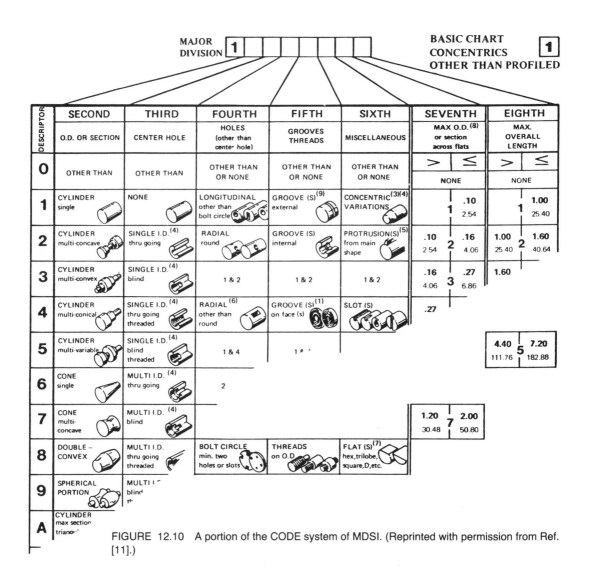

FIGURE 12.10 A portion of the CODE system of MDSI. (Reprinted with permission from Ref. [11].)

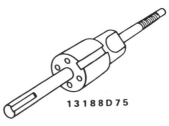

13188D75

FIGURE 12.11 Workpart of Example 12.3. (Reprinted with permission from Ref. [11].)

Although these advantages exist in the GT machine cell, it is a matter of considerable inconvenience and disruption for the shop to make the conversion from a conventional process type layout (Figure 12.3) to the GT cell layout (Figure 12.4). This is surely one of the reasons group technology has not been more widely applied in the manufacturing industries. Today, many practitioners argue that it is possible to achieve a good share of the benefits of GT without physically rearranging the machines into cells. Notwithstanding these arguments, we consider in the current section some of the important aspects of group technology machine cells.

The composite part concept

Part families are defined by the fact that their members have similar design and manufacturing attributes. The composite part concept takes this part family definition to its logical conclusion. It conceives of a hypothetical part that represents all of the design and corresponding manufacturing attributes possessed by the various individuals in the family. Such a hypothetical part is illustrated in Figure 12.12. To produce one of the members of the part family, operations are added and deleted corresponding to the attributes of the particular part design. For example, the composite part in Figure 12.12 is a rotational part made up of seven separate design and manufacturing features. These features are listed in Table 12.2.

A machine cell would be designed to provide all seven machining capabilities. The machine, fixtures, and tools would be set up for efficient flow of workparts through the cell. A part with all seven attributes, such as the composite part of Figure 12.12, would go through all seven processing steps. For part designs without all seven features, unneeded operations would simply be canceled.

In practice, the number of design and manufacturing attributes would be greater than seven, and allowances would have to be made for variations in overall

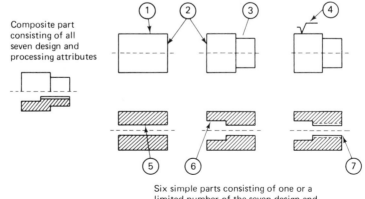

FIGURE 12.12   Composite part concept.

TABLE 12.2   Design and Manufacturing Attributes of the
Composite Part in Figure 12.12

Number	Design and manufacturing attribute
1	Turning operation for external cylindrical shape
2	Facing operation for ends
3	Turning operation to produce step
4	External cylindrical grinding to achieve specified surface finish
5	Drilling operation to create through-hole
6	Counterbore
7	Tapping operation to produce internal threads

size and shape of parts in the part family. Nevertheless, the composite part concept is useful for visualizing the machine cell design problem.

Types of GT machine cells

The organization of machines into cells can follow one of three general patterns:

1.   Single machine cell
2.   Group machine layout
3.   Flow line design

The single machine approach can be used for workparts whose attributes allow them to be made on basically one type of process, such as turning or milling. For example, the composite part of Figure 12.12 could be produced on a conventional turret lathe with the exception of the cylindrical grinding operation (operation 4 in Table 12.2). Even the grinding operation could be set up on the lathe with a little trouble.

The group machine layout is a cell design in which several machines are used together, with no provision for conveyorized parts movement between the machines. The cell contains the machines needed to produce a certain family of parts, and the machines are organized with the proper fixtures, tools, and operators to efficiently produce the parts family.

The flow line cell design is a group of machines connected by a conveyor system. Although this design approaches the efficiency of an automated transfer line, the limitation of the flow line layout is that all the parts in the family must be processed through the machines in the same sequence. Certain of the processing steps can be omitted, but the flow of work through the system must be in one direction. Reversal of work flow is accommodated in the more flexible group machine layout, but not conveniently in the flow line configuration. One possible

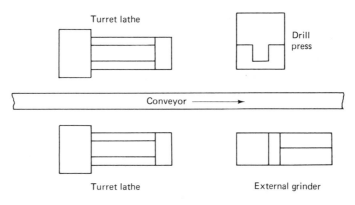

FIGURE 12.13   Flow line cell design.

flow line cell design for producing the parts family of Figure 12.12 (the composite part) is illustrated in Figure 12.13.

12.6   BENEFITS OF GROUP TECHNOLOGY

Although group technology is expected to be an important principle in future production plants, it has not yet achieved the widespread application which might be expected. There are several reasons for this. First, as we have already indicated, there is the problem of rearranging the machines in the plant into GT cells. Many companies have been inhibited from adopting group technology because of the expense and disruption associated with this transition to GT machine cells. Second, there is the problem of identifying part families among the many components produced in the plant. Usually associated with this problem is the expense of parts classification and coding. Not only is this procedure expensive, but it also requires a considerable investment in time and personnel resources. Managers often feel that these limited resources can better be allocated to other projects than group technology with its uncertain future benefits. Finally, it is common for companies to encounter a general resistance among its operating personnel when changeover to a new system is contemplated.

When these problems are solved and group technology is applied, the company will typically realize benefits in the following areas:

Product design
Tooling and setups
Materials handling
Production and inventory control
Employee satisfaction
Process planning procedures

Product design benefits

In the area of product design, improvements and benefits are derived from the use of a parts classification and coding system, together with a computerized design-retrieval system. When a new part design is required, the engineer or draftsman can devote a few minutes to figure the code of the required part. Then the existing part designs that match the code can be retrieved to see if one of them will serve the function desired. The few minutes spent searching the design file with the aid of the coding system may save several hours of the designer's time. If the exact part design cannot be found, perhaps a small alteration of the existing design will satisfy the function. Use of the automated design-retrieval system helps to eliminate design duplication and proliferation of new part designs.

Other benefits of GT in design are that it improves cost estimating procedures and helps to promote design standardization. Design features such as inside corner radii, chamfers, and tolerances are more likely to become standardized with group technology. According to DeVries et al. [5], a 10% reduction in the number of drawings can be expected through standardization with GT.

Tooling and setups

Group technology also tends to promote standardization of several areas of manufacturing. Two of these areas are tooling and setups.

In tooling, an effort is made to design group jigs and fixtures that will accommodate every member of a parts family. Workholding devices are designed to use special adapters which convert the general fixture into one that can accept each part family member.

The machine tools in a GT cell do not require drastic changeovers in setup because of the similarity in the workparts processed on them. Hence, setup time is saved, and it becomes more feasible to try to process parts in an order so as to achieve a bare minimum of setup changeovers. It has been estimated that the use of group technology can result in a 69% reduction in setup time [5].

Materials handling

Another advantage in manufacturing is a reduction in the workpart move and waiting time. The group technology machine layouts lend themselves to efficient flow of materials through the shop. The contrast is sharpest when the flow line cell design is compared to the conventional process-type layout (Figures 1.3 and 1.4).

Production and inventory control

Several benefits accrue to a company's production and inventory control function as a consequence of group technology.

Production scheduling is simplified with group technology. In effect, grouping of machines into cells reduces the number of production centers that must be scheduled. Grouping of parts into families reduces the complexity and size of the parts scheduling problem. And for those workparts that cannot be processed through any of the machine cells, more attention can be devoted to the control of these parts. Because of the reduced setups and more efficient materials handling with machine cells, production lead times, work-in-process, and late deliveries can all be reduced. Estimates on what can be expected are provided by DeVries et al. [5]:

70% reduction in production times

62% reduction in work-in-process inventories

82% reduction in overdue orders

Employee satisfaction

The machine cell often allows parts to be processed from raw material to finished state by a small group of workers. The workers are able to visualize their contributions to the firm more clearly. This tends to cultivate an improved worker attitude and a higher level of job satisfaction.

Another employee-related benefit of GT is that more attention tends to be given to product quality. Workpart quality is more easily traced to a particular machine cell in group technology. Consequently, workers are more responsible for the quality of work they accomplish. Traceability of part defects is sometimes very difficult in a conventional process-type layout, and quality control suffers as a result.

Process planning procedures

The time and cost of the process planning function can be reduced through standardization associated with group technology. A new part design is identified by its code number as belonging to a certain parts family, for which the general process routing is known. The logic of this procedure can be written into computer software to form a computer-automated process planning system. This topic will be discussed in detail in the following chapter.

REFERENCES

[1] ABOU-ZEID, M. R., "Group Technology," *Industrial Engineering*, May, 1975, pp. 32–39.

[2] THE BRISCH ORGANIZATION, *The Modern Approach to Industrial Classification and Coding,* Brisch, Birn & Partners Ltd., Inc., Fort Lauderdale, Fla.

[3] BURBIDGE, J. L., *The Introduction of Group Technology,* Heinemann, London, 1975.

[4] DESAI, D. T., "Parts Coding Using Group Technology," *Industrial Engineering,* November, 1981, pp. 78–86.

[5] DEVRIES, M. F., HARVEY, S. M., AND TIPNIS, V. A., *Group Technology,* Publication MDC 76-601, Machinability Data Center, Cincinnati, Ohio, 1976.

[6] GALLAGHER, C. C., AND KNIGHT, W. A., *Group Technology,* Butterworth & Company Ltd., Kent, England, 1973.

[7] GROOVER, M. P., *Automation, Production Systems, and Computer-Aided Manufacturing,* Prentice-Hall, Inc., Englewood Cliffs, N.J., 1980, Chapter 18.

[8] HAM, I., "Introduction to Group Technology," *Technical Report MMR76-03,* Society of Manufacturing Engineers, Dearborn, Mich., 1976.

[9] HOUTZEEL, A., *Classification and Coding,* TNO, Organization for Industrial Research, Inc., Waltham, Mass.

[10] HOUTZEEL, A., "The Many Faces of Group Technology," *American Machinist,* January, 1979, pp. 115–120.

[11] MANUFACTURING DATA SYSTEMS, INC., *CODE: The Parts Classification Data Retrieval System for Computer-Aided Manufacturing,* Product Information PI-30-6000-0, 1977.

[12] OPITZ, H., *A Classification System to Describe Workpieces,* Pergamon Press Ltd., Oxford, England.

[13] OPITZ, H., AND WIENDAHL, H. P., "Group Technology and Manufacturing Systems for Medium Quantity Production," *International Journal of Production Research,* Vol. 9, No. 1, 1971, pp. 181–203.

[14] ORGANIZATION FOR INDUSTRIAL RESEARCH, *Miclass, Migroup, Miplan, Migraphics* (marketing brochure), Waltham, Mass.

[15] PHILLIPS, R. H., AND ELGOMAYEL, J., "Group Technology Applied to Product Design," *Educational Module* Manufacturing Productivity Educational Committee, Purdue Research Foundation, West Lafayette, Inc., 1977.

[16] SCHAFFER, G. H., "Implementing CIM," *Special Report 736, American Machinist,* August, 1981, pp. 151–174.

[17] STAUFFER, R. N., "The Rewards of Classification and Coding," *Manufacturing Engineering,* May, 1979, pp. 48–52.

[18] TNO, *An Introduction to MICLASS,* Organization for Industrial Research, Inc., Waltham, Mass.

PROBLEMS

12.1. Determine the form code (first five digits) in the Opitz system for the part shown in Figure 12.1. Assume the following dimensions for the part: Overall length = 2.5 in., large outside diameter (O.D.) = 2.2 in., small O.D. = 1.5 in., length of large O.D. = 1.0 in., hole diameter = 0.75 in.

12.2. Determine the form code (five digits) in the Opitz system for the composite part shown in Figure 12.12. Assume that the part is of the following dimensions: Length

= 3.0 in., large O.D. = 2.375 in., small O.D. = 1.75 in., length of large diameter = 1.75 in., hole size = 0.5 in., counterbore diameter = 1.25 in., counterbore length = 1.25 in.; the thread is 13 UNC.

12.3. Develop a simple parts classification and coding scheme of no more than four digits which can be used to uniquely code the various possible combinations of part attributes for the composite part shown in Figure 12.12. The coding scheme should consider shape features and size, but not material. What is the structure of your coding scheme (hierarchical, chain-type, or hybrid)? What is the meaning of each digit in the code number? How many different values can each digit take? (*Note*: This problem will require a considerable amount of thought and deliberation and will probably require a significant amount of time to complete.)

chapter 13

Computer-Aided Process Planning

13.1   THE PLANNING FUNCTION

This chapter examines several process planning functions which can be implemented by computer systems. Process planning is conerned with determining the sequence of individual manufacturing operations needed to produce a given part or product. The resulting operation sequence is documented on a form typically referred to as a route sheet. The route sheet is a listing of the production operations and associated machine tools for a workpart or assembly.

Closely related to process planning are the functions of determining appropriate cutting conditions for the machining operations and setting the time standards for the operations. All three functions—planning the process, determining the cutting conditions, and setting the time standards—have traditionally been carried out as tasks with a very high manual and clerical content. They are also typically routine tasks in which similar or even identical decisions are repeated over and over. Today, these kinds of decisions are being made with the aid of computers. In the first four sections of this chapter we consider the process planning function and how computers can be used to perform this function. In Sections 13.5

and 13.6 we study machinability data systems and computer-generated time standards, respectively.

Traditional process planning

There are variations in the level of detail found in route sheets among different companies and industries. In the one extreme, process planning is accomplished by releasing the part print to the production shop with the instructions "make to drawing." Most firms provide a more detailed list of steps describing each operation and identifying each work center. In any case, it is traditionally the task of the manufacturing engineers or industrial engineers in an organization to write these process plans for new part designs to be produced by the shop. The process planning procedure is very much dependent on the experience and judgment of the planner. It is the manufacturing engineer's responsibility to determine an optimal routing for each new part design. However, individual engineers each have their own opinions about what constitutes the best routing. Accordingly, there are differences among the operation sequences developed by various planners. We can illustrate rather dramatically these differences by means of an example.

In one case cited in Ref. [21], a total of 42 different routings were developed for various sizes of a relatively simple part called an "expander sleeve." There were a total of 64 different sizes and styles, each with its own part number. The 42 routings included 20 different machine tools in the shop. The reason for this absence of process standardization was that many different individuals had worked on the parts: 8 or 9 manufacturing engineers, 2 planners, and 25 NC part programmers. Upon analysis, it was determined that only two different routings through four machines were needed to process the 64 part numbers. It is clear that there are potentially great differences in the perceptions among process planners as to what constitutes the "optimal" method of production.

In addition to this problem of variability among planners, there are often difficulties in the conventional process planning procedure. New machine tools in the factory render old routings less than optimal. Machine breakdowns force shop personnel to use temporary routings, and these become the documented routings even after the machine is repaired. For these reasons and others, a significant proportion of the total number of process plans used in manufacturing are not optimal.

Automated process planning

Because of the problems encountered with manual process planning, attempts have been made in recent years to capture the logic, judgment, and experience required for this important function and incorporate them into computer programs. Based on the characteristics of a given part, the program automatically generates the manufacturing operation sequence. A computer-aided process planning (CAPP) system offers the potential for reducing the routine clerical work of manufacturing engineers. At the same time, it provides the opportunity to generate production routings which are rational, consistent, and perhaps even optimal. Two alternative approaches to computer-aided process planning have been developed. These are:

1. Retrieval-type CAPP systems (also called variant systems)
2. Generative CAPP systems

The two types are described in the following two sections.

13.2 RETRIEVAL-TYPE PROCESS PLANNING SYSTEMS

Retrieval-type CAPP systems use parts classification and coding and group technology as a foundation. In this approach, the parts produced in the plant are grouped into part families, distinguished according to their manufacturing characteristics. For each part family, a standard process plan is established. The standard process plan is stored in computer files and then retrieved for new workparts which belong to that family. Some form of parts classification and coding system is required to organize the computer files and to permit efficient retrieval of the appropriate process plan for a new workpart. For some new parts, editing of the existing process plan may be required. This is done when the manufacturing requirements of the new part are slightly different from the standard. The machine routing may be the same for the new part, but the specific operations required at each machine may be different. The complete process plan must document the operations as well as the sequence of machines through which the part must be routed. Because of the alterations that are made in the retrieved process plan, these CAPP systems are sometimes also called by the name "variant system."

Figure 13.1 will help to explain the procedure used in a retrieval process planning system. The user would initiate the procedure by entering the part code number at a computer terminal. The CAPP program then searches the part family matrix file to determine if a match exists. If the file contains an identical code number, the standard machine routing and operation sequence are retrieved from the respective computer files for display to the user. The standard process plan is examined by the user to permit any necessary editing of the plan to make it compatible with the new part design. After editing, the process plan formatter prepares the paper document in the proper form.

If an exact match cannot be found between the code numbers in the computer file and the code number for the new part, the user may search the machine routing file and the operation sequence file for similar parts that could be used to develop the plan for the new part. Once the process plan for a new part code number has

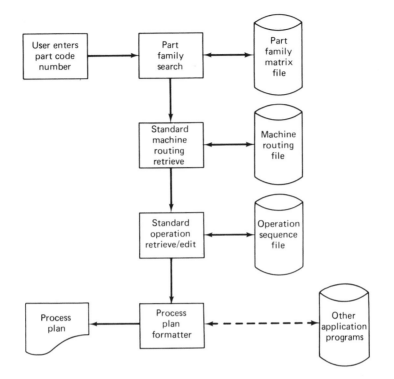

FIGURE 13.1 Information flow in a retrieval-type computer-aided process planning system.

been entered, it becomes the standard process for future parts of the same classification.

In Figure 13.1 the machine routing file is distinguished from the operation sequence file to emphasize that the machine routing may apply to a range of different part families and code numbers. It would be easier to find a match in the machine routing file than in the operation sequence file. Some CAPP retrieval systems would use only one such file which would be a combination of operation sequence file and machine routing file.

The process plan formatter may use other application programs. These could include programs to compute machining conditions, work standards, and standard costs. Standard cost programs can be used to determine total product costs for pricing purposes.

A number of retrieval-type computer-aided process planning systems have been developed. These include MIPLAN, one of the MICLASS modules [6,20], the CAPP system developed by Computer-Aided Manufacturing—International [1], COMCAPP V by MDSI, and systems by individual companies [10]. We will use MIPLAN as an example to illustrate these industrial systems.

EXAMPLE 13.1

MIPLAN is a computer-aided process planning package available from the Organization for Industrial Research, Inc., of Waltham, Massachusetts. It is basically a retrieval-type CAPP system with some additional features. The MIPLAN system consists of several modules which are used in an interactive, conversational mode.

To operate the system, the user can select any of four different options to create the process plan for a new part:

1. The first option is a retrieval approach in which the user inputs a part code number and a standard process plan is retrieved from the computer file for possible editing. To generate the part code number, the planner may elect to use the MICLASS interactive parts classification and coding system (see Section 12.4).

2. In the second option, a process plan is retrieved from the computer files by entering an existing part number (rather than a part code number). Again, the existing process plan can be edited by the user, if required.

3. A process plan can be created from scratch, using standard text material stored in computer files. This option is basically a specialized word-processing system in which the planner selects from a menu of text related to machines and processes. The process plan is assembled from text passages subject to editing for the particular requirements of the new part.

4. The user can call up an incomplete process plan from the computer file. This may occur when the user is unable to complete the process plan for a new part at one sitting. For example, the planner may be interrupted in the middle of the procedure to solve some emergency problem. When the procedure is resumed, the incomplete plan can be retrieved and finished.

After the process plan has been completed using one of the four MIPLAN options, the user can have a paper document printed out by the computer. A typical MIPLAN output is shown in Figure 13.2. It is also possible for the user to store the completed process plan (or the partially completed plan as with option 3) in the computer files, or to purge an existing plan from the files. This might be done, for example, when an old machine tool is replaced by a more productive machine, and this necessitates changes in some of the standard process plans.

Computer graphics can be utilized to enhance the MIPLAN output. This possibility is illustrated in Figure 13.3, which shows a tooling setup for the machining operation described. With this kind of pictorial process planning, drawings of workpart details, tool paths, and other information can be presented visually to facilitate communication to the manufacturing shops.

13.3 GENERATIVE PROCESS PLANNING SYSTEMS

Generative process planning involves the use of the computer to create an individual process plan from scratch, automatically and without human assistance. The computer would employ a set of algorithms to progress through the various techni-

ORGANIZATION FOR INDUSTRIAL RESEARCH, INC.					FACILITY — F1	
PART NUMBER: A63799		S/O #	PRJ #	ORDER QTY	MINIMUM QTY	DUE DATES	PRI #
PART NAME: SHAFT, ARM							

ORGANIZATION FOR INDUSTRIAL RESEARCH, INC.						FACILITY — F1	
PART NUMBER: A63799		S/O #	PRJ #	ORDER QTY	MINIMUM QTY	DUE DATES	PRI #
PART NAME: SHAFT, ARM		A34UB	45D3	1000	935	249	2
PLNG REV: 02	DWG REV: 0						
PLANNER: ADAMS							

INSPECTIONS				CODE #: 1310-1181-2111-0000-0100-0000-0000-00		
	#1	#2	#3	SPECIAL INSTRUCTIONS/HANDLING:		
MFG ENG Q/A		FHB PC AH		1/2″ DIA MS-5000 H.R. STEEL (2″ LGTHS)		

OPER NO	MACH TOOL	OPERATION DESCRIPTION — ASSY INSTRUCTIONS	SETUP TIMES	PIECE TIMES	OPR

0010	5145	S/U COLLET
		ROUGH TURN MACHINE PER TAPE NO. LS982A
		   0.440 DIA. BY 1.750 LENGTH
		   0.300 DIA. BY 0.8120 LENGTH
		   0.275 DIA. BY 0.4375 LENGTH
		FINISH 3/64 GROOVES (TYP) AND CHAMFERS
		   0.270 DIA. BY 0.375 LENGTH
		CHAMFER CUTOFF TO 1.906
			2.00	0.173	

0015	1026	#2 CENTERS BOTH ENDS	0.25	0.004	
0020	9401	CARBURIZE AND HARDEN	0.50		
0030	4063	S/U BETWEEN CENTERS GRIND OD HOLD CONCENTRICITY HOLD 0.4200 DIM. HOLD 0.2600 DIM. HOLD 0.2815 DIM. HOLD 0.2712 DIM.	1.25	0.0983	
0040	9501	BLAST TO CLEAN		0.001	
0050	9201	CHROME PLATE PER PRINT	0.38		
0060	9805	FINAL INSPECT			

FIGURE 13.2   Route sheet generated by MIPLAN. (Reprinted with permission from Ref. [15].)

cal and logical decisions toward a final plan for manufacturing. Inputs to the system would include a comprehensive description of the workpart. This may involve the use of some form of part code number to summarize the workpart data, but it does not involve the retrieval of existing standard plans. Instead, the generative CAPP system synthesizes the design of the optimum process sequence, based on an analysis of part geometry, material, and other factors which would influence manufacturing decisions.

In the ideal generative process planning package, any part design could be presented to the system for creation of the optimal plan. In practice, current generative-type systems are far from universal in their applicability. They tend to fall short of a truly generative capability, and they are developed for a somewhat limited range of manufacturing processes.

We will illustrate the generative process planning approach by means of a system called GENPLAN developed at Lockheed-Georgia Company [5,22].

PART NO: 190105		PART NAME: FRONT PLATE		FORMAT:

PLNG REV: 1   DWG REV: A   PLANNER: ADAMS

CODE #: 8798-3711-1189-3433-1400-0000-0000-00

0040	5002 MACHINE PER TAPE #1
	SET UP IN FIXTURE #1
	WITH STD ANGLE PLATE
	#A123 PER SKETCH
	ROUGH AND FINISH FACE
	− HOLD 0.25 + − 0.02 DIM.
	USE 4″ DIA FACE MILL
	C-DRILL (3) HOLES
	DRILL (1) HOLE 3/8 DIA.
	DRILL (2) TOOLING HOLES
	5/16 DIA.
	DRILL (2) TOOLING HOLES
	0.365/0.370 DIA.
	AND REAM TO 0.376/0.370 DIA.
	SET UP = 2.50
	PIECE TIME = 0.350

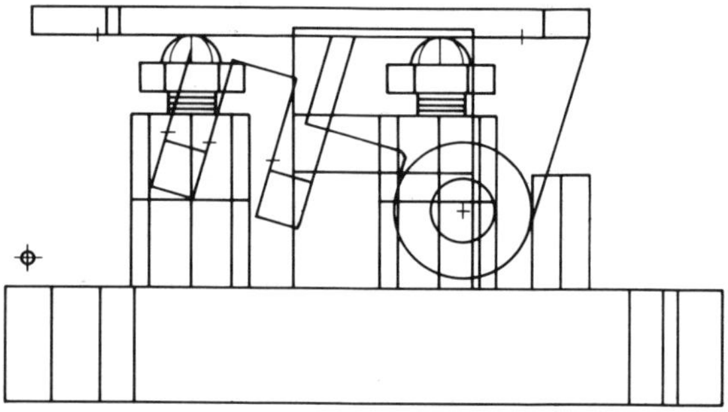

FIGURE 13.3 Pictorial process planning. (Reprinted with permission from Ref. [15].)

EXAMPLE 13.2

GENPLAN is close to a generative process planning system, but it requires a human planner to assist with some of the manufacturing decisions. Also, there are several versions of GEN-PLAN (one for parts fabrication, and another for assembly), which means that it is not a system of universal applicability.

To operate the system, the planner enters a part classification code using a coding scheme developed at Lockheed. GENPLAN then analyzes the characteristics of the part based on the code number (e.g, part geometry, workpiece material, and other manufacturing-related features) to synthesize an optimum process plan. It does not store stand-ard manufacturing plans. Rather, it stores machine tool capabilities and it employs the logic and technological science of manufacturing. The output is a document specifying the sequence of operations, machine tools, and calculated process times. An example of a computer-generated route sheet produced by GENPLAN is shown in Figure 13.4. Process plans that previously required several hours to accomplish manually are now done typically by GENPLAN in 15 minutes.

PART NUMBER			OPP	PROD	SPARES	OTHER	EDITION	TYPE MFG	TOOLING	DCC	METHOD SERIAL	MODEL	DATE	PAGE
3313793-3			X				303A					C-130		1

PART NAME	STIFFENER						GROUP	06.00		PLANNER	KEYS 2932

MAKE FROM NUMBER		OPP	CLAD (X)	GAUGE / WALL	WIDTH / O.D.	LGTH/BLANK/1st STRIP	LGTH E.A.P. (STRIP)	PTS/BLANK	INSP.
LS10133-1402-5						9			

MATL. DESCRIPTION	TYPE MATERIAL	ALLOY	CONDITION	SPECIFICATION	AUTHORITY
EXTRU	ALUM	7075	T6511	QQ-A-200/11	62672

GA FORM 8857.1   DESCRIPTION

BC	CC	LC	OPER	TOOL CODE	TOOL IDENTIFICATION	DESCRIPTION	T.S. CODE	SET UP	RUN/100	REJ.	ACPT.
18	35	640	006	ATT		SHAPE SHORT	260	40	10		
					FLANGE IN MUL. LENGTHS						
18	35	391	010	ATT	SAW TO LENGTH	SAW0391	254A	15	35		
18	35	627	020	MPS7253	IDENTIFY M/T A PCT.		2237	08	00		
57			030		INSPECT						
							CC-TOT	63	45		
18	67	166	040	ATT		SCRIBE FOR TRIM	265	05	100		
18	67	169	050		TRIM TO SCRIBE LINE		4012	03	140		
18	67	212	060	ATT	DRILL	DRIL212	249	15	100		
18	67	500	070	MPS7302	BURR-BREAK SHARP EDGES		500M	05	217		
18	67	696	080	MPS7257	IDENTIFY TAG AND SEAL		2237	08	00		

CONTINUE ON PAGE 2

3313793-3   P   303A   OLD   GENPLAN OPSHEET

PART NUMBER			OPP	PROD	SPARES	OTHER	EDITION	TYPE MFG	TOOLING	DCC	METHOD SERIAL	MODEL	DATE	PAGE
3313793-3			X				303A					C-130		2

PART NAME	STIFFENER						GROUP	06.00		PLANNER	KEYS 2932

MAKE FROM NUMBER		OPP	CLAD (X)	GAUGE / WALL	WIDTH / O.D.	LGTH/BLANK/1st STRIP	LGTH E.A.P. (STRIP)	PTS/BLANK	INSP.
LS10133-1402-5						9			

MATL. DESCRIPTION	TYPE MATERIAL	ALLOY	CONDITION	SPECIFICATION	AUTHORITY
EXTRU	ALUM	7075	T6511	QQ-A-200/11	62672

GA. FORM 8857.1   DESCRIPTION

BC	CC	LC	OPER.	TOOL CODE	TOOL IDENTIFICATION	DESCRIPTION	T.S. CODE	SET UP	RUN/100	REJ.	ACPT.
57			090		INSPECT						
							CC-TOT	36	557		
18	42	104	100	MPS1322F			002	00	00		
18	42	140	110	MPS10121-1			002	00	00		
18	42	140	120	MPS101229	COLOR 24424		002	00	00		
18	42	668	130	MPS7253	IDENTIFY		916E	00	657		
57			140		INSPECT HARDNESS						
57			150		INSPECT						
							CC-TOT	00	657		
			160		STOCK						
			RFR		TO ADD BREAK SHARP EDGES						

FIGURE 13.4   Route sheet generate by GENPLAN. (Reprinted with permission from Ref. [5].)

13.4   BENEFITS OF CAPP

Whether it is a retrieval system or a generative system, computer-aided process planning offers a number of potential advantages over manually oriented process planning.

1. *Process rationalization.*   Computer-automated preparation of operation routings is more likely to be consistent, logical, and optimal than its manual counterpart. The process plans will be consistent because the same computer software is being used by all planners. We avoid the tendency for drastically different process plans from different planners, as described in Section 13.1. The process plans tend to be more logical and optimal because the company has presumably incorporated the experience and judgment of its best manufacturing people into the process planning computer software.

2. *Increased productivity of process planners.*   With computer-aided process planning, there is reduced clerical effort, fewer errors are made, and the planners have immediate access to the process planning data base. These benefits translate into higher productivity of the process planners. One system was reported to increase productivity by 600% in the process planning function [10].

3. *Reduced turnaround time.*   Working with the CAPP system, the process planner is able to prepare a route sheet for a new part in less time compared to manual preparation. This leads to an overall reduction in manufacturing lead time.

4. *Improved legibility.*   The computer-prepared document is neater and easier to read than manually written route sheets. CAPP systems employ standard text, which facilitates interpretation of the process plan in the factory.

5. *Incorporation of other application programs.*   The process planning system can be designed to operate in conjunction with other software packages to automate many of the time-consuming manufacturing support functions. We discuss two of these related planning functions, machinability data systems and computerized work standards, in the following sections.

13.5   MACHINABILITY DATA SYSTEMS

In a machine shop, process planning should include selection of the cutting conditions that are to be used in the various machining operations. The cutting conditions consist of the speed, feed, and depth of cut. Depth of cut is usually predetermined by the workpiece geometry and operation sequence. Therefore, the problem reduces to one of determining the proper speed and feed combination. Machinability data systems are basically intended to solve this problem.

Definition of the problem

Stated precisely, the objective of a machinability data system is to select cutting speed and feed rate given that the following characteristics of the operation have been defined:

1. Type of machining operation
2. Machine tool
3. Cutting tool
4. Workpart
5. Operating parameters other than feed and speed

The magnitude of the problem can best be appreciated by contemplating the multitude of different parameters that are included within these five operation characteristics. A partial list of the parameters is presented in Table 13.1.

The methods of solving the speed/feed selection problem are:

1. Experience and judgment of process planner, foreman, or machine operator

TABLE 13.1   Characteristics of a Machining Operation

1. Type of Machining Operation
   a. Process type—turning, facing, drilling, tapping, milling, boring, grinding, etc.
   b. Roughing operation versus finishing operation.

2. Machine Tool Parameters
   a. Size and rigidity
   b. Horsepower
   c. Spindle speed and feed rate levels
   d. Conventional or NC
   e. Accuracy and precision capabilities
   f. Operating time data

3. Cutting Tool Parameters
   a. Tool material type (high-speed steel, cemented carbide, ceramic, etc.)
   b. Tool material chemistry or composition
   c. Physical and mechanical properties (hardness, wear resistance, etc.)
   d. Type of tool (single point, drill, milling cutter, etc.)
   e. Geometry (nose radius, rake angles, relief angles, number of teeth, etc.)
   f. Tool cost data

4. Workpart Characteristics
   a. Material—basic type and specific grade
   b. Hardness and strength of work material
   c. Geometric size and shape
   d. Tolerances
   e. Surface finish
   f. Initial surface condition of workpiece

5. Operating Parameters Other Than Feed and Speed
   a. Depth of cut
   b. Cutting fluid, if any
   c. Workpiece rigidity
   d. Fixtures and jigs used

Source: From Ref. [3].

2. Handbook recommendations
3. Computerized machinability data systems

Relying on the experience and judgment of any individual is the least systematic approach and carries the greatest risk. The risk lies in the potential loss of the individual who has acquired the experience and judgment over many years in the shop. Personal judgment is also undesirable because it usually has no scientific foundation. Cutting conditions derived from personal experience are not based on economic criteria.

Handbook recommendations are compiled from the experiences of more than one person. Handbooks of machinability data are generally developed from a systematic analysis of large quantities of machining data. The cutting recommendations are often based on laboratory experiments whose objective is to determine speeds and feeds. The best known of these handbooks is the *Machining Data Handbook* [9].

Although the handbook approach represents a definite improvement over personal judgment, it often suffers from several drawbacks when applied to a particular company's machining environment. First, handbook recommendations tend to be conservative, meaning that the suggested feeds and speeds are based on worst-case conditions. Second, handbooks must be considered as general guides and may not coincide with the particular product line and machine tools of a given shop. Third, the use of handbooks is not compatible with the automation of the process planning function using a computerized data base.

Computerized machinability data system

To overcome these difficulties, efforts have been directed to the development of computerized machinability data systems. These efforts date back to the early 1960s and are continuing today. Some of the systems have been developed by individual firms to meet their own specific requirements. The importance of these systems has grown with the increase in the use of NC machines and the economic need to operate these machines as efficiently as possible. The importance of computerized machinability data systems will continue to grow with the development of integrated manufacturing data bases.

Computerized machinability data systems have been classified into two general types by Pressman and Williams [17]:

1. Data base systems
2. Mathematical model systems

DATA BASE SYSTEMS. These systems require the collection and storage of large quantities of data from laboratory experiments and shop experience. The data base is maintained on a computerized storage file that can be accessed either by a remote terminal or in a batch mode for a more permanent printout of cutting recommendations. An example of a typical printout is shown in Figure 13.5.

COST AND PRODUCTION RATE FOR TURNING
THROWAWAY CARBIDE TOOLS

DATA SET NO	WORK MATERIAL	HARD NESS	TOOL MATL	CUT SPD F/M	FEED IN/REV	TOOL LIFE MIN	FEED COST $	RAPD TRAV $	LOAD UNLD $	SET UP $	INDX INST $	HLDR DEPR $	INSERT COST $	TOTAL COST $/PC	PROD RATE PC/HR
16	AISI 4340	515	C-8	300	0.0100	8	0.87	0.04	0.34	0.15	0.04	0.00	0.11	1.58	6.1
17	AISI 4340	515	C-8	200	0.0100	30	1.30	0.04	0.34	0.15	0.01	0.00	0.04	1.91	4.8
18	AISI 4340	515	C-8	150	0.0100	48	1.74	0.04	0.34	0.15	0.01	0.00	0.03	2.33	3.9
19	VASJET1000	52RC	C-8	162	0.0100	15	1.61	0.04	0.34	0.15	0.04	0.00	0.11	2.32	4.0
20	VASJET1000	52RC	C-8	133	0.0100	30	1.96	0.04	0.34	0.15	0.02	0.00	0.07	2.60	3.5
21	VASJET1000	52RC	C-8	110	0.0100	45	2.37	0.04	0.34	0.15	0.02	0.00	0.05	2.99	3.0
22	VASJET1000	52RC	C-8	95	0.0100	60	2.74	0.04	0.34	0.15	0.01	0.00	0.04	3.36	2.7
23	250 MARAGE	53RC	C-3	345	0.0100	5	0.75	0.04	0.34	0.15	0.06	0.00	0.16	1.53	6.6
24	250 MARAGE	53RC	C-3	315	0.0100	15	0.82	0.04	0.34	0.15	0.02	0.00	0.05	1.45	6.4
25	250 MARAGE	53RC	C-3	275	0.0100	35	0.94	0.04	0.34	0.15	0.01	0.00	0.02	1.53	5.9

FIGURE 13.5 Typical output format of a computerized machinability data system. (Reprinted with permission from Ref. [2].)

To collect the machinability data for a data base system, cutting experiments are performed over a range of feasible conditions. These experiments are commonly conducted in the laboratory. However, many data base machinability systems allow for shop data to be entered into the files also. For each set of conditions, computations are made to determine the cost of the operation. Not only is the total cost per piece calculated, but the cost components that make up the total are also calculated. Examples of these costs are illustrated in Figure 13.5.

The computations are based on the traditional concept in machining economics that the total cost per piece is composed of elements as given in the following equation:

$$C_{pc} = C_o T_m + C_o T_h + \frac{T_m}{T}(C_t + C_o T_{tc}) \tag{13.1}$$

where C_{pc} = cost per workpiece, \$/piece

     C_o = cost to operate the machine tool (labor, machine, and applicable overhead), \$/min

     T_m = machining time, min

     T_h = workpiece handling time, min

      T = tool life, min

     C_t = cost of tooling, \$/cutting edge

     T_{tc} = tool change time, min

The organization of a typical data base system consists of files of data not unlike the arrangement of Table 13.1. To access these files, the user would have to enter certain descriptive data that would identify the type of machining operation, work material, tooling, and so on. The printout would consist of a listing of the machining recommendations corresponding to the input data.

MATHEMATICAL MODEL SYSTEMS. These systems go one step beyond the data base systems. Instead of simply retrieving cost information on operations that have already been performed, the mathematical model systems attempt to predict the optimum cutting conditions for an operation. The prediction is generally limited to optimum cutting speed, given a certain feed rate. The definition of optimal is based on either the objective of minimizing cost or maximizing production rate.

A common mathematical model to predict optimum cutting speed relies on the familiar Taylor equation for tool life,

$$VT^n = C \tag{13.2}$$

where $\qquad V =$ surface speed, ft/min or m/s

$T =$ tool life, min

C and $n =$ constants

By combining Eqs. (13.1) and (13.2), and accounting for the fact that machining time T_m is inversely proportional to cutting speed, V, the equation for minimum cost cutting speed can be derived.

$$V_{min} = \frac{C}{\left[\dfrac{1-n}{n} \dfrac{C_o T_{tc} + C_t}{C_o} \right]^n} \qquad (13.3)$$

In a similar way, the cutting speed that yields maximum production rate can also be derived:

$$V_{max} = \frac{C}{\left[\dfrac{1-n}{n} T_{tc} \right]^n} \qquad (13.4)$$

Equations (13.3) and (13.4), or equations similar to these, are used in the predictive-type machinability data systems to determine recommendations that approximate optimal cutting conditions.

The potential weakness of the mathematical model systems lies in the validity of the Taylor tool-life equation. Equation (13.2) is an empirical equation derived from experimental data that contain random errors. These random variations tend to distort the accuracy of the minimum cost and maximum production equations. Also, there are dangers in extrapolating the Taylor equation beyond the range over which the experimental data were collected.

EXAMPLE 13.3

A computerized machinability data base system was developed for the Abex Corporation by Zimmers [16,24]. It is principally a data base system.

Five data files are used in the Abex system to store the required machinability data needed in the computations.

1. *Machine tool data (DB100):* machine identification, horsepower, and available speeds and feeds.
2. *Base cutting conditions (DB200):* base cutting conditions for certain types of materials for HSS and cemented carbide tools.

MACHINABILITY ANALYSIS – COST SUMMARY AND CUTTING CONDITION DETAILS

PART NO. 03189-T   OPER NO. 100   DATE 03/

MATERIAL DESCRIPTION – 4340B   0.86 HP/CU. IN.

OPERATION NAME – TURN

CUT DETAILS – DIAMETER – 3.5000   LENGTH – 19.0000   DEPTH – 0.100

FINISH – 125. RMS

CALCULATIONS USING SHOP DATA

CUT SPEED RPM	CUT SPEED SFPM	FEED RATE IPR	FEED RATE IPM	TOOL LIFE MIN.	WORK DIAM IN.	TOOL DESCRIPTION OP****TH**RAK** **MATL**NR***LA	SIZE TOLER IN.	SURF FIN URMS	TOOL COST $/100	TIP COST $/100	MACH COST $/100	TOTAL COST $/100	PROD RATE HRS/100	PARTS PER T.L.
448.	470.	0.010	4.48	15.	4.000	11 78 0103P0600	0.005	125.	0.19	5.35	63.54	125.61	9.15	3.54
381.	400.	0.010	3.81	30.	4.000	11 78 0103P0600	0.005	125.	0.11	3.14	74.66	128.33	9.71	6.02
343.	360.	0.010	3.43	45.	4.000	11 78 0103P0600	0.005	125.	0.08	2.32	82.96	133.52	10.38	8.13
310.	325.	0.010	3.10	60.	4.000	11 78 0103P0600	0.005	125.	0.07	1.93	91.90	140.96	11.25	9.79

MINIMUM COST CONDITIONS,

512.	470.	0.010	5.12	15.	3.500	11 78 0103P0600	0.005	125.	0.19	5.35	63.54	125.61	9.15	3.54

SHOP DATA

MACH OVHD $/MIN	APPR TO WRK IN.	RPDTRV COST $/100	NO. CUTS	LOAD TIME MIN.	LOAD COST $/100	SETUP TIME MIN.	SETUP COST $/100
0.15	4.0	4.05	1.	2.3	34.49	21.0	3.14

MACHINE GROUPS – GROUP NO.

GROUP NO.	MACHINE DESCRIPTION	MACH HP
126-0	WARNER AND SWASEY NO. 2A TURRET LATHE 15	10.500

1	RPM— 321.	FPR— 0.0110	IPM— 3.53	HRS/100— 9.49
2	RPM— 439.	FPR— 0.0110	IPM— 4.82	HRS/100— 7.00
3	RPM— 600.	FPR— 0.0110	IPM— 6.59	HRS/100— 5.18

FIGURE 13.6 Output format of Abex computerized machinability data system. (Reprinted courtesy of the Society of Manufacturing Engineers.)

3. *Experimental data (DB300):* results of laboratory tests to supplement file DB200. Data contained in the experimental data bank include operation type, tooling, work material, cutting conditions, observed surface finish, and tool-life values.

4. *Shop-generated data (DB400):* results of successful experience in production. Data contained in DB400 are similar to those of DB300 except that they originated in the shop rather than the laboratory.

5. *Plant cost data (DB500):* cost parameters needed to compute cost per piece and optimum conditions.

Input to the program consists of data on the workpart, depth of cut, surface finish requirements, and other data related to the operation. (Development of the process sequence must precede the use of the machinability data program.) During operation of the program, the computer performs a series of data file searches and calculations to determine the recommended cutting conditions. A typical printout is shown in Figure 13.6. If shop data (DB400) are available, the recommendations are based on this file. Otherwise, experimental data (DB300) are used, and if no applicable data are available in DB300, recommended speeds and feeds are based on the DB200 file.

13.6   COMPUTER-GENERATED TIME STANDARDS

Work measurement can be defined as the development of a time standard to indicate the value of a work task. According to a survey conducted by *Industrial Engineering* magazine and Patton Consultants, Inc., 95% of the manufacturing firms responding to the survey use work measurement [18]. There are various purposes for which work measurement is used. Among these are:

Wage incentives
Estimating and job costing
Production scheduling and capacity planning
Measurement of worker performance

Work measurement is related to process planning because in a well-organized process planning system, a time standard must be determined for each of the operations listed on the route sheet.

The techniques used to determine time standards in manufacturing include the following:

1. Direct time study
2. The use of standard data
3. Predetermined time standard systems
4. Estimates based on previous experience
5. Work sampling

It is not the purpose of this book to explain each of these techniques. The reader is referred to one of the standard texts on work measurement, such as Mundel [11] or Niebel [12].

By far the most often used work measurement technique is direct time study. According to the survey, 46% of all standards are set by this method. Direct time study involves the direct observation of the task, timing the elements of the work cycle with a stopwatch, rating the performance of the operation (also called "leveling"), applying the necessary allowances (for personal delay, machine cycle, etc.), and calculating the standard time for the job. Among the disadvantages of the direct time study method are:

> The performance rating is often disputed by the worker.
>
> The time standard cannot be set until after the job is in production. This often means wasted time in the shop until after the standard has been set.
>
> There tends to be variability in the standards among time study analysts.
>
> Much time is required by the time study analyst in taking the study and calculating the standard.
>
> Even though many of the jobs done in a given plant are similar, the stopwatch time study procedure must nevertheless be applied to all new jobs. (This, of course, depends on the terms of the labor—management agreement.)

In recent years, a number of computer packages have been commercially introduced for setting time standards. The objective of these computer packages is to reduce the time required in the development of work standards, and to overcome some of the problems associated with direct time study. In addition, these systems provide a manufacturing data base on operation time standards, standard cost data, tooling information, job instructions, and so on.

The computerized systems are based on the use of standard data stored in computer files. The term *standard data*, when applied in the context of work measurement, refers to previously determined time values corresponding to particular work elements or groups of work elements. These elements of various type comprise all manual production activities in the factory. For example, in a turret lathe operation, the workpiece must be loaded onto the machine and unloaded. The time required to load (or unload) the piece depends on the weight and configuration of the part, how it is to be fixtured in the turret lathe, and so on. Standard data can be developed to indicate the time required for the loading element as well as other work elements in the operation cycle. The individual element times depend on the attributes of the job (workpiece weight, size, machine tool, etc.). The computer stores these elemental times either in a data file or in the form of a mathematical formula. To use the package, the time study analyst must analyze the job to be timed by dividing it into its elements and specifying the attributes of the job for each element. The computer then retrieves from the file or calculates the element times, sums the times, and applies the necessary allowances to determine the standard time for the total cycle.

Among the advantages of using a computerized system for generating time standards are the following:

1. Reduction in time required by the time study analyst to set the standard. In application, a 50% reduction in clerical effort was reported compared to manual methods of applying standard data [23].
2. Greater accuracy and uniformity in the time standards.
3. Ease of maintaining the methods and standards file when engineering and methods changes occur.
4. Elimination of the controversial performance rating (leveling) step.
5. Time standards can often be set before the job gets into production.
6. Improved manufacturing data base for production planning, scheduling, forecasting labor requirements, tool control, and so on.

There are several available commercial systems for computerized work standards. We will use the 4M DATA system as an example of these packages. It is available from the MTM Association in Fair Lawn, New Jersey. 4M stands for "Micromatic Methods and Measurement."

EXAMPLE 13.4[1]

The 4M DATA system is a computerized work measurement system for applying MTM-1 predetermined time standards. The current version is MOD II. The MTM-1 motions are elementary body motions that are used in all manual production operations. About 90% of the manual work performed in the typical industrial situation consists of the motions REACH, GRASP, MOVE, POSITION, and RELEASE. These motions can be combined into GET and PLACE motion groups. Special symbols for these two motion groups convey the necessary data to the 4M computer program for calculating the proper standard time values. In addition to the GET and PLACE motion groups, the computer program also recognizes all other MTM-1 motions with their corresponding variables. The methods/time study analyst can develop the time standard for a given manual operation cycle by calling the required elements in the 4M DATA package.

Figures 13.7 and 13.8 illustrate the GET and PLACE notations. Figures 13.9 and 13.10 show two of the output reports. Figure 13.9 shows the 4M Element Analysis form generated by the 4M DATA system for a particular operation. This report indicates the right- and left-hand activity for the sequence of elements in the operation. The time units (MUs) are 0.00001 h, which is typically used in the MTM system. Figure 13.10 shows an MTM analysis report (this is for a different operation than that shown in Figure 13.9). The various indices shown at the bottom

[1]This example is based on marketing literature furnished by the MTM Association, Refs. [7] and [8].

GET

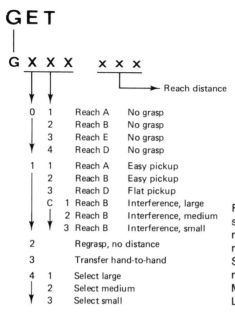

FIGURE 13.7 GET symbol in 4M DATA system. (Reprinted from Ref. [8] by permission of MTM Association. Copyrighted by the MTM Association for Standards and Research. No reprint permission without written consent from the MTM Association, 16-01 Broadway, Fair Lawn, NJ 07410.)

PLACE

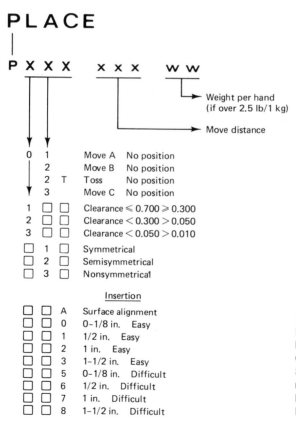

FIGURE 13.8 PLACE symbol in 4M DATA system. (Reprinted from Ref. [8] by permission of MTM Association. Copyrighted by the MTM Association for Standards and Research. No reprint permission without written consent from the MTM Association, 16-01 Broadway, Fair Lawn, NJ 07410.)

4M ELEMENT ANALYSIS

REQUESTED BY JPO   TIME 11.05.25   DATE 02/21/80   PAGE 1

ETS, PLACE NEXT COIL AND IRON IN FIXTURE

LEARNING LEVEL 100

UNITY   UPDATED   TOTAL MU = MANUAL 943
                                                             = PROCESS 190

			Action	RH OR BODY MOTIONS	FREQ.	PROCESS TIME	LH	RH	NET MANUAL
N	G01 = 9	G01 = 9	REACH	CONTROL BUTTON			83	83	83
	P01 = 1	P01 = 1	MOVE	BUTTON			25	25	25
			PROCESS	HOLD BUTTONS		190			
			PROCESS	COMPLETE PRESS ACTION		250*			
		G12 = 14	GET	COIL, IRON IN TRAY				173P	173
	G3 = 12	P01 = 26	MOVE	PART TO LH			152	239P	295
		G11 = 8	GET	COIL, IRON IN FIXTURE				109P	109
RE	P121 = 6	P03 = 24	MOVE	TO TRAY POCKET			238	227	258
				TOTAL MU		190	498	856	943

PERCENT GRA 9 PERCENT POS 10 PERCENT PROC 17 PERCENT

FIGURE 13.9   4M analysis output form generated by the 4M DATA system. (Reprinted from Ref. [8] by permission of MTM Association. Copyrighted by the MTM Association for Standards and Research. No reprint permission without written consent from the MTM Association, 16-01 Broadway, Fair Lawn, NJ 07410.)

317

4M DATA SYSTEM   R0860   MTM ANALYSIS REPORT   REQUESTED BY RE   TIME 09.05.50   DATE 04/11/80   PAGE 1

PART 2476SP8X-5   WATT-HOUR METER   ORIGINATED 04/11/80 BY MT
OPERATION 09   ASSEMBLE BRACKETS TO RING   REVISION   / / BY
DEPT. 605G   TOOLING

PRESS DIE
ALLOWANCE - MANUAL 0.080   STANDARD HOURS = SET UP   0.0000 UNITS PER   RUN 0.00625
          - PROCESS 0.080   HOUR 160.08

LH MOTIONS	RH OR BODY MOTIONS	FREQ.	PROCESS TIME	LH	RH	NET MANUAL
010 112-01 PLACE RING ON FIXTURE	--- RUN ---					
001 R128 G1A RL	GET RING ON BAR G12 = 12					B
002 RECEIVE SIDE OF RING G3 = 24 R24A G3	MOVE RING TO LH P01 = 10			205	271	327B
003 R3A G3	RECEIVE RING FOR BETTER GRASP G3 = 3				109	109
004 REACH TO CONTACT RING G02 = 4 R4B	MOVE RING TOWARD FIXTURE P02 = 4			64	69	69
005 ASSIST RIGHT HAND SP M1C P22NS4	PLACE RING ON FIXTURE P232 = 1			294	294	294
006 ASSIST RIGHT HAND SP APA	PRESS RING IN FIXTURE APA			106	106	106

ELEMENT SUBTOTAL   467

TOTAL RUN MU - MANUAL   5784   WITH ALLOWANCE   6247
             - PROCESS   0   WITH ALLOWANCE   0
                                  TOTAL STANDARD RUN TIME   6247 MU
                                  CYCLE QUANTITY   1.00
                                  UNIT STANDARD RUN TIME   6247 MU
                                                             0.00625 HOURS
                                                             0.375 MINUTES
                                  UNITS PER HOUR   160.08

MAI 71 PERCENT   RMB 43 PERCENT   GRA 22 PERCENT   POS 34 PERCENT   PROC 0 PERCENT

FIGURE 13.10   MTM analysis report generated by the 4M DATA system. (Reprinted from Ref. [8] by permission of MTM Association. Copyrighted by the MTM Association for Standards and Research. No reprint permission without written consent from the MTM Association, 16-01 Broadway, Fair Lawn, NJ 07410.)

of the report (MAI, RMB, GRA, and POS) represent various indices and ratios that might indicate possible methods improvements.

REFERENCES

[1] COMPUTER AIDED MANUFACTURING—INTERNATIONAL, INC, *Process Planning Program*, PR-81-ASPP-01.3, Arlington, Tex., 1982.

[2] FIELD, M., AND ACKENHAUSEN, A. R., *Determination and Analysis of Machining Costs and Production Rates Using Computer Techniques*, Report AFMDC 68-1, Machinability Data Center, Cincinnati, Ohio, 1968.

[3] GROOVER, M. P., *Automation, Production Systems, and Computer-Aided Manufacturing*, Prentice-Hall, Inc., Englewood Cliffs, N.J., 1980, Chapters 16, 18.

[4] HALEVI, G., *The Role of Computers in Manufacturing Processes*, John Wiley & Sons, New York, 1980, Chapters 11−14.

[5] HEGLAND, D. E., "Out in Front with CAD/CAM at Lockheed−Georgia," *Production Engineering*, November, 1981, pp. 44−48.

[6] HOUTZEEL, A., "Computer-Assisted Process Planning Minimizes Design and Manufacturing Cost," *Industrial Engineering*, November, 1981, pp. 60−64.

[7] MTM ASSOCIATION, "4M DATA: Computerized Work Measurement System," Publication 3233-77, Fairlawn, N.J.

[8] MTM ASSOCIATION, "4M DATA—MOD II: A Computerized Work Measurement System," Publication 3233-80B, Fairlawn, N.J.

[9] MACHINABILITY DATA CENTER, *Machining Data Handbook*, 2nd ed., Cincinnati, Ohio, 1972.

[10] MCNEELY, R. A., AND MALSTROM, E. M., "Computer Generates Process Routings," *Industrial Engineering*, July, 1977, pp. 32−35.

[11] MUNDEL, M. E., *Motion and Time Study*, 5th ed., Prentice-Hall, Inc., Englewood Cliffs, N.J., 1978.

[12] NIEBEL, B. W., *Motion and Time Study*, 7th ed., Richard D. Irwin, Inc., Homewood, Ill., 1982.

[13] O'NEAL, M. H., AND MOORE, C., "Multi-plant Computer System for Standards Provides Tool for Overall Manufacturing Control," *Industrial Engineering*, August, 1980, pp. 54−62.

[14] ONITIRI, T. A., "The Development of a Simulator for Manufacturing Methods Decisions," *Ph.D. dissertation,* Lehigh University, Bethlehem, Pa., 1981.

[15] ORGANIZATION FOR INDUSTRIAL RESEARCH, Inc., *Miclass, Migroup, Miplan, Migraphics* (marketing brochure), Waltham, Mass.

[16] PARSONS, N. R. (Ed.), *N/C Machinability Data Systems*, Society of Manufacturing Engineers, Dearborn, Mich., 1971.

[17] PRESSMAN, R. S., AND WILLIAMS, J. E., *Numerical Control and Computer-Aided Manufacturing*, John Wiley & Sons, New York, 1977, Chapter 9.

[18] RICE, R. S., "Survey of Work Measurement and Wage Incentives," *Industrial Engineering*, July, 1977, pp. 18−31.

[19] SCHAFFER, G., "GT via Automated Process Planning," *American Machinist*, May, 1980, pp. 119–122.

[20] SCHAFFER, G., "Implementing CIM," Special Report 736, *American Machinist*, August, 1981, pp. 151–174.

[21] TNO, *The 1978 MICLASS User's Meeting*, Organization for Industrial Research, Inc., Waltham, Mass., 1978.

[22] TULKOFF, J., "Lockheed's GENPLAN," *Proceedings, Eighteenth Annual Meeting and Technical Conference*, Numerical Control Society, Dallas, Tex., May, 1981, pp. 417–421.

[23] WEAVER, R. F., KOLLMAR, J. J., AND BOEPPLE, E. A., "Developing Standards by Computer," *Industrial Engineering*, January, 1978, pp. 26–31.

[24] ZIMMERS, E. W., JR., "Practical Applications of Computer Augmented Systems for Determination of Metal Removal Parameters and Production Rate Standards in a Job Shop," *Tech Paper MS71-136*, Society of Manufacturing Engineers, Dearborn, Mich., 1971.

PROBLEMS

13.1. Develop a simple parts classification and coding scheme of no more than four digits which can be used to uniquely code the various possible combinations of part attributes for the composite part of Figure 12.12. (The reader should refer back to Problem 12.3.) Make a list of all of the separate manufacturing operations needed to accomplish the machining of each part attribute. Using the coding scheme and the list of operations, write a computer program to accomplish retrieval-type automated process planning. The program should be written for interactive use. The user would enter the part code number, and the program would generate the list of operations in the correct sequence to machine that part. The program should not allow for any editing by the user.

13.2. Same as Problem 13.1 except that the program should allow for user editing of the retrieved plan for such purposes as inserting dimensions, changing text, and so on.

13.3. For a certain tool and work material, the parameters of the Taylor tool life equation, Eq. (13.2), are $C = 1000$ and $n = 0.25$. These parameters apply to a turning operation using a feed of 0.012 in per revolution and a depth of cut of 0.075. The following data also apply to the operation:

> Machine and labor cost = \$0.40/min
>
> Tool cost = \$0.60/cutting edge
>
> Tool change time = 1.5 min
>
> Work diameter = 4.0 in.
>
> Work length = 17.5 in.
>
> Workpiece handling time = 4.0 min

Write a computer program to perform the necessary calculations and print out the results according to the general format illustrated in Figure 13.5. The work material is C4330 and the Rockwell hardness is C20. Tool grade is Carboloy 350. The following speeds should be used to construct the output table: 400, 500, 550, 600, 650, and 700 ft/min. The individual cost components should be separated as indicated by the terms of Eq. (13.1).

PART VI

Computer-Integrated Production Management Systems

chapter 14

Production Planning and Control

14.1 INTRODUCTION

This part of the book is concerned with the use of computers to manage the production function. The subject has traditionally been referred to as production planning and control. This function has been practiced for many years. Attempts to use the computer in production planning date from the late 1950s and early 1960s. The early attempts were directed toward computerizing the same clerical procedures which had been done by hand for years. These procedures included preparation of schedules, shortage lists, inventory lists, and similar documents. During the late 1960s and early 1970s, a few individuals began to recognize the tremendous opportunities provided by the computer to make fundamental changes in the procedures and organization of production planning and control. Joseph Orlicky [8], George Plossl [9], and Oliver Wight [10,11] stand out as some of the principal pioneers in these efforts to modernize and computerize the production management function. MRP (material requirements planning) was one of the first computerized procedures which significantly improved the way things were done. Since MRP was first implemented, many additional improvements in production planning and control have been introduced by taking advantage of the data processing and computational powers of the computer. Several of these computerized procedures are examined in this and the next two chapters.

Our objective in these three chapters is not to present a textbook treatment on the theory and practice of production planning and control. Instead, our purpose is to describe how computers are utilized to carry out the production management function in the CAD/CAM age.

In this first chapter we review the way in which production planning and control has traditionally been organized and carried out (pre-computer). We then examine some of the problems which have resulted from these manual methods. Attempts are being made in industry to solve these problems by installing computer-integrated production management systems (CIPMS). The components and structure of these systems are examined in this chapter. Two of the components we consider important enough to rate separate chapters. Chapter 15 deals with MRP, both the original material requirements planning and the more recent MRP II, which stands for manufacturing resource planning. Chapter 16 covers shop floor control and computerized production monitoring systems.

14.2   TRADITIONAL PRODUCTION PLANNING AND CONTROL

At least a dozen separate functions can be identified as constituting the cycle of activities in traditional production planning and control. Organizationally, some of these functions are performed by departments in the firm other than the production control department. The functions are described in the following sections.

Forecasting

The forecasting function is concerned with projecting or predicting the future sales activity of the firm's products. Sales forecasts are often classified according to the time horizon over which they attempt to estimate. Long-range forecasts look ahead five years or more and are used to guide decisions about plant construction and equipment acquisition. Intermediate-range forecasts estimate one or two years in advance and would be used to plan for long-lead-time materials and components. Short-term forecasts are concerned with a three- to six-month future. Decisions on personnel (e.g., new hiring), purchasing, and production scheduling would be based on the short-term forecast.

Production planning

This is sometimes called aggregate production planning and its objective is to establish general production levels for product groups over the next year or so. It is based on the sales forecast and is used to raise or lower inventories, stabilize production over the planning horizon, and allow for the launching of new products into the company's product line. Aggregate production planning is a function that precedes the detailed master production schedule.

Process planning

As we described in Chapter 13, process planning involves determining the sequence of manufacturing operations required to produce a certain product and/or its components. Process planning has traditionally been carried out by manufacturing engineers as a very manual and clerical procedure. The resulting document, prepared by hand, is called a route sheet and is a listing of the operations and machine tools through which the part or product must be routed. The term "routing" is sometimes applied to describe the process planning function.

Estimating

For purposes of determining prices, predicting costs, and preparing schedules, the firm will determine estimates of the manufacturing lead times and production costs for its products. The manufacturing lead time is the total time required to process a workpart through the factory. The production costs are the sum of the material costs, labor, and applicable overhead costs needed to produce the part. These estimates of lead times and costs are based on data contained in the route sheets, purchasing files, and accounting records.

Master scheduling

The aggregate production plan must be translated into a *master schedule* which specifies how many units of each product are to be delivered and when. In turn, this master schedule must be converted into purchase orders for raw materials, orders for components from outside vendors, and production schedules for parts made in the shop. These events must be timed and coordinated to allow delivery of the final product according to the master schedule.

Specifically, the master schedule or master production schedule is a listing of the products to be produced, when they are to be delivered, and in what quantities. The scheduling periods in the master schedule are typically months, weeks, or dates. The master schedule must be consistent with the plant's production capacity. It should not list more quantities of products than the plant is capable of producing with its given resources of machines and labor.

Requirements planning

Based on the master schedule, the individual components and subassemblies that make up each product must be planned. Raw materials must be ordered to make the various components. Purchased parts must be ordered. And all of these items must be planned so that the components and assemblies are available when needed.

This whole task is called requirements planning or material requirements planning. The term MRP (for material requirements planning) has come into common usage since the introduction of computerized procedures to perform the massive data processing required to accomplish this function. However, the function itself had to be accomplished manually by clerical workers before computers were used.

Purchasing

The firm will elect to manufacture some components for its products in its own plants. Other components will be purchased. Deciding between these alternatives is the familiar ''make-or-buy'' decision. For the components made in-house, raw materials have to be acquired. Ordering the raw materials and purchased components is the function of the purchasing department. Materials will be ordered and the receipt of these items will be scheduled according to the timetable defined during the requirements planning procedure.

Machine loading and scheduling

Also based on the requirements planning activity is production scheduling. This involves the assignment of start dates and due dates for the components to be processed through the factory. Several factors make scheduling complex. First, the number of individual parts and orders to be scheduled may run into the thousands. Second, each part has its own individual process routing to be followed. Some parts may have to be routed through dozens of separate machines. Third, the number of machines in the shop is limited, and the machines are different. They perform different operations and have different features and capacities.

     The total number of jobs to be processed through the factory will typically exceed the number of machines by a substantial margin. Accordingly, each machine, or work center, will have a queue of jobs waiting to be processed. Allocating the jobs to work centers is referred to as machine loading. Allocating the jobs to the entire shop is called shop loading.

Dispatching

Based on the production schedule, the dispatching function is concerned with issuing the individual orders to the machine operators. This involves giving out order tickets, route sheets, part drawings, and job instructions. The dispatching function in some shops is performed by the shop foremen, in other shops by a person called a dispatcher.

Expediting

Even with the best plans and schedules, things go wrong. It is the expediter's job to compare the actual progress of the order against the production schedule. For orders that fall behind schedule, the expediter recommends corrective action. This may involve rearranging the sequence in which orders are to be done on a certain machine, coaxing the foreman to tear down one setup so that another order can be run, or hand-carrying parts from one department to the next just to keep production going. There are many reasons why things go wrong in production: parts-in-process have not yet arrived from the previous department, machine breakdowns, proper tooling not available, quality problems, and so forth.

Quality control

The quality control department is responsible for assuring that the quality of the product and its components meets the standards specified by the designer. This function must be accomplished at various points throughout the manufacturing cycle. Materials and parts purchased from outside suppliers must be inspected when they are received. Parts fabricated inside the company must be inspected, usually several times during processing. Final inspection of the finished product is performed to test its overall functional and appearance quality.

Shipping and inventory control

The final step in the production control cycle involves shipping the product directly to the customer or stocking the item in inventory. The purpose of inventory control is to ensure that enough products of each type are available to satisfy customer demand. Competing with this objective is the desire that the company's financial investment in inventory be kept at a minimum. Inventory control interfaces with production control since there must be coordination between the various product's sales, production, and inventory level. Inventory control is often included within the production control department.

The inventory control function applies not only to the company's final products. It also applies to raw materials, purchased components, and work-in-process within the factory. In each case, planning and control are required to achieve a balance between the danger of too little inventory (with possible stockouts) and the expense of too much inventory.

The block diagram of Figure 14.1 depicts the relationships among the production planning and control functions as well as various other functions of the firm, customers, and outside suppliers. In the diagram, the production planning and control functions are highlighted in bold blocks.

14.3 PROBLEMS WITH TRADITIONAL PRODUCTION PLANNING AND CONTROL

There are many problems that occur during the cycle of activities in the traditional approach to production planning and control. Many of these problems result directly from the inability of the traditional approach to deal with the complex and

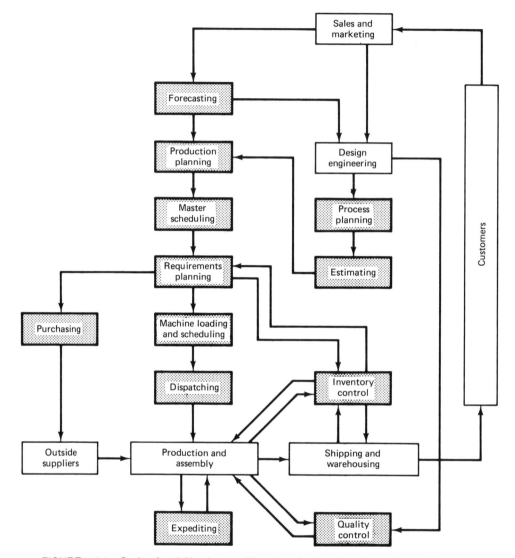

FIGURE 14.1 Cycle of activities in a traditional production planning and control system.

ever-changing nature of manufacturing. The types of problems commonly encountered in the planning and control of production are the following:

1. *Plant capacity problems.* Production falls behind schedule due to a lack of labor and equipment. This results in excessive overtime, delays in meeting delivery schedules, customer complaints, backordering, and other similar problems.

2. *Suboptimal production scheduling.* The wrong jobs are scheduled because of a lack of clear order priorities, inefficient scheduling rules, and the ever-changing status of jobs in the shop. As a consequence, production runs are interrupted by jobs whose priorities have suddenly increased, machine setups are increased, and jobs that are on schedule fall behind.

3. *Long manufacturing lead times.* In an attempt to compensate for problems 1 and 2, production planners allow extra time to produce an order. The shop becomes overloaded, order priorities become confused, and the result is excessively long manufacturing lead times.

4. *Inefficient inventory control.* At the same time that total inventories are too high for raw materials, work-in-progress, and finished products, there are stockouts that occur on individual items needed for production. High total inventories mean high carrying costs, while raw material stockouts mean delays in meeting production schedules.

5. *Low work center utilization.* This problem results in part from poor scheduling (excessive product changeovers and job interruptions), and from other factors over which plant management has limited control (e.g., equipment breakdowns, strikes, reduced demand for products).

6. *Process planning not followed.* This is the situation in which the regular planned routing is superseded by an ad hoc process sequence. It occurs, for instance, because of bottlenecks at work centers in the planned sequence. The consequences are longer setups, improper tooling, and less efficient processes.

7. *Errors in engineering and manufacturing records.* Bills of materials are not current, route sheets are not up to date with respect to the latest engineering changes, inventory records are inaccurate, and production piece counts are incorrect.

8. *Quality problems.* Quality defects are encountered in manufactured components and assembled products, resulting in rework or scrapped parts, thus causing delays in the shipping schedule.

14.4 COMPUTER-INTEGRATED PRODUCTION MANAGEMENT SYSTEM

There have been several factors working over the last several decades to cause the evolution of a more modern and effective approach to the problems of production planning and control cited above. The most obvious of these factors was the development of the computer, a powerful tool to help accomplish the vast data processing and routine decision-making chores in production planning that had previously been done by human beings.

In addition to the computer, there were other factors which were perhaps less dramatic but equally important. One of these was the increase in the level of professionalism brought to the field of production planning and control. Production planning has been gradually transformed from what was largely a clerical function into a recognized profession requiring specialized knowledge and academic training. Systems, methodologies, and even a terminology have developed to deal with the problems of this professional field.

Important among the methodologies of production planning and control, and another significant factor in the development of the field, is operations research. The computer became the important tool in production planning, but many of the decision-making procedures and software programs were based on the analytical models provided by operations research. Linear programming, inventory models, queueing theory, and a host of other techniques have been effectively applied to problems in production planning and control.

Another factor that has acted as a driving force in the development of better production planning is increased competition from abroad. Many American firms have lost their competitive edge in international and even domestic markets. Increasing U.S. productivity is seen as one important way to improve our competitive position. Better management of the production function is certainly a key element in productivity improvement.

Finally, a fifth factor is the increase in the complexity of both the products manufactured and the markets that buy these products. The number of different products has proliferated, tolerances and specifications are more stringent, and customers are more particular in their requirements and expectations. These changes have placed greater demands on manufacturing firms to manage their operations more efficiently and responsively.

As a consequence of these factors, companies are gradually abandoning the traditional approach in favor of what we are calling computer-integrated production management systems. There are other terms which are used to describe these systems and their major components. IBM uses the term "communications-oriented production information and control system—COPICS" [6]—to identify the group of system elements. George Plossl integrates the various system concepts under the name "manufacturing control" [9]. Computer-Aided Manufacturing—International, calls its development effort in this area the "factory management project" [2]. Oliver Wight refers to the use of MRP II, or manufacturing resource planning, to consolidate the manufacturing, engineering, and financial functions of the firm into one operating system [11]. All of these terms refer to computerized information systems designed to integrate the various functions of production planning and control and to reduce the problems described in Section 14.3.

Figure 14.2 presents a block diagram illustrating the functions and their relationships in a computer-integrated production management system. Many of these functions are nearly identical to their counterparts in traditional production planning and control. For example, forecasting, production planning, the development of the master schedule, purchasing, and other functions appear the same in Figures 14.1 and 14.2. To be sure, modern computerized systems have been developed to perform these functions, but the functions themselves remain relatively unchanged. More significant changes have occurred in the organization and execution of production planning and control through the implementation of such schemes as MRP, capacity planning, and shop floor control. What follows is a brief description of

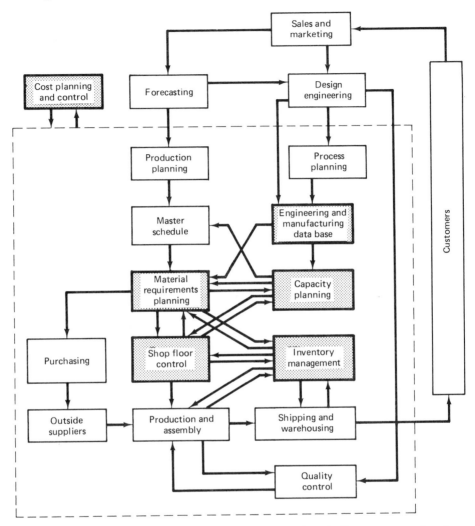

FIGURE 14.2 Cycle of activities in a computer-integrated production management system.

some of the recently developed functions in a CIPMS. We will neglect those functions which are nearly the same as their conventional counterparts. The newer functions are highlighted in Figure 14.2 by bold blocks.

Engineering and manufacturing data base

This data base comprises all the information needed to fabricate the components and assemble the products. It includes the bills of material (assembly lists), part design data (either as engineering drawings or some other suitable format), process route sheets, and so on. Ideally, these data should be contained in some master file to avoid duplication of records and to facilitate update of the files when design engineering changes are made or route sheets are updated. As shown in Figure 14.2, the design engineering and process planning functions provide the inputs for the engineering and manufacturing data base.

Material requirements planning (MRP)

MRP involves determining when to order raw materials and components for assembled products. It can also be used to reschedule orders in response to changing production priorities and demand conditions. The term *priority planning* is now widely used in describing computer-based systems for time-phased planning of raw materials, work-in-progress, and finished goods.

We will devote most of the following chapter (Chapter 15) to the subject of material requirements planning.

Capacity planning

MRP is concerned with the planning of materials and components. Capacity planning, on the other hand, is concerned with determining the labor and equipment resources needed to meet the production schedule.

Capacity planning will often necessitate a revision in the master production schedule. It would be infeasible, and counterproductive in all likelihood, to develop a master schedule that exceeds plant capacity. Therefore, the master schedule is checked against available plant capacity to make sure that the schedule can be realized. If not, either the schedule or plant capacity must be adjusted to be brought into balance. Capacity planning has always been of concern in traditional production planning and control. However, it is an area of planning whose recognition has been growing in recent years due to its impact on the ability to achieve the master production schedule.

The term ''plant capacity'' is used to define the maximum rate of output that the plant can produce under a given set of assumed operating conditions. The assumed operating conditions refer to the number of shifts (one, two, or three shifts per day), number of days of plant operation per week, employment levels, and whether or not overtime is included in the definition of plant capacity. Capacity for a production plant is traditionally measured in terms of output units of the plant. Examples would be tons of steel for a steel mill, number of automobiles for a car assembly plant, and barrels of oil for a refinery. When the output units of a plant are nonhomogeneous, input units may be more appropriate for measuring plant capacity. A job shop, for instance, may use labor hours or available machine hours to measure capacity.

Capacity planning is concerned with determining what labor and equipment capacity is required to meet the current master production schedule as well as the long-term future production needs of the firm. Capacity planning is typically performed in terms of labor and/or machine hours available.

The function of capacity planning in the overall production planning and control system is shown in Figure 14.2. The master schedule is transformed into material and component requirements using MRP. Then these requirements are compared with available plant capacity over the planning horizon. If the schedule is incompatible with capacity, adjustments must be made either in the master schedule or in plant capacity. The possibility of adjustments in the master schedule is indicated by the arrow in Figure 14.2 leading from capacity planning to the master schedule.

Capacity adjustments can be accomplished in either the short term or the long term. Capacity planning for short-term adjustments would include decisions on such factors as the following:

1. *Employment levels.* Employment in the plant can be increased or decreased in response to changes in capacity requirements.
2. *Number of work shifts.* Increasing or decreasing the number of shifts per week.
3. *Labor overtime hours or reduced workweek.*
4. *Inventory stockpiling.* This would be used to maintain steady employment during temporary slack periods.
5. *Order backlogs.* Deliveries of product to customers would be delayed during busy periods.
6. *Subcontracting.* Letting of jobs to other shops during busy periods, or taking in extra work during slack periods.

Capacity planning to meet long-term capacity requirements would include the following types of decisions:

1. Investing in more productive machines or new types of machines to manufacture new product designs
2. New plant construction
3. Purchase of existing plants from other companies
4. Closing down or selling off existing facilities which will not be needed in the future

Inventory management

In the manufacturing environment, inventory management is closely tied to material requirements planning. The objectives are simple—to keep the investment in inventory low while maintaining good customer service. The use of computer systems has provided opportunities to accomplish these objectives more effectively. We discuss some of these computer systems in Chapter 15.

Shop floor control

The term "shop floor control" refers to a system for monitoring the status of production activity in the plant and reporting the status to management so that effective control can be exercised. We examine shop floor control and the use of computers to monitor production in Chapter 16.

Cost planning and control

The cost planning and control system consists of the data base to determine expected costs to manufacture each of the firm's products. It also consists of the cost collection and analysis software to determine what the actual costs of manufacturing are and how these actual costs compare with the expected costs. The following section is devoted to this important area in the operations of a computer-integrated production management system.

14.5   COST PLANNING AND CONTROL

As indicated in Figure 14.2, the cost planning and control function encompasses most of the other functions within the computer-integrated production management system. It receives data from all of the other CIPMS modules and reduces them to a lowest common denominator: money. The objectives of the cost planning and control system are to help answer the following questions:

1.   What are the *expected* costs to manufacture and sell each of the company's products?
2.   What are the *actual* costs to manufacture and sell each of the company's products?
3.   What are the differences between what it should cost and what it does cost, and how are these differences explained?

The underlying basis for attempting to answer these questions is the objective of minimizing the costs of manufacturing the firm's products.

Cost planning

Cost planning is concerned with the first of the three questions: What are the expected costs of manufacturing a product? An attempt is made to answer the question by determining the standard cost for the product. The standard cost for the product is the aggregate cost of labor, materials, and allocated overhead costs. The standard costs are compiled from various data sources and other modules in the CIPMS. The following list includes several standard data sources:

1. *The bill of materials* gives the components and materials used in the product.
2. *Process route sheets* list the manufacturing operations used for each component in the product.
3. *Time standards* specify the operation times for each operation listed on the route sheets.
4. *Labor and machines rates* allow the time standards to be converted into dollar costs for each operation.
5. *Material quotations from purchasing* provide information on material costs, based on catalog price data or direct quotes from potential vendors.
6. *Accounting data* determine appropriate overhead rates.

With so much data collected from many different sources, the computation of a standard cost for a product is not an insignificant task. To accomplish the task and determine a meaningful cost value, the current approach involves use of a data base which is common for engineering, manufacturing, and accounting. In this way, all departments have access as needed to the same information files and there is greater consistency and accuracy of computations based on data in these files.

Development of standard costs for all of the company's products provides a yardstick against which the actual production cost performance can be measured. Determining the actual costs is the function of cost control.

Cost control

Cost control is concerned with the second and third questions: What are the actual costs of manufacturing? And what are the differences between actual costs and expected costs?

In any manufacturing activity, there will be differences between the standard costs computed in cost planning and the actual costs that occurred during production. The reasons why these differences happen comprise a never-ending list. Actual prices of raw materials increase above quoted prices, machines break down, differing lot sizes influence production costs, actual process sequences deviate from the planned route sheets, and a vast collection of other reasons result in variances between actual costs and standard costs.

Cost control involves the collection of data from which the actual costs of the product can be calculated. Data on material costs can be compiled through the purchasing department. Data on labor costs can be collected by means of the shop floor control system, and we defer discussion of this topic until Chapter 16. Overhead costs are usually excluded from consideration because they do not represent an actual expense of the product but rather an allocation of general factory and corporate expenses. Included within the cost control function is the preparation of reports that document actual product costs and variances from standard costs.

REFERENCES

[1] CHASE, R. B., AND AQUILANI, N. J., *Production and Operations Management*, 3rd ed., Richard D. Irwin, Inc., Homewood, Ill., 1981, Chapters 13–16.

[2] COMPUTER-AIDED MANUFACTURING—INTERNATIONAL, INC., *Functional Specifications for an Advanced Factory Management System*, Report R-79-JSIG-01, Arlington, Tex., March, 1979.

[3] GROOVER, M. P., *Automation, Production Systems, and Computer-Aided Manufacturing*, Prentice-Hall, Inc., Englewood Cliffs, N.J., 1980, Chapter 17.

[4] HALEVI, G., *The Role of Computers in Manufacturing Processes*, John Wiley & Sons, New York, 1980, Chapters 7, 8, 10.

[5] INTERNATIONAL BUSINESS MACHINES CORPORATION, *Production Information and Control System,* Publication GEZO-0280-2, White Plains, NY, 1968.

[6] INTERNATIONAL BUSINESS MACHINES CORPORATION, *Communications Oriented Production Information and Control System,* Publication G320-1974, White Plains, NY, 1972.

[7] MONKS, J. G., *Operations Management, Theory and Problems* McGraw-Hill Book Company, New York, 1977, Chapter 7, 8, 9.

[8] ORLICKY, J., *Material Requirements Planning*, McGraw-Hill Book Company, New York, 1975.

[9] PLOSSL, G. W., *Manufacturing Control*, Reston Publishing Company, Inc., Reston, Va., 1973.

[10] WIGHT, O. W., *Production and Inventory Management in the Computer Age*, CBI Publishing Company, Inc., Boston, 1974.

[11] WIGHT, O. W., *MRP II: Unlocking America's Productivity Potential*, Oliver Wight Limited Publications, Inc., Williston, Vt., 1981.

chapter 15

Inventory Management and MRP

15.1 INTRODUCTION

Chapter 14 provided a general description of a modern computer-integrated production management system (CIPMS). One of the most important components of the CIPMS is the material requirements planning system. MRP is sometimes thought of as an inventory management system. In this chapter we describe material requirements planning, its underlying concepts, how it works, and its benefits. Entire books have been written on this subject, for example, [5]. The interested reader is invited to explore the many facets of MRP in that volume, details of which are beyond the scope of our book on CAD/CAM.

MRP has expanded to mean more than material requirements planning. Wight [10] is using the term "MRP II" to represent manufacturing resource planning, a system for planning and controlling the operational, engineering, and financial resources of a manufacturing firm. We shall examine MRP II later. We begin the chapter with a general introduction to inventory management.

15.2 INVENTORY MANAGEMENT

Inventory control is concerned with achieving an optimum balance between two competing objectives. The objectives are first, to minimize investment in inventory, and second, to maximize the service levels to the firm's customers and its own operating departments. The purpose of this section is not to provide a comprehensive discussion of the theory and methods by which these objectives are achieved. For that purpose, the reader is referred to any of several books in the field, for example, Peterson and Silver [6]. Instead, our purpose is to describe how inventory control is accomplished in a modern computer-integrated production management system.

Inventory types and general control procedures

There are four types of inventory with which a manufacturing firm must concern itself:

1. *Raw materials and purchased components.* These are the basic materials out of which the product and components are made. Purchased components are similar to raw materials. The difference is that the supplier does all of or a portion of the processing on the component.

2. *In-process inventory.* After processing begins on the raw materials or purchased components, there is a time lapse (the manufacturing lead time) before the processing has been completed. During that time, the materials are in-process.

3. *Finished product.* The finished product may be stored in the factory or warehouse before shipment to the customer. This category includes service parts as well as end product.

4. *Maintenance, repair, and tooling inventories.* These are the cutting tools and fixtures used to operate the machines in the factory and the repair parts needed to fix the machines when they break down. Electrical spare parts and plumbing supplies also fit into this category.

To manage these various kinds of inventories, two alternative control procedures can be used:

1. *Order point systems.* This has been the traditional approach to inventory control. In these systems, the items are restocked when the inventory levels become low.

2. *Material requirements planning.* MRP is sometimes thought of as an inventory control procedure. It is really more than that, as we shall see.

It is important that the proper control procedure be applied to each of the four types of inventory. In general, MRP is the appropriate control procedure for inventory types 1 and 2 (raw materials, purchased components, and in-process inventory). Order point systems are often considered the appropriate procedure to control inventory types 3 and 4 (finished product, maintenance and repair parts, and tooling). In some cases, order point systems may be applied for certain widely used raw materials and purchased components.

Order point systems

Order point systems are concerned with the two related problems of determining when to order and how much to order. Determining when to order is often accomplished by establishing a "reorder point." When the inventory level for a particular item falls to the reorder point, it is time to restock the item. A computerized inventory control system can be programmed to track the inventory level continuously as transactions are entered against existing stocks. It automatically indicates when it is time to reorder, perhaps even generating a purchase requisition to do so.

The problem of determining how much to order is typically based on the use of the familiar economic order quantity (EOQ) formula, which states

$$EOQ = \sqrt{\frac{2DS}{H}} \qquad (15.1)$$

where EOQ = quantity of the item to be ordered

$\quad\quad\quad D$ = annual demand rate for the item

$\quad\quad\quad S$ = setup cost or ordering cost per order

$\quad\quad\quad H$ = annual holding cost for the item to be carried in inventory

EXAMPLE 15.1

The EOQ formula might be quite appropriate for controlling inventory on certain tooling items. Suppose that we wanted to use the formula [Eq. (15.1)], to determine the order quantity for a particular drill bit that is used in the shop. The annual usage rate (demand rate) for the drill is 8000 units per year. The cost to place an order, including the vendor's setup charge, is $100. And the holding cost for us to carry the drill in inventory is figured at $1.60 per drill each year. The holding cost would typically be calculated by applying an interest rate to the unit cost of the item. For example, if the interest rate were 20% and the drill cost $8.00 per unit, the holding cost would be $0.20 \times 8.00 = \$1.60$ per drill per year.

Using these values in the EOQ equation, we get

$$EOQ = \sqrt{\frac{2(8000)(100)}{1.60}}$$

$$EOQ = 1000 \text{ units per order}$$

Each time an order is placed, it is most economical to order 1000 units.

Computer software can be written to perform this economic order quantity calculation for all the items stocked in inventory. The program keeps track of usage or demand rates for the items in stock and computes the term D in Eq. (15.1). Up-to-date files are maintained by the system on current stock levels so that when inventory falls below the reorder point, a new order for the optimum order quantity can be issued. Problem 15.2 at the end of the chapter deals with such a program.

The inventory management module

The inventory management module of the computer-integrated production management system would be organized to accomplsih two major functions:

1. Inventory accounting
2. Inventory planning and control

Both of these functions apply to all types of inventories (raw materials, in-process inventories, finished product, and maintenance and tooling items).

INVENTORY ACCOUNTING. Inventory accounting is concerned with inventory transactions and inventory records. The accuracy and completeness of these transactions and records is critical to the success of the CIPMS. Material requirements planning, for example, depends on accurate inventory records to perform its planning function.

Inventory transactions include receipts, disbursements or issues, returns, and loans (e.g., the loaning of a tool or fixture that would be returned after use). Inventory transactions would also allow for adjustments to be made in the records as a result of a physical inventory count. The various changes would be entered by means of a terminal located at the site of the transaction (in the receiving department, raw material stores, tool crib, etc.).

The purpose of entering the various transactions is to maintain accurate inventory records. Additions and subtractions are made to the inventory balance as a result of the transactions. The term ''item master file'' is sometimes used to describe the computerized inventory record file. The types of data contained in the

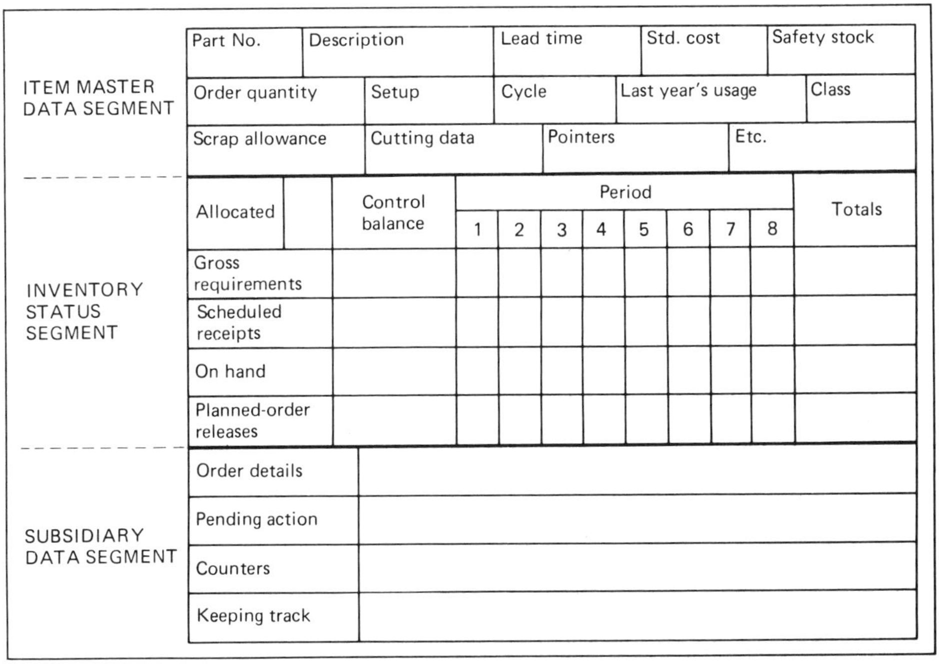

ITEM MASTER DATA SEGMENT	Part No.		Description		Lead time		Std. cost		Safety stock			
	Order quantity		Setup		Cycle		Last year's usage		Class			
	Scrap allowance		Cutting data		Pointers			Etc.				
INVENTORY STATUS SEGMENT	Allocated		Control balance		Period								Totals
				1	2	3	4	5	6	7	8	
	Gross requirements											
	Scheduled receipts											
	On hand											
	Planned-order releases											
SUBSIDIARY DATA SEGMENT	Order details											
	Pending action											
	Counters											
	Keeping track											

FIGURE 15.1   Record of an inventory item. (Reprinted with permission from Orlicky [5].)

record for a given item would typically include the categories shown in Figure 15.1. The file contains three segments:

1. Item master data segment
2. Inventory status segment
3. Subsidiary data segment

The first segment gives the item's identification (by part number) and other data, such as lead time, cost, and order quantity. The second segment (inventory status segment) provides a time-phased record of inventory status. In MRP it is important to know not only the current level of inventory, but also the future changes that will occur against the inventory status. Therefore, the inventory status segment lists the gross requirements for the item, scheduled receipts, on-hand status, and planned-order releases. The third file segment (subsidiary data segment) contains miscellaneous information pertaining to purchase orders, scrap or rejects, engineering change actions, and so on.

INVENTORY PLANNING AND CONTROL. In addition to maintaining an accurate item master file, the inventory management module must also be concerned with the planning and control of inventories. Many of these planning and control activities are accomplished by the MRP module. Other functions include:

Determining economic lot sizes
Determining safety stock levels
Determining ordering policies and reorder points
ABC inventory analysis
Analysis of usage rates for lot size calculations and other purposes
Automatic generation of requisitions for purchasing

In these and other inventory planning functions, the purpose is to keep inventory investment as low as possible while achieving a satisfactory service level. The inventory management module would accomplish some of these functions automatically, without human intervention. In other cases where human judgment is required, the proper information would be provided to the decision maker for appropriate action.

15.3 MATERIAL REQUIREMENTS PLANNING

Material requirements planning is a computational technique that converts the master schedule for end products into a detailed schedule for the raw materials and components used in the end products. The detailed schedule identifies the quantities of each raw material and component item. It also tells when each item must be ordered and delivered so as to meet the master schedule for the final products.

MRP is often considered to be a subset of inventory control. While it is an effective tool for minimizing unnecessary inventory investment, MRP is also useful in production scheduling and purchasing of materials.

The concept of MRP is relatively straightforward. What complicates the application of the technique is the sheer magnitude of the data to be processed. The master schedule provides the overall production plan for final products in terms of month-by-month or week-by-week delivery requirements. Each of the products may contain hundreds of individual components. These components are produced out of raw materials, some of which are common among the components. For example, several parts may be produced out of the same sheet steel. The components are assembled into simple subassemblies. Then these subassemblies are put together into more complex assemblies—and so forth, until the final product is assembled together. Each production and assembly step takes time. All of these factors must be incorporated into the MRP computations. Although each separate computation is uncomplicated, the magnitude of all the data to be processed is so large that the application of MRP is virtually impossible unless carried out on a digital computer.

15.4   BASIC MRP CONCEPTS

Material requirements planning is based on several basic concepts which are implicit in the preceding description but not explicitly defined. These concepts are:

1. Independent versus dependent demand
2. Lumpy demand
3. Lead times
4. Common use items

Independent versus dependent demand

This distinction is fundamental to MRP. *Independent* demand means that demand for a product is unrelated to demand for other items. End products and spare parts are examples of items whose demand is independent. Independent demand patterns must usually be forecasted.

Dependent demand means that demand for the item is related directly to the demand for some other product. The dependency usually derives from the fact that the item is a component of the other product. Not only component parts, but also raw materials and subassemblies, are examples of items that are subject to dependent demand.

Whereas demand for the firm's end products must often be forecasted, the raw materials and component parts should not be forecasted. Once the delivery schedule for the end products is established, the requirements for components and raw materials can be calculated directly. For example, even though the demand for automobiles in a given month can only be forecasted, once that quantity is established we know that four regular tires will be needed to deliver the car plus one spare tire.

MRP is the appropriate technique for determining quantities of dependent demand items. These items constitute the inventory of manufacturing: raw materials, work-in-progress, component parts, and subassemblies. Accordingly, MRP is a very powerful tool in the planning and control of manufacturing inventories.

Lumpy demand

In an order point system, the assumption is generally made that the demand for the item in inventory will occur at a gradual, continuous rate. This assumption is important for developing the mathematical model to derive the economic lot size formula. In a manufacturing situation, demand for the raw materials and components of a product will occur in large increments rather than in small, almost continuous units. The large increments correspond to the quantities needed to make a certain batch of the final product. When the demand occurs in these large steps, it is referred to by the term "lumpy demand." MRP is the appropriate approach for dealing with inventory situations characterized by lumpy demand.

Lead times

The lead time for a job is the time that must be allowed to complete the job from start to finish. In manufacturing there are two kinds of lead times: ordering lead times and manufacturing lead times. An *ordering* lead time for an item is the time required from initiation of the purchase requisition to receipt of the item from the vendor. If the item is a raw material that is stocked by the vendor, the ordering lead time should be relatively short, perhaps a few weeks. If the item must be fabricated by the vendor, the lead time may be substantial, perhaps several months.

Manufacturing lead time is the time needed to process the part through the sequence of machines specified on the route sheet. It includes not only the operation times but also the nonproductive time that must be allowed.

In MRP, lead times are used to determine starting dates for assembling final products and subassemblies, for producing component parts, and for ordering raw materials.

Common use items

In manufacturing, the basic raw materials are often used to produce more than one component type. Also, a given component may be used on more than one final product. For example, the same type of steel rod stock may be used to produce screws on an automatic screw machine. Each of the screw types may then be used on several different products. MRP collects these common-use items from different products to effect economies in ordering the raw materials and manufacturing the components.

15.5   INPUTS TO MRP

MRP converts the master production schedule into the detailed schedule for raw materials and components. For the MRP program to perform this function, it must operate on the data contained in the master schedule. However, this is only one of three sources of input data on which MRP relies. The three inputs to MRP are:

1.   The master production schedule and other order data
2.   The bill-of-materials file, which defines the product structure
3.   The inventory record file

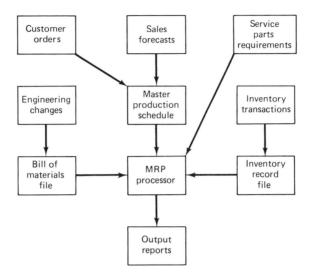

FIGURE 15.2 Structure of a material requirements planning (MRP) system.

Figure 15.2 presents a diagram showing the flow of data into the MRP processor and its conversion into useful output reports. The three inputs are described in the sections below.

Master production schedule

As explained in Chapter 14, the master production schedule is a list of what end products are to be produced, how many of each product is to be produced, and when the products are to be ready for shipment. The general format of a master production schedule is illustrated in Figure 15.3. Manufacturing firms generally work toward monthly delivery schedules. However, in Figure 15.3, the master schedule uses weeks as the time periods (for purposes of an example that will be developed later). The master schedule must be based on an accurate estimate of demand for the firm's product, together with a realistic assessment of its production capacity.

Week number		6	7	8	9	10
Product P1				50		100
Product P2			70	80	25	
etc.						

FIGURE 15.3 Master production schedule for products P1 and P2, showing week delivery quantities.

Product demand that makes up the master schedule can be separated into three categories. The first consists of firm customer orders for specific products. These orders usually include a specific delivery date which has been promised to the customer by the sales department. The second category is forecasted demand. Based on statistical techniques applied to past demand, estimates provided by the sales staff, and other sources, the firm will generate a forecast of demand for its various product lines. This forecast may constitute the major portion of the master schedule. The third category is demand for individual component parts. These components will be used as repair parts and are stocked by the firm's service department. This third category is often excluded from the master schedule since it does not represent demand for end products. In Figure 15.2 we show it as feeding directly into the MRP processor.

Bill of materials file

In order to compute the raw material and component requirements for end products listed in the master schedule, the product structure must be known. This is specified by the bill of materials, which is a listing of component parts and subassemblies that make up each product. Putting all these assembly lists together, we have the bill-of-materials file (BOM).

The structure of an assembled product can be pictured as shown in Figure 15.4. This is a relatively simple product in which a group of individual components make up two subassemblies, which in turn make up the product. The product structure is in the form of a pyramid, with lower levels feeding into the levels above. We can envision one level below that shown in Figure 15.4. This would consist of the raw materials used to make the individual components. The items at each successively higher level are called the parents of the items in the level directly below. For example, subassembly S1 is the parent of components C1, C2, and C3. Product P1 is the parent of subassemblies S1 and S2.

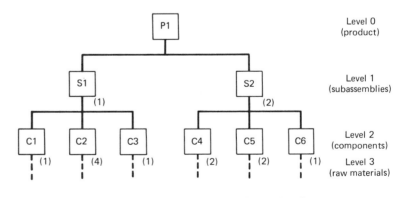

FIGURE 15.4 Product structure for product P1.

The product structure must also specify how many of each item are included in its parent. This is accomplished in Figure 15.4 by the number in parentheses to the right and below each block. For example, subassembly S1 contains four of component C2 and one each of components C1 and C3.

Inventory record file

It is mandatory in material requirements planning to have accurate current data on inventory status. This is accomplished by utilizing a computerized inventory system which maintains the inventory record file or item master file. The features of such a system were discussed in Section 15.2 and illustrated in Figure 15.1.

A definition of the lead time for the raw materials, components, and assemblies must be established in the inventory record file. The ordering lead time can be determined from purchasing records. The manufacturing lead time can be determined from the process route sheets (or routing file).

It is important that the inputs to the MRP processor be kept current. The bill-of-materials file must be maintained by feeding any engineering changes that affect the product structure into the BOM. Similarly, the inventory record file is maintained by inputing the inventory transactions to the file.

15.6   HOW MRP WORKS

The material requirements planning processor operates on the data contained in the master schedule, the bill-of-materials file, and the inventory record file. The master schedule specifies a period-by-period list of final products required. The BOM defines what materials and components are needed for each product. The inventory record file contains information on the current and future inventory status of each component. The MRP program computes how many of each component and raw material are needed by "exploding" the end-product requirements into successively lower levels in the product structure. Referring to the master schedule in Figure 15.3, 50 units of product P1 are specified in the master schedule for week 8. Now referring to the product structure in Figure 15.4, 50 units of P1 explode into 50 units of subassembly S1 and 100 units of S2, and the following numbers of units for the components:

C1:   50 units
C2: 200 units
C3:   50 units
C4: 200 units
C5: 200 units
C6: 100 units

The quantities of raw materials for these components would be determined in a similar manner.

There are several factors that must be considered in the MRP parts and materials explosion. First, the component and subassembly quantities given above are gross requirements. Quantities of some of the components and subassemblies may already be in stock or on order. Hence the quantities that are in inventory or scheduled for delivery in the near future must be subtracted from gross requirements to determine net requirements for meeting the master schedule.

A second factor in the MRP computations is manifested in the form of lead times: ordering lead times and manufacturing lead times. The MRP processor must determine when to start assembling the subassemblies by offsetting the due dates for these items by their respective manufacturing lead times. Similarly, the component due dates must be offset by their manufacturing lead times. Finally, the raw materials for the components must be offset by their respective ordering lead times. The material requirements planning program performs this lead-time-offset calculation from data contained in the inventory record file and from route sheet data.

A third factor that complicates MRP is common-use items. Some components and many raw materials are common to several products. The MRP processor must collect these common use items during the parts explosion. The total quantities for each common use item are then combined into a single net requirement for the item.

Finally, a feature of MRP that should be emphasized is that the master production schedule provides time-phased delivery requirements for the end products, and this time phasing must be carried through the calculations of the individual component and raw material requirements.

EXAMPLE 15.2

To illustrate how MRP works, let us consider the requirements planning procedure for one of the components of product P1. The component we will consider is C4. This part happens to be used also on one other product: P2 (see the master schedule of Figure 15.3). However, only one of item C4 is used on each P2 produced. The product structure of P2 is given in Figure 15.5. Component C4 is made out of raw material M4. One unit of M4 is needed to produce a unit of C4. The ordering and manufacturing lead times needed to make the MRP computations are as follows:

P1: assembly lead time = 1 week
P2: assembly lead time = 1 week
S2: assembly lead time = 1 week
S3: assembly lead time = 1 week
C4: manufacturing lead time = 2 weeks
M4: ordering lead time = 3 weeks

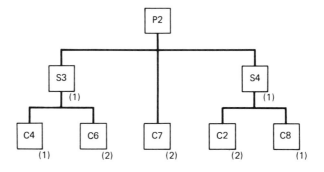

FIGURE 15.5   Product structure for product P2.

Period		1	2	3	4	5	6	7
Item: RAW MATL. M4								
Gross Requirements								
Scheduled Receipts				40				
On Hand	50			90				
Net Requirements								
Planned Order Releases								

FIGURE 15.6   Initial inventory status of material M4 in Example 15.2.

The current inventory and order status of item M4 is shown in Figure 15.6. There are no stocks or orders for any of the other items listed above.

The solution is presented in Figure 15.7. The delivery requirements for products P1 and P2 must be offset by the 1-week assembly lead time to obtain the planned order releases. Since the subassemblies to make the products must be ready, these order release quantities are "exploded" into requirements for subassemblies S2 (for P1) and S3 (for P2). These net requirements are then offset by the 1-week lead time and combined (in week 6) to obtain the gross requirements for part C4. Net requirements are equal to gross requirements for P1, P2, S2, S3, and C4 because of no on-hand inventory and no planned orders. We see the effect of current and planned stocks in the time-phased inventory picture for M4. The on-hand stock of 50 plus the scheduled receipts of 40 are used to meet the gross requirements of 70 units of M4 in week 3. Twenty units remain after meeting these requirements, which can be applied to the gross requirements of 280 units of M4 in week 4. Net requirements in week 4 are therefore $280 - 20 = 260$ units. With an ordering lead time of 3 weeks, the order release for the 260 units must be planned for week 1.

Period		1	2	3	4	5	6	7	8	9	10
Item: PRODUCT P1											
Gross Requirements									50		100
Scheduled Receipts											
On Hand	0										
Net Requirements									50		100
Planned Order Releases								50		100	
Item: PRODUCT P2											
Gross Requirements								70	80	25	
Scheduled Receipts											
On Hand	0										
Net Requirements								70	80	25	
Planned Order Releases							70	80	25		
Item: SUBASSBY S2											
Gross Requirements								100		200	
Scheduled Receipts											
On Hand											
Net Requirements								100		200	
Planned Order Releases							100		200		
Item: SUBASSBY S3											
Gross Requirements								70	80	25	
Scheduled Receipts											
On Hand											
Net Requirements								70	80	25	
Planned Order Releases							70	80	25		
Item: COMPONENT C4											
Gross Requirements							70	280	25	400	
Scheduled Receipts											
On Hand											
Net Requirements							70	280	25	400	
Planned Order Releases				70	280	25	400				
Item: RAW MATL. M4											
Gross Requirements				70	280	25	400				
Scheduled Receipts				40							
On Hand	50			90	20						
Net Requirements				-20	260	25	400				
Planned Order Release		260	25	400							

FIGURE 15.7 MRP solution for Example 15.2.

15.7 MRP OUTPUT REPORTS

The material requirements planning program generates a variety of outputs that can be used in the planning and management of plant operations. These outputs include:

1. Order release notice, to place orders that have been planned by the MRP system
2. Reports showing planned orders to be released in future periods
3. Rescheduling notices, indicating changes in due dates for open orders
4. Cancellation notices, indicating cancellation of open orders because of changes in the master schedule
5. Reports on inventory status

The outputs of the MRP system listed above are called primary outputs by Orlicky [5]. In addition, secondary output reports can be generated by the MRP system at the user's option. These reports include:

1. Performance reports of various types, indicating costs, item usage, actual versus planned lead times, and other measures of performance
2. Exception reports, showing deviations from schedule, orders that are overdue, scrap, and so on
3. Inventory forecasts, indicating projected inventory levels (both aggregate inventory as well as item inventory) in future periods

15.8 BENEFITS OF MRP

There are many advantages claimed for a well-designed, well-managed material requirements planning system. Among these benefits reported by MRP users are the following. The statistical data are compiled from Schaffer [8].

> *Reduction in inventory.* MRP mainly affects raw materials, purchased components, and work-in-process inventories. Users claim a 30 to 50% reduction in work-in-process.
>
> *Improved customer service.* Some MRP proponents claim that late orders are reduced 90%.
>
> *Quicker response to changes in demand and in the master schedule.*
>
> *Greater productivity.* Claims are that productivity can be increased by 5 to 30% through MRP. Labor requirements are reduced correspondingly.
>
> *Reduced setup and product changeover costs.*
>
> *Better machine utilization.*

Increased sales and reductions in sales price. These are also claimed as MRP benefits by some users.

As a result of the recognition given to these benefits in trade journals and professional societies, applications of material requirements planning have grown dramatically since the late 1960s. Today, more and more manufacturing firms of any significant size either use a computerized MRP system or are planning to implement one.

15.9 MRP II: MANUFACTURING RESOURCE PLANNING

In his book on manufacturing resource planning, Wight [10] defines four classes of MRP users. The characteristics of the four classes are listed in Table 15.1. The lowest level is class D, in which the potential utility of material requirements planning is hardly realized at all. MRP is used basically as a data processing system with many of the traditional production control procedures (e.g., shortage lists, expediting) still being used. At the top of the list is the class A MRP user. This is a company that uses material requirements planning together with capacity planning, shop floor control, and other components of a computer-integrated production management system. The next step beyond the class A user is MRP II. To describe MRP II, let us first examine the progressive evolution of material requirements planning into manufacturing resource planning.

TABLE 15.1   Four Classes of MRP Users

Class of User	*Characteristics*
Class A	Uses closed-loop MRP
	Integrated system has MRP, capacity planning, shop floor control, vendor scheduling, etc.
	MRP system used to help plan sales, engineering, production, purchasing, etc.
	No shortage lists to override the production schedules
Class B	System has MRP, capacity planning, shop floor control, but no vendor scheduling
	Not used much to help manage the business; used as a production control system
	Needs help from shortage list
	Inventory is higher than need be
Class C	System used for inventory ordering rather than scheduling
	Scheduling by shortage list
	Master schedule is overloaded
Class D	MRP working in the data processing department only
	Inventory records are poor
	Master schedule, if it exists at all, is overstated and mismanaged
	Relies on shortage list and expediting rather than MRP

Four steps of MRP

Material requirements planning has changed significantly over the years. Four steps can be identified in the evolution of MRP [10]:

1. An improved ordering method
2. Priority planning
3. Closed-loop MRP
4. MRP II

The first step was implemented when initial use of the computer was made to perform the requirements planning calculations. Before the computer, this task was performed manually and consumed tremendous amounts of time and manpower to accomplish. Computerized MRP systems represented a tremendous improvement in the ordering of raw materials and components because of the speed and accuracy with which the requirements planning task could be performed.

The need for the second step in MRP evolution grew out of attempts to implement step one MRP in conjunction with an unrealistic master schedule. This was a master schedule that ignored the limitation imposed by plant capacity and other constraints. It caused the MRP processor to generate schedules and requirements that could not be accomplished by the factory. As a result, the use of shortage lists continued. To overcome these problems, the MRP systems began to incorporate priority planning into their computations. The term ''priority planning'' connotes an MRP system which determines not only what materials should be ordered but also when those materials will be required. The planning of material requirements can be phased into time periods (weeks, or even days). Priority planning not only provides a means for dealing with rush jobs by increasing their priorities; it also helps to unexpedite jobs whose priorities have been reduced.

Step three MRP represents the level of achievement of the class A MRP user. Closed-loop MRP is an improvement over step two MRP because it not only plans the priorities but also provides feedback information relative to executing the priority plan. Closed-loop MRP means that the various functions in production planning and control (capacity planning, inventory management, shop floor control, and MRP in Figure 14.2) have been integrated into a single system. It also means that there is feedback from vendors, the production shop, and so on, when problems arise in implementing the production plan.

MRP II

Closed-loop MRP represents a significant achievement in terms of tieing together the various separate functions of a production planning and control system. However, there is one final step in the evolution of MRP (at least, as it is currently conceived). This fourth step involves a link-up between the closed-loop MRP system and the financial systems of the company. Manufacturing resource planning is the name given to this combination.

MRP II possesses two basic characteristics which go beyond closed-loop MRP:

1. It is an operational and financial system.
2. It is a simulator.

The operational and financial system makes MRP II a company-wide system, concerned with all facets of the business, including sales, production, engineering, inventories, and cash flows. In all cases, the operations of the individual departments are reduced to the same common denominator: financial data. This common base provides the company management with the information needed to manage it successfully. For example, raw materials on hand can be converted into their equivalent cost and summed over all stocks in inventory. Work-in-process can be evaluated by adding raw material costs to the cost of labor turned in against the particular part numbers and orders. Other operating data can be expressed in money terms by a similar calculation procedure.

MRP II is also a simulator which is intended to answer ''what if'' questions. The simulator can be used to simulate the probable outcomes of alternative production plans and management decisions which are under consideration.

In essence, manufacturing resource planning is quite similar to the general model of a computer-integrated production management system described in Chapter 14. The CIPMS includes not only the operational system (MRP, inventory management, capacity planning, etc.) but also the cost planning and control module. This cost planning and control module is the connection with the company's accounting and financial systems.

REFERENCES

[1] CHASE, R., AND AQUILANO, N., *Production and Operations Management,* 3rd ed., Richard D. Irwin, Inc., Homewood, Ill., 1981, Chapter 16.

[2] GROOVER, M. P., *Automation, Production Systems, and Computer-Aided Manufacturing,* Prentice-Hall, Inc., Englewood Cliffs, N.J., 1980, Chapter 17.

[3] HALEVI, G., *The Role of Computers in Manufacturing Processes,* John Wiley & Sons, Inc., New York, 1980.

[4] INTERNATIONAL BUSINESS MACHINES CORPORATION, *Communications Oriented Production Information and Control System,* Publication G320-1974.

[5] ORLICKY, J., *Material Requirements Planning,* McGraw-Hill Book Company, New York, 1975.

[6] PETERSON, R., AND SILVER, E. A., *Decision Systems for Inventory Management and Production Planning,* John Wiley & Sons, Inc., New York, 1979.

[7]   PLOSSL, G. W., "MRP Yesterday, Today, and Tomorrow," *Production and Inventory Management*, Third Quarter, 1980, pp. 1–10.

[8]   SCHAFFER, G. H., "Implementing CIM," *American Machinist*, August, 1981, pp. 151–174.

[9]   WIGHT, O. W., *Production and Inventory Management in the Computer Age*, Cahners Books, Boston, 1974.

[10] WIGHT, O. W. *MRP II: Unlocking America's Productivity Potential*, Oliver Wight Limited Publications, Inc., Williston, Vt., 1981.

PROBLEMS

15.1. The monthly usage rate for a certain type of cemented carbide insert used in the machine shop is averaging 1100 units. The inserts cost $4.36 each when ordered in quantities over 500. It is estimated that it costs an average of $60 each time a batch of inserts is ordered. The holding cost for this shop is based on using 25% per year of the cost of the item in inventory. What is the economic order quantity for this insert?

15.2. Develop a computer program for an automated inventory control and reordering system. The program should have the following capabilities and specifications:
(a)   It should maintain records for 200 items in inventory.
(b)   It must have a reorder point for each item.
(c)   It should calculate the usage rate (demand rate) for each item when the reorder point is reached.
(d)   It should compute the EOQ by Eq. (15.1). Assume that the annual holding cost and ordering cost are known for each item.

15.3. Using the master schedule of Figure 15.2 and the product structures in Figures 15.3 and 15.5, determine the time-phased requirements for component C6. The raw material used in component C6 is M6. Two units of C6 are obtained from every unit of M6. Lead times are as follows:

P1:   assembly lead time = 1 week
P2:   assembly lead time = 1 week
S2:   assembly lead time = 1 week
S3:   assembly lead time = 1 week
C6:   manufacturing lead time = 2 weeks
M6:   ordering lead time = 2 weeks

Assume that the current inventory status for all of the items above is: units on hand = 0, units on order = 0. The format of the solution should be similar to that presented in Example 15.1.

15.4. Solve Problem 15.3 if the current inventory and order status for S3, C6, and M6 is:

S3:   inventory on hand = 2, on order = 0
C6:   inventory on hand = 5, on order = 10 due for delivery in week 2
M6:   inventory on hand = 10, on order = 50 due for delivery in week 2

15.5. Product P4 is assembled out of 2 units of S5 and 1 unit of S6. Both S5 and S6 are subassemblies. S5 is made of 1 unit of C2, 2 units of C3, and 4 units of C4. S6 is made of 3 units of C3 and 1 unit of C5. Draw a diagram of the product P4 structure similar to the format of Figure 15.3.

15.6. This problem makes use of the data for product P4 in Problem 15.5. The master schedule specifies that 100 units of P4 are to be delivered in week 10, 150 units in week 11, and 200 units in weeks 12, 13, and 14. The lead times for each item are given as follows:

P4:   assembly lead time = 2 weeks
S5:   assembly lead time = 1 week
S6:   assembly lead time = 1 week
C2:   ordering lead time = 3 weeks
C3:   ordering lead time = 4 weeks
C4:   ordering lead time = 2 weeks
C5:   ordering lead time = 3 weeks

Assume that the current inventory status for these items is: units on hand = 0, units on order = 0. Also assume that these components and subassemblies are not common-use items on any other products.

Develop a computer program to compute the time-phased gross and net requirements and planned order release dates for the data of this problem. The output format should be similar to that shown in Figure 15.6. The planning horizon should cover weeks 1 through 14 for all components, subassemblies, and the final product P4.

chapter 16

Shop Floor Control
and Computer Process Monitoring

16.1 INTRODUCTION

Production management systems are concerned with two related objectives: planning, and control of the manufacturing operations. The functions of production planning, development of the master schedule, capacity planning, and MRP all deal with the planning objective. In this chapter we consider the control objective. Systems that accomplish this objective are often referred to by the term "shop floor control" (SFC).

At the time of this writing, the techniques of shop floor control are in the midst of a transition from manual methods to computerized methods. Shop floor control has been recognized as a distinct problem area in production management for many years. In 1973, the American Production and Inventory Control Society published a book entitled *Shop Floor Controls* [1]. At that time computer and data collection hardware for SFC was not nearly as well developed as it is today. Even today, this area must be considered as one which is rapidly changing both in terms of hardware and in terms of theory. What we attempt to present is the latest practice in computerized shop floor control systems. This includes systems that establish a direct connection between the computer and the manufacturing process for the purpose of monitoring the operation.

16.2   FUNCTIONS OF SHOP FLOOR CONTROL

Production managers are faced with the problem of acquiring up-to-date information on the progress of orders in the factory and making use of that information to control factory operations. This is the problem addressed by a shop floor control system. The functions of a shop floor control system are classifed by Raffish [17] as follows:

1. Priority control and assignment of shop orders.
2. Maintain information on work-in-process for MRP.
3. Monitor shop order status information.
4. Provide production output data for capacity control purposes.

These functions are explained in the following subsections. The reader should refer back to Figure 14.2 to recall the relationships and lines of communication between shop floor control and the other functions (MRP, capacity planning, etc.) in a computer-integrated production management system.

Priority control and assigment of shop orders

MRP and priority planning are concerned with the time-phased planning of materials, work-in-process, and assembly of final product. Priority control can be considered as the execution phase of this planning process. Priority control is concerned with maintaining the appropriate priorities for work-in-process in response to changes in job order status. Suppose that the delivery date requirement for one batch of product was moved forward because demand for the product had increased such that current inventory levels were running low. Suppose also that the delivery date on another order had been pushed back due to low demand for that product. Priority control would be concerned with increasing the priority for the first job and decreasing the priority for the second job. The existence of priority control in the production management system recognizes that job priorities might change after the job order is originally issued to the shop.

   The priorities for the jobs in the shop might be redetermined on a weekly or even daily basis. Once these priorities are established, the assignment of work to the work centers in the factory must be made. The assignment of shop orders is basically a problem in operation scheduling and we will consider this topic in Section 16.4.

Maintain information on work-in-process

Shop floor control is sometimes defined as a method of controlling the work-in-process in the factory. All of the functions and objectives of SFC boil down to this one goal of managing the parts and assemblies that are currently being processed in the shop. Information relating to quantities and completion dates for the various steps in the production sequence are compared against the plan generated in MRP. Any discrepancies, due, for example, to parts scrapped in production, might require additional raw materials to be ordered and adjustments made in the priority plan for other components in that product.

Monitor shop order status

This is similar to the work-in-process function and relies on the same basic data. One of the principal reporting documents in SFC is the Work Order Status Report. As its name suggests, it provides information on the status of the orders in the shop.

This report should be updated several times per week, depending on the nature of the product and the processes in the shop. It might be sufficient to display the report on a CRT rather than in hard-copy form. However, an exception report should be printed periodically in document form. This would indicate the orders that were behind schedule as well as other noteworthy exceptions that happened during the period (e.g., significant machine breakdowns).

The accuracy and currentness of the work order status report are dependent on the correctness and timeliness of the basic data collected in the shop. These data deal with shop order transactions such as job completions, material movement, time turned in against an order, and so on. The method of collecting these data is critical to the success and value of the shop floor control system. The design of this factory data collection (FDC) system can take on various forms and we consider these systems in Sections 16.5 and 16.6.

Production output data for capacity control

Capacity planning is the function of a computer-integrated production management system which determines the labor and equipment resources that will be required to accomplish the master production schedule. It is a planning function. On the other hand, capacity control is concerned with making adjustments in labor and equipment usage to meet the production schedule. To make these adjustments effectively, the capacity control function must have up-to-date information on production rates and order status from the factory data collection system.

16.3 THE SHOP FLOOR CONTROL SYSTEM

The shop floor control system should be designed to accomplish the four functions efficiently and effectively. There are various ways in which the SFC system can be configured, ranging in degrees of computer involvement. At the current level of development in manufacturing technology, none of the shop floor control systems exclude human participation. In other words, SFC systems are not control systems in the sense of automatic feedback control or computer process control.[1] Even the most computerized and modern shop floor control systems today require human beings as a vital link in the control loop. The purpose of the computer system in SFC is to generate information by which the human beings can make good decisions on effective factory management and implementation of the master schedule.

The organization of a computerized shop floor control system is illustrated in Figure 16.1. The diagram differentiates between those portions of SFC which are computer driven and those which require human participation. The computer generates various documents which are utilized by people to control production in the factory.

The shop floor control system consists of three steps. These steps are manifested as three computer software modules which are linked together in the SFC system. The three steps or modules are:

1. Order release
2. Order scheduling
3. Order progress

We describe these shop floor control steps in the following subsections.

Order release

The purpose of the order release module is to provide the necessary documentation that accompanies an order as it is processed through the shop. These documents are referred to collectively as the shop packet and their release to the factory constitutes the release of the order. The shop packet for an order consists of:

1. *Route sheet*—listing the operation sequence and tools needed
2. *Material requisitions*—to draw the necessary raw materials (or components for assemblies) from stock
3. *Job cards*—enough job cards to report the labor for each operation on the route sheet
4. *Move tickets*—to move the parts between work centers
5. *Parts list*—for assembly jobs

---

[1]We consider computer process control in Chapter 18.

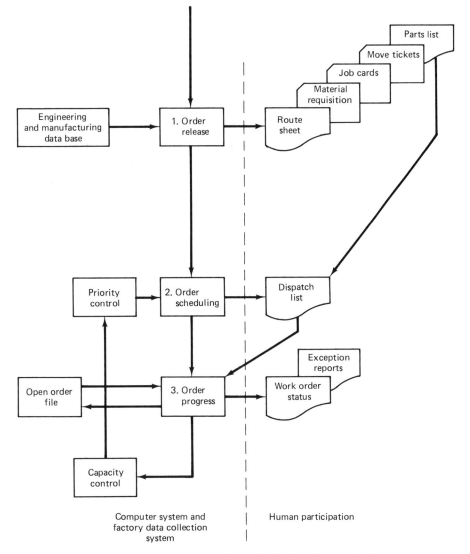

FIGURE 16.1 Flow of information in shop floor control.

The shop packet moves with the job through the sequence of processing or assembly operations. It comprises the documentation necessary to complete the job and to monitor the status of the job in the shop.

There are two inputs to the order release module. First, the basic information, which indicates that this module should release a particular order, is provided by MRP, capacity planning, and other elements of the computer-integrated production management system. (These elements are shown in Figure 14.2, but not in Figure 16.1.) The second input is the engineering and manufacturing data base, which contains the standard routing file and product structure file. The standard routing file contains the sequence of operations and work centers needed to process the order. This is used to prepare the route sheet in the shop packet. The product structure file contains the bill of materials for material requisitions and parts lists.

Order scheduling

The purpose of order scheduling is to make assignments of the orders to the various machines in the factory. Inputs to this module consist of the order release and priority control. Each job order has a certain priority determined by its due data and other factors. It is the function of priority control to schedule the orders for production according to their priorities. Section 16.4 examines several methods of assigning priorities for production scheduling.

The basic document prepared by the order scheduling module is the dispatch list. It reports the jobs that should be done at each work center and certain details about the routing of the part. One possible format of this dispatch list is shown in Figure 16.2. The dispatch list is generated each day or each shift. In an effective shop floor control system, this document represents to the foreman or production supervisor the best way for accomplishing the master schedule. Certain events may occur, such as a machine breakdown or poor raw materials, which prevent the supervisor from following the dispatch list completely. These exceptions will be reported, and, through subsequent adjustments in priority control, an attempt will be made to put the order back on schedule.

Order progress

The order scheduling module satisfies the first function of shop floor control: priority control and assignment of work orders. The purpose of the order progress module is to accomplish the remaining three functions of SFC: to provide data relative to work-in-process, shop order status, and capacity control. The order pro-

DISPATCH LIST

WORK CENTER 372   DATE   1-17-83

WORK ORDER	LOT QUANTITY	PRIOR WORK CENTER	NEXT WORK CENTER	START DATE	DUE DATE	REMAINING HOURS
1573	25	370	400	1–14	1–25	27.6
1688	300	370	400	1–14	1–28	40.5
1692	107	265	375	1–17	1–31	13.6
1693	14	370	342	1–17	2–4	12.0
2053	27	240	375	1–20	1–28	16.2

FIGURE 16.2 Dispatch list.

WORK ORDER STATUS REPORT

DEPARTMENT __2__   DATE __1–17–83__

WORK ORDER	LOT QUANTITY	DUE DATE	CURRENT WORK CENTER	REMAINING HOURS	STATUS	PRIORITY CLASS
1688	300	1–28	372	40.5	Late	Rush
1689	25	2–4	265	27.2	On time	Med
1690	105	2–4	265	20.5	On time	Med
1691	60	—	—	—	Cancel	—
1692	107	1–31	372	13.6	On time	Hi
1693	14	2–4	372	12.0	On time	Med
1694	10	2–28	370	40.8	On time	Lo

FIGURE 16.3 Work order status report.

gress module is designed to accept data collected from the shop floor and to generate reports that can be used to assist production management.

The various documents contained in the shop packet provide the means to identify the job. When a machine operator completes a particular process specified in the route sheet, relevant data to indicate the completion are entered into the order progress module. The data would include such items as piece count, scrappage (if any), operator identification number, operation number, work centers, and time of completion. The order progress module maintains a file of the transactions reported on each of the uncompleted jobs. This file is called the *open order file*, and it contains the latest status of each job order in the factory. The types of reports that can be generated from the open order file include the following:

1. *Work order status reports.* These are detailed reports that show the progress of each job through the shop. One possible format for the work order status report is illustrated in Figure 16.3.
2. *Exception reports.* These are designed to pinpoint deviations from the production schedule, overdue jobs, inconsistent piece counts (e.g., an increase in the number of pieces in the order), and other exception information.

These reports can be generated daily to achieve better control over jobs in the plant. Depending on the design of the shop floor control system, the information in the open order file might also be obtained on an inquiry basis.

16.4 OPERATION SCHEDULING

An overview of the scheduling problem

The master production schedule gives the timetable for end-product deliveries. This is translated into material and component requirements by using MRP, and checked against plant production capacity by means of capacity planning. The next link in this planning chain is operation scheduling, which is performed by the order scheduling module in shop floor control.

Operation scheduling is concerned with the problem of assigning specific jobs to specific work centers on a weekly, daily, or hourly basis. The end products specified in the master schedule consist of components, each of which is manufactured by a sequence of processing operations. Operation scheduling involves the assignment of start dates and completion dates to the batches of individual components and the designation of work centers on which the work is to be performed. The scheduling problem is complicated by the fact that there may be hundreds or thousands of individual jobs competing for time on a limited number of work centers. These complications are compounded by unforeseen interruptions and delays such as machine breakdowns, changes in job priority, work absenteeism, and strikes.

The objectives of an operation scheduling system are to assign jobs to work centers so as to:

1. Meet the required delivery dates for completion of all work on the jobs.
2. Minimize in-process inventory. This is accomplished by minimizing the aggregate manufacturing lead time.
3. Maximize utilization of machines and labor resources.

There are a variety of scheduling methods used in production. Different methods are appropriate, depending on whether the factory is engaged in job shop operations, batch production, or mass production. To describe the operation scheduling problem, let us consider the typical job shop or batch production situation. Jobs are to be processed through the appropriate sequence of work centers. A *job* is defined as a single part or batch of parts. A job could also consist of groups of components to be assembled. Operation scheduling can be described as consisting of the following two steps:

1. Machine loading
2. Job sequencing

To process the jobs through the factory, the jobs must be assigned to work centers. Since the total number of jobs exceeds the number of work centers, each work center will have a queue of jobs waiting to be processed. Allocating the jobs to the work centers is referred to as *machine loading*. Ten jobs may be the loading for a particular work center. The unanswered question is: In what sequence will the 10 jobs be processed? Answering this question is the problem in job sequencing.

Job sequencing involves determining the order in which to process the jobs through a given work center. To accomplish this, priorities are established among the jobs in the queue. Then the jobs are processed in the order of their relative priorities. When a job is completed at one work center, it enters the queue at the next work center in its process routing. That is, it becomes part of the machine loading for the next work center, and the priority rule determines its sequence of processing among those jobs.

Priority rules for job sequencing

A priority rule in operation scheduling is a guide for determining the sequence in which jobs will be processed through a given work center. Some of the priority rules used in industry are the following:

> Highest priority is given to jobs with the "earliest due date."
>
> Highest priority goes to the job with the "shortest processing time."
>
> Jobs are processed on a "first-come-first-serve" basis.
>
> Highest priority is given to the job with the "least slack" in its schedule, where slack is defined as follows:

$$\text{slack} = (\text{time remaining until due date}) - (\text{process time remaining}) \quad (16.1)$$

> Highest priority is given to the job with the lowest "critical ratio" where the critical ratio is defined as follows:

$$\text{critical ratio} = \frac{\text{time remaining until due date}}{\text{process time remaining}} \quad (16.2)$$

> The "red tag" method, where rush jobs are identified with a red ticket by the expediter.

An example will help to illustrate how these priority rules are used to determine the sequence in which jobs would be processed.

EXAMPLE 16.1

Suppose that we are presently at day 15 on the production scheduling calendar for the G & Z Machine Company and that there are three jobs (shop orders A, B, and C) in the queue for a particular work center. The jobs arrived at the work center in the order A, then B, then C. The following table gives the parameters of the scheduling problem for each job:

Job	Remaining process time (days)	Due date
A	5	25
B	16	34
C	7	24

Using the "earliest due date" we would schedule the jobs in the order C−A−B. With the "shortest processing time" priority rule, the jobs would be schedule A−C−B. If jobs were scheduled first-come-first-serve, the sequence would be A−B−C.

Using the "least slack" rule, we would determine the slack for each job according to Eq. (16.1). For job A, the slack is $(25 - 15) - 5 = 5$ days. For job B, the slack is $(34 - 15) - 16 = 3$ days. And for job C, slack $= (24 - 15) - 7 = 2$ days. The jobs would be processed in the order C−B−A.

Finally, using the "critical ratio" to establish order priority, Eq. (16.2) would be used for each job. For job A, the critical ratio $= (25 - 15)/5 = 2.0$ For job B, the critical ratio $= (34 - 15)/16 = 1.19$. For job C the critical ratio $= (24 - 15)/7 = 1.29$. The processing sequence would be B—C—A.

The results are summarized in the following table

Priority rule	Job sequence
Earliest due date	C−A−B
Shortest processing time	A−C−B
First-come-first-serve	A−B−C
Least slack	C−B−A
Critical ratio	B−C−A

The reader will note that, for the data of this example, five different priority rules have yielded five different job sequences.

The question of which among the solutions is the best depends on one's criteria for defining what is best. The shortest processing time rule will usually result in the lowest average manufacturing lead time and therefore the lowest in-process inventory. However, this may result in customers whose jobs have long processing times to be disappointed. First-come-first-serve seems like the fairest criteria, but it denies the opportunity to deal with differences in due dates among customers and genuine rush jobs. The earliest due date, least slack, and critical ratio rules address this issue of relative urgency among jobs. Then there is the "red tag" priority rule (not considered in Example 16.1), the method used by expediters to indicate which jobs should be processed on a "rush" basis. The trouble with the red tag method is that shops often end up with more jobs red tagged than not.

Let us evaluate the five priority rules used in Example 16.1 in terms of manufacturing lead time and job lateness.

EXAMPLE 16.2

For example 16.1, assume that the work center to which the data apply is the final work center in the routing for each of the three jobs. Accordingly, the remaining process time refers to that work center. The problem is to determine the average manufacturing lead time for the three jobs and the aggregate job lateness for the three jobs.

The manufacturing lead time for each job is the remaining process time plus time spent in the queue waiting to be processed at the work center. The average manufacturing lead time is the average for the three jobs.

Job lateness for each job is defined as the number of days the job is completed after the due date. If it is completed before the due date, it is not late. Therefore, its lateness is zero. The aggregate lateness is the sum of the lateness times for the individual jobs.

Results of the processing through the last work center are displayed on the time chart of Figure 16.4 for each of the five priority rules considered in Example 16.1.

Using the earliest due date rule to illustrate the arithmetic, the lead time for job C is its processing time of 7 days. The lead time of job A is its waiting time plus processing time $= 7 + 5 = 12$ days. The lead time of job B is $7 + 5 + 16 = 28$ days. The average is $(7 + 12 + 28)/3 = 15\frac{2}{3}$ days.

Job lateness for job C under the earliest due date rule is zero since it is completed before its due date. Job A is 2 days late and job B is completed 9 days behind schedule. The aggregate job lateness is 11 days.

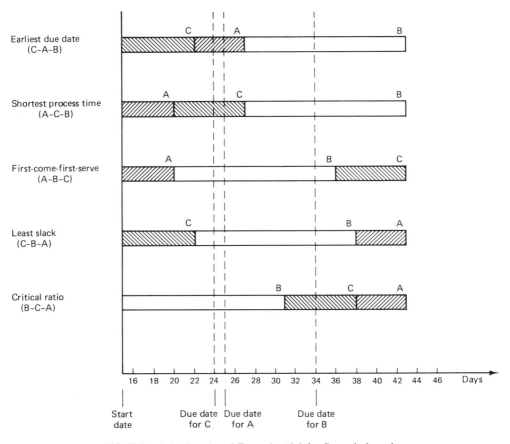

FIGURE 16.4   Results of Example 16.2 for five priority rules.

The results for all five rules are summarized in the following table:

Priority rule	Sequence	Avg mfg. lead time (days)	Aggregate job lateness (days)
Earliest due date	C−A−B	15.67	11
Shortest process time	A−C−B	15.0	12
First-come-first-serve	A−B−C	18.0	21
Least slack	C−B−A	19.33	22
Critical ratio	B−C−A	22.33	32

The relatively simple earliest due date and shortest process time rules seem to perform best for the data of this example. Any of the common priority rules can be programmed into the order scheduling module of a shop floor control system. As Example 16.2 illustrates, the firm must be careful to select a priority rule or combination of rules that accomplish its desired scheduling objectives.

16.5   THE FACTORY DATA COLLECTION SYSTEM

The purpose of the factory data collection (FDC) system in shop floor control is to provide basic data for monitoring order progress. In a computerized SFC system these data are submitted to the order progress module for analysis and generation of work order status reports and exception reports. The types of SFC data that would be collected by the FDC system include:

    Piece counts
    Count on scrapped parts or parts needing rework
    Completion of operations in the routing sequence
    Machine breakdowns
    Labor time turned in against a job

These parameters of factory activity represent the basic data from which status information on shop orders can be determined.

Another purpose for which a factory data collection system may be used is time and attendance reporting for accounting and payroll departments. Many FDC systems include terminals for employees to clock in and clock out at the beginning and end of the shift. These types of records are often incorporated into the shop floor control system.

Kasarda [14] lists five methods that have been used to collect data from the shop floor:

1. Job traveler
2. Employee time sheet
3. Operation tear strips or punched cards included with shop packet
4. Centralized shop floor terminals
5. Individual work center terminals

All of these methods require the coorperation of the shop workers in recording the data on a paper document or entering the data at a computer terminal. Job travelers and time sheets are the most manual and clerical approaches, while individual work center terminals represent the least manual (and highest initial investment) approach. Our interest is more with the computerized FDC systems (central terminals and work center terminals), but we will describe the manual approaches first.

Job traveler

The job traveler is basically a log sheet that is included with the shop packet moving with the job through the various work centers specified on the route sheet. Each employee who works on the job is required to record the time spent, piece counts, and other data onto the log sheet. When the job is completed, the sheet contains the complete history for that shop order. It can be reviewed, analyzed, and summarized by production control, accounting, and others.

From the standpoint of shop floor control, there are several important disadvantages with this method and little that can be said in its favor. It is perhaps the easiest method to implement and does not require use of a computer. Among its disadvantages, one is the problem that no data are collected on the order until the shop has completed all processing. Therefore, it is not possible to determine the work status of the order during process, except by physically locating the order in the factory and visually determining its status. This hardly satisfies the functions of shop floor control. Another problem is the accuracy and reliability of the data recorded on the log sheet. It is not possible to detect errors in the log sheet as they are being recorded. Because of its significant disadvantages, use of the job traveler does not lend itself to a computerized shop floor control system.

Employee time sheet

A second method of shop data collection is the employee time sheet. A time sheet is prepared by each worker, detailing information about the order. The employee must copy data (e.g., order number, operation number, etc.) onto the time sheet from the shop packet that accompanies the job. The time spent on the order, quantities produced, and similar data must also be recorded.

One big advantage of the employee time sheet method over the job traveler is that the data are submitted in a much more timely manner. Time sheets can be turned in each shift, thus allowing the status of an order to be determined on a more current basis. This does not happen automatically. It requires a clerical employee to summarize the data and an alert foreman to track down any problems or discrepanices which are indicated by the time sheets. The clerical procedure involves a time lag which delays the reporting on order status.

This delay is one of the disadvantages of the time sheet method. The order status report may not be prepared until several days after the work is reported. This is probably too late for factory managment to react to a problem effectively. A second disadvantage is that the time sheet method still relies on the employee to transcribe the data. Clerical errors may result from this manual procedure.

Operation tear strips

This represents an attempt to reduce the amount of clerical detail work involved with the job traveler and time sheet methods. Instead, when the shop packet is prepared at the time of order release, a set of preprinted tear strips or prepunched cards are also generated. These tear strips or cards follow the job through the factory together with the shop packet. When a worker finishes a particular operation on the job, one of the tear strips or cards is turned in. The employee must record only a minimum of data, such as piece count completed, since all the other information has been preprinted or prepunched.

The job tickets can be submitted daily or at the completion of an operation on the job which may occur during the shift. With the assistance of a clerical employee working at a computer terminal, job status data can be turned in to the SFC order progress module in a fairly timely fashion.

The disadvantages of the tear strip (or prepunched job ticket) method are that the tickets may get lost (of course, this could happen to the entire shop packet) or that an insufficient quantity of tickets will be prepared. Also, there is a time delay between when the tickets are turned in and when the clerk reports them into the order progress module.

Centralized shop terminal

To overcome the disadvantages of the previous clerical methods, the use of data entry terminals in the shop has been introduced. This approach ranges from the installation of one central terminal to the use of individual workstation terminals. We discuss the individual workstation terminals in the next subsection.

The centralized terminal approach may consist of a single terminal located near the center of the plant. If the plant is large, this may require some employees to walk great distances to enter the order data. A variation of the central terminal approach is to have a limited number of satellite terminals strategically located at convenient and accessible points throughtout the shop.

The use of central terminals requires that human operators enter the data manually. Workpiece counts, operation number, machine downtime, and product quality data are examples of data that can be entered by means of human operators located in the shop. In discrete-parts manufacturing, manual entry stations constitute a common form of factory data collection system in use today. The terminals range in complexity from simple pushbutton devices to sophisticated typewriter keyboard units with built-in microprocessor and data storage capability. Factory workers must be trained to key in the proper data, including operator identification number, the type of data being reported, and the data themselves in correct format.

Shop floor data entered from manual entry stations can be communicated directly to the computer. This is sometimes referred to as an ''on-line'' system. The alternative is to use an ''off-line'' configuration, in which the data are collected and stored in a data storage device for subsequent processing by the computer. An example of a system that can operate in either mode is the IBM 3630 Plant Communication System [13], which consists of various types of data-entry terminals, output devices, communication links, and a controller. The data-entry stations capture information from the factory, such as production piece counts, inventory counts, material and job status data, quality inspection data, job times (for costing purposes), and tool usage data. Factory management defines the type of information to be collected by the system. The time entry stations are located at entrance and exit doors and serve the function of time clocks to punch in and out. The controller unit collects and stores the data from the various entry stations. The shop records can be stored on diskettes for processing by the plant computer to generate the desired reports. Figures 16.5, 16.6, and 16.7 show a variety of data-entry terminals available on the IBM 3630 system.

The disadvantage of the centralized FDC system is the inconvenience to the operator, who may have to waste time walking back and forth to the one central

FIGURE 16.5   IBM 3647 Time and Attendance Terminal with magnetic card reader port. (Courtesy of IBM Corp.)

FIGURE 16.6 IBM 3641 Reporting Terminal used by workers to input work order data in the factory. (Courtesy of IBM Corp.)

FIGURE 16.7 IBM 3643 Keyboard Display, an interactive terminal for plant use. (Courtesy of IBM Corp.)

input terminal. At the end of the shift there will probably be other workers waiting to use the same terminal.

Another potential disadvantage arises in the case of the off-line entry configuration. In this system, the data entered may be stored until the end of the shift and then batch processed on the computer. This results in a delay on the reporting of order status which may be no less than in the tear strip method of factory data collection. The company may have invested thousands of dollars in a central terminal installation which provides results that are little better than the manual method.

Individual work center terminals

The ultimate FDC system for convenience and responsiveness is the individual work center entry terminal. The convenience to the operator results from the fact that the terminal is located in the immediate work area, perhaps attached to the machine tool or other processing equipment. The system should be more responsive than other methods because the operator can report status on the order more frequently; hence status reports are more current.

It is possible with the work center terminal to customize the front panel and other features of the entry device to facilitate the types of inputs which are likely to be made at that workstation. The order progress module can be programmed to properly interpret abbreviated inputs when entered at a particular work center. All of this results in simpler and faster data entry for the operator.

Cost is an important factor in the workstation terminal FDC systems. To be economically justified, the cost of each terminal must be low enough to make the installation worthwhile. Several hundred dollars is probably the upper limit on price per terminal installation.

The next logical step beyond the individual workstation terminal is to establish a direct connection between the workstation and the computer, uninterrupted by the necessity for a human operator to enter the data manually on a data entry device. This is the subject of computer process monitoring which we discuss in Section 16.6. One final manual input method might be mentioned before discussing this computer-automated technique.

Voice data input

In Section 8.9 we discussed voice numerical control, a method for programming numerically controlled machine tools. Another use of voice−computer communication is data input for manufacturing information systems. This technology is sometimes referred to as automatic speech recognition and it represents an attempt to simplify the human−machine interface. Under the proper circumstances, voice entry of information to the computer represents the easiest and quickest method of input. Examples of applications in manufacturing include quality inspection, inventory control, and part identification. There is potential for applying these systems in shop floor control data collection.

The characteristics that make a good industrial application of an automatic speech recognition system are described in Ref. [10]. First, there must be an existing computerized reporting system. Second, the data entry steps must be repetitive. Third, use of voice input requires a limited number of operators using the system. Most voice systems are speaker dependent, which means that the speech recognition processor recognizes only the unique voice patterns of the individuals who use the system. Fourth, a reasonable vocabulary is required to enter the data. This usually means a limitation of several dozen words. Fifth, it is desirable to capture the production data at the source.

EXAMPLE 16.3

An example of a factory data collection system and factory management system is reported in Ref. [6]. The system is installed at Hughes Aircraft Company's Microelectronics Technology Division in Fullerton, California, and is called FAMIS (FActory Management Information System). It was developed as a turnkey hardware and software system by Digital Datacom, Inc. The system consists of the central computer, disk memory and file storage, video display terminals, printers, and several card readers. A job traveler FDC system is used, where the traveler tickets are prepared on the printers as needed. These tickets contain all the data needed to report on order status. The job tickets are mark-sensitive tab cards, which allows work entries to be made with an ordinary number 2 pencil. These penciled entries consist of checks in boxes printed on the tab card. The job card is then inserted into the card reader to enter data. FAMIS is used to generate a variety of management reports on production and quality performance.

16.6 COMPUTER PROCESS MONITORING

All the factory data collection systems described in the preceding section required some form of human participation. Computer process monitoring (also sometimes called computer production monitoring) is a data collection system in which the computer is connected directly to the workstation and associated equipment for the purpose of observing the operation. The monitoring function has no direct effect on the mode of operation except that the data provided by monitoring may result in improved supervision of the process. The industrial process is not regulated by commands from the computer. Any use that is made of the computer to improve process performance is indirect, with human operators acting on information from the computer to make changes in the plant operations. The flow of data between the process and the computer is in one direction only—from process to computer.

The components used to build a computer process monitoring system include transducers and sensors, analog-to-digital converters (ADC), multiplexers, real-time clocks, and other electronic devices. Many of these components are described in Chapter 17. These components are assembled into various configurations for process monitoring. We discuss three such configurations:

1. Data logging systems
2. Data acquisition systems
3. Multilevel scanning

A particular computer process monitoring system is highly custom designed and may consist of a combination of these possible configurations. For example, a data acquisition system may include multilevel scanning.

Data logging systems

A *data logger* (DL) is a device that automatically collects and stores data for off-line analysis. Strictly speaking, the data could be analyzed by a person without the aid of computer. Our interest here is in data logging systems that operate in conjunction with computers. Data loggers can be classified into three types [4]:

1. Analog input/analog output
2. Analog input/analog and digital output
3. Analog and digital input/analog and digital output

Type 1 can be a simple one-channel strip chart recording potentiometer for tracking temperature values using a thermocouple as the sensing device. Types 2 and 3 are more sophisticated instruments which have multiple input channels and make use of multiplexers and ADCs to collect process or experimental test data from several sources. The DL can be interfaced with tape punches, magnetic tape units, teletypes and printers, plotters, and so on. They can also be interfaced with the computer for periodic transfer of data.

A *programmable data logger* (PDL) is a device that incorporates a microprocessor as part of the system. The microprocessor serves as a controller to the data logger and can be programmed by means of a keyboard. The programmable data logger can easily accommodate changes in rate or sequence of scanning the inputs. The PDL can also be programmed to perform such functions as data scaling, limit checking (making certain the input variables conform to prespecified upper and lower bounds), sounding alarms, and formatting the data to be in a compatible and desirable format with the interface devices.

Data acquisition systems

The term *data acquisition system* (DAS) normally implies a system that collects data for direct communication to a central computer. It is therefore an on-line system, whereas the data logger is an off-line system. However, the distinction between the DL and the DAS has become somewhat blurred as data loggers have become directly connected to computers.

Data acquisition systems gather data from the various production operations for processing by the central computer. The basic data can be analog or digital data which are collected automatically (transducers, ADC, multiplexers, etc.). It is a factory-wide system, as compared with data loggers, which are often used locally within the plant. The number of input channels in the DAS is therefore typically greater than in the DL system. For the data logger the number of input channels might range between 1 and 100, while the data acquisition system might have as many as 1000 channels or more. The rate of data entry into the DL system might

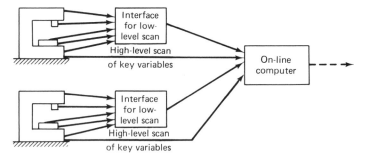

FIGURE 16.8   Multilevel scanning in computer process monitoring.

be 10 readings per second for multiple-channel applications. By contrast, the DAS would have to be capable of a data sampling rate of up to 1000 per second. Because of these differences, the data acquisition system would be typically more expensive than the data logger.

Multilevel scanning

In the data acquisition system, it is possible for the total number of monitored variables to become quite large. Although it is technically feasible for all these variables to be monitored through multiplexing, some of the signals would not be needed under normal operating conditions. In such a situation it is convenient to utilize a multilevel scan configuration, as illustrated schematically in Figure 16.8. With multilevel scanning, there would be two (or more) process scanning levels, a high-level scan and a low-level scan. When the process is running normally, only the key variables and status data would be monitored. This is the high-level scan. When abnormal operation is indicated by the incoming data, the computer switches to the low-level scan, which involves a more complete data logging and analysis to ascertain the source of the malfunction. The low-level scan would sample all the process data or perform an intensive sampling for a certain portion of the process that might be operating out of tolerance.

REFERENCES

[1]   AMERICAN PRODUCTION and INVENTORY CONTROL SOCIETY, *Shop Floor Controls*, 1973.

[2]   BAILEY, S. J., "Data Acquisition Systems Mature into Control Loop Decision Makers," *Control Engineering*, February, 1982, pp. 66−70.

[3]   BOEING COMPUTER SERVICES COMPANY, *PMS (Production Management System)*, The Boeing Company, 1979.

[4]   BROWN, J., "Choosing between a Data Logger (DL) and Data Acquisition System (DAS)," *Instruments and Control Systems*, September, 1976, pp. 33−38.

[5] CHASE, R. B., and AQUILANO, N. J., *Production and Operations Management*, 3rd ed., Richard D. Irwin, Inc., Homewood, Ill., 1981, Chapter 14.

[6] "FAMIS SPELLS SAVINGS," *Assembly Engineering*, October 1981, pp. 42−44.

[7] GREEN, A. M., "Shop Floor Data Closes the Loop in Factory Management," *Iron Age*, September 28, 1981, pp. 75−79.

[8] GROOVER, M. P., *Automation, Production Systems and Computer-Aided Manufacturing,* Prentice-Hall, Inc., Englewood Cliffs, N.J., 1980, Chapters 12, 17.

[9] HALEVI, G., *The Role of Computers in Manufacturing Processes*, John Wiley & Sons, Inc., New York, 1980, Chapter 9.

[10] HUBER, R. F., "Tell It to Your Machines," *Production*, June, 1980.

[11] INTERNATIONAL BUSINESS MACHINES CORPORATION, *Production Information and Control System, Publication GE20-0280-2.*

[12] INTERNATIONAL BUSINESS MACHINES CORPORATION, *Communications Oriented Production Information and Control System*, Vol. I, Publication G320-1974.

[13] INTERNATIONAL BUSINESS MACHINES CORPORATION, *IBM 3630 Plant Communication System, System Description*, Publication GA24-3652-2, Endicott, N.Y., 1979.

[14] KASARDA, J. B., "Shop Floor Control Must Be Able to Provide Real-Time Status and Control," *Industrial Engineering*, November, 1980, pp. 74−78, 96.

[15] MAY, N., "Shop Floor Controls—Principles and Use," *Proceedings, Twenty-fourth Annual International Conference*, American Production and Inventory Control Society, Boston, October, 1981, pp. 170−174.

[16] MONKS, J. G., *Operations Management: Theory and Problems*, McGraw-Hill Book Company, New York, 1977, Chapter 9.

[17] RAFFISH, N., "Let's Help Shop Floor Control," *Production and Inventory Management Review and APICs News*, American Production and Inventory Control Society, July, 1981, pp. 17−19.

[18] RAFFISH, N., "The Engineer's Impact on Shop Floor Control," *Manufacturing Engineering*, February, 1982, pp. 59−61.

[19] WIGHT, OLIVER W., *Production and Inventory Management in the Computer Age*, Cahners Books, Boston, 1974, Chapter 8.

PROBLEMS

16.1. Four jobs are to be scheduled through a certain work center. The following table gives data regarding the due date (values given are in terms of the production calendar day) and remaining process time.

Job	Remaining process time (days)	Due date
A	12	39
B	7	26
C	9	37
D	6	45

In the shop calendar, the current date is day 10.
(a) Use the "earliest due date" priority rule to set the production sequence for these jobs.
(b) Use the "shortest processing time" rule to establish the sequence.
(c) Use the "least slack" rule to sequence the jobs.
(d) Use the "critical ratio" to sequence the jobs.

16.2. For the solutions of Problem 16.1, evaluate the results using the criteria of average manufacturing lead time and aggregate job lateness. It is suggested that the reader plot the results on a time chart similar to Figure 16.4.

16.3. Three jobs are currently waiting in the queue to be processed through a certain milling machine. Following the milling operation, each of the jobs must be processed through a surface grinder. There is no queue of work at the grinder nor is it anticipated that there will be between now and when these jobs are processed. The processing times at the two machines for each of the three jobs are given in the following table, together with the due dates.

Job	Milling process time (days)	Grinding process time (days)	Due date
A	5	12	30
B	8	4	37
C	6	7	28

Assume that the present date is day zero.
(a) Use the earliest-due-date rule to schedule the jobs through the milling machine.
(b) On a piece of graph paper, devise a method of plotting the progress of the three jobs as they are each processed through the milling machine and the grinding machine.
(c) Determine the average manufacturing lead time and the aggregate job lateness for this schedule.

16.4. Solve Problem 16.3 except use the shortest-processing-time rule instead of the earliest-due-date rule.

16.5. Solve Problem 16.3 except use the least-slack rule instead of the earliest-due-date rule.

16.6. Solve Problem 16.3 except use the critical ratio to schedule the jobs instead of the earliest-due-date rule.

PART VII

Computer Control

chapter 17

Computer—Process Interfacing

17.1 INTRODUCTION

In this part of the book we consider how computers are used to control manufacturing processes. In this first chapter, the problem of interfacing the computer with the process is examined. In Chapter 18 we survey the various types of computer process control systems, such as direct digital control (DDC) and supervisory computer control. Chapter 19 is concerned with the technology of quality inspection and quality control by means of computers. Finally, Chapter 20 presents a good example of computer process control in manufacturing by examining computer-integrated manufacturing systems.

To be useful, the computer must be capable of communicating with its environment. In a data processing system, this communication is accomplished by the various input/output devices discussed in Chapter 2, such as card readers, printers, and CRT consoles. In computer-aided manufacturing, the environment of the computer includes not only these devices but also includes one or more manufacturing processes. Functioning as a process control system, the computer must be capable of sensing the important process variables from the operation and providing the necessary responses to maintain effective control over the process. In the following sections we examine some of the important components of a computer process control system.

17.2 MANUFACTURING PROCESS DATA

Let us begin by considering the nature of the data that must be communicated between the manufacturing process and the computer. These data can be classified into three categories:

1. Continuous analog signals
2. Discrete binary data
3. Pulse data

Continuous analog data can be represented by a variable that assumes a continuum of values over time. During the manufacturing process cycle, it remains uninterrupted, and the values it can assume are restricted to a finite range. Examples of analog variables include temperature, pressure, liquid flow rate, and velocity. Each of these phenomena are continuous functions over time and are capable of taking on an infinite number of values in a certain range. The number of values permitted is a function of the ability of measuring instruments to distinguish between different signal levels.

Discrete binary data can take on either of two possible values, such as on and off, or open and closed. Switches, motors, valves, and lights are all devices whose status at any time may be a binary function. Binary-valued process variables can represent many situations, such as:

The state of a machine tool (automatic or controlled cycle, operational or down, workpart in place or station empty, etc.).

The state of special sensing switches which permit the process to continue. For example, some industrial robot installations include several switch devices under a floor mat which close under the weight of an intruding worker. The closed state of the switch is interpreted as a hazard condition under which robot motion is stopped and alarm devices are activated.

Binary data are represented in electronic digital systems as typically one of two voltage levels whose values depend on the specific devices that make up the system. Typical voltage levels are 0 and +5V.

Pulse data consist of a train of pulsed electrical signals from devices called pulse generators. The pulse train can be used to drive devices such as stepping motors. The magnitude of each pulse is fixed; the magnitude of the pulse train is the number of pulses it contains. Since the number of pulses in a series can be counted over a period of time, that number can be represented as digital data. Conversely, digital data can be used to produce a pulse train of a given magnitude. The three general types of manufacturing process data are illustrated in Figure 17.1

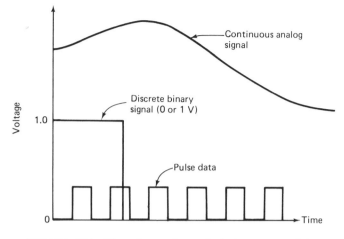

FIGURE 17.1   Three types of manufacturing process data.

17.3   SYSTEM INTERPRETATION OF PROCESS DATA

The three categories of manufacturing process data must be capable of interacting with the computer. For monitoring the process, input data must be entered into the computer. For controlling the process, output data must be generated by the computer and converted into signals understandable by the manufacturing process. There are six categories of computer–process interface representing the inputs and outputs for the three types of process data. These categories are:

1.   Analog to digital
2.   Contact input
3.   Pulse counters
4.   Digital to analog
5.   Contact output
6.   Pulse generators

The general configuration of the computer process interface is illustrated in Figure 17.2.

Analog-to-digital interfacing involves transforming real-valued signals into digital representations of their magnitude. A number of different steps must be accomplished to effect this conversion process. These steps involve the following hardware:

1.   Transducers, which convert a measurable process characteristic (flow rate, temperature, process, etc.) into electrical voltage levels corresponding in magnitude to the state of the characteristic of the pro-

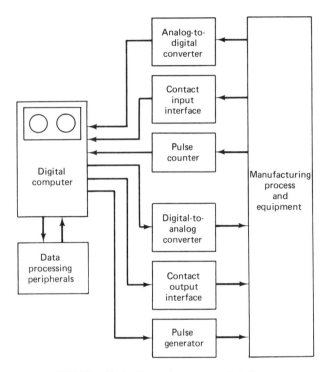

FIGURE 17.2 Computer–process interface.

cess measured. A thermocouple is an example of a device in this category. It converts temperature into a small voltage level.

2. Signal conditioners, which filter random electrical noise and smooth the analog signal emanating from transducing devices.

3. Multiplexers, which connect several process-monitoring devices to the analog-to-digital converter. Each process signal is sampled at periodic intervals and passed on to the converter.

4. Amplifiers scale the incoming signal up or down to the range of the analog-to-digital converter being used.

5. The analog-to-digital converter (ADC) transforms the incoming real-valued process signals into their digital equivalents.

6. The digital computer's I/O section accepts the digital signals from the ADC. A limit comparator is often connected between the I/O sections and the ADC to prevent out-of-limit signals from distracting the CPU.

The contact input interface is a set of simple contacts that can be opened and closed to indicate the status of limit switches, button positions, and other binary-type data. It serves as the intermediary between discrete process data and the computer, which periodically scans the signal status and compares it with prepro-grammed values.

Digital transducers belong to a class of electronic measuring instruments that generate as output a series of electrical pulses of uniform magnitude. A pulse counter is used to convert the pulse trains into a digital representation, which is then applied to the computer's input channel. The last three devices discussed are concerned with processing data from the computer to the process.

The digital-to-analog converter takes digital data generated by the computer and transforms it into a pseudoanalog signal. The signal is considered pseudoana-log because the computer is only capable of a limited-precision digital word, so that an infinite number of analog signal levels cannot be generated.

Contact output interfaces, like their input counterparts, are sets of contacts that can be opened or closed. The output word of the computer is used to turn on indicator lights, alarms, and even equipment functions such as cutting oil pumps.

Pulse generators are frequently used as output devices which convert digital words from the computer into pulse trains that are used to drive devices such as stepping motors.

17.4   INTERFACE HARDWARE DEVICES

Hardware components are required to accomplish the computer–process interface described in the preceding section. In this section we consider some of these hardware devices.

Transducers and sensors

A transducer converts one type of physical quantity into another which can be more conveniently used and evaluated. When transducers are used to measure the value of a physical quantity, they are referred to as sensors. There are two types of transducers: analog and digital.

An analog transducer generates a continuous signal, such as an electrical vol-tage or current, that can be interpreted as the value of a measured process variable. A calibration between the measured variable and the output signal is required to determine the exact nature of their correspondence. Signals generated by these transducers are classified as low-level (measured in millivolts) and high-level (greater than 1 V).

A digital transducer generates a digital signal in the form of a set of parallel bits or a series of countable pulses. The output of the transducer represents the quantity measured by it. Because of their compatibility with digital computers and the ease with which they can be read when used as stand-alone instruments, digital transducers are finding increased use in industrial applications.

Analog-to-digital converters

An analog-to-digital converter (ADC) is a device that transforms a continuous analog signal into digital form. The process of converting the analog signal into a digital value occurs in three phases:

1. The continuous signal is periodically sampled to form a series of discrete-time analog signals.
2. Each discrete analog value falls within one of a finite number of predefined amplitude levels called quantization levels, which consist of discrete voltages over the range of the device.
3. The amplitude levels are converted into digital form. This phase is sometimes called encoding.

In selecting an analog-to-digital converter for a particular manufacturing process application, consideration must be given to the operating characteristics of the ADC device. Among these characteristics are the following:

Sampling rate. This is the rate at which the analog signal is sampled. The higher this rate, the more closely approximated the signal will be. High sampling rates are also very important for multiplexed systems, in which all signals must be sampled in a short period of time. The upper limit on the sampling rate is imposed by the conversion time of the ADC.

Conversion time for an applied analog signal to be transformed into a digital output.

Resolution of the ADC. Resolution refers to the precision with which an analog signal can be represented in digital form. It is determined by the number of quantization levels, which depend on the number of bits used by the ADC. The number of quantization levels available is given by:

$$\text{number of quantization levels} = 2^N \tag{17.1}$$

where N is the number of bits used by the ADC. The resolution of the device would be the reciprocal of the number of quantization levels. The spacing between adjacent quantization levels would be the full-scale value for the incoming analog signal divided by the number of quantization levels. The incoming analog signal is typically amplified to provide a full scale of 0 to 10 V. The following example will illustrate these concepts of quantization levels, resolution, and spacing between quantization levels.

EXAMPLE 17.1

A continuous-voltage signal is to be converted to digital form by an ADC. The actual full-scale range of this voltage is zero to 2.0 V, but an amplifier is used to magnify this range to zero to 10 V. A 10-bit analog-to-digital converter will be used. It is desired to determine the number of quantization levels, the resolution, and the spacing between adjacent quantization levels.

According to Eq. (17.1), the number of quantization levels is

$$2^{10} = 1024 \text{ levels}$$

The resolution would be the reciprocal of the number of quantization levels:

$$\text{resolution} = (1024)^{-1} = 0.00097 = 0.097\%$$

Finally, the spacing between adjacent quantization levels is the full-scale value of 10 V, divided by the number of quantization levels:

$$\text{spacing} = \frac{10 \text{ V}}{1024} = 0.00977 \text{ V}$$

There are a variety of methods used to convert an analog signal into digital form. Two methods that we will describe are:

1. Successive approximation method
2. Integrating ADC method

In the successive approximation method, trial voltages are compared to the incoming analog signal. Special circuitry yields a "1" when the input voltage exceeds the trial voltage, and a "0" when the input voltage is less than the trial voltage. After each successive iteration, the trial voltage is decremented to half of its previous value. For 6-bit precision conversion time is typically 9 μs for this method.

EXAMPLE 17.2

It is desired to use a successive-approximation-type ADC to encode an incoming signal whose full scale value is 10 V. Let us suppose that during the moment of conversion the signal value is 6.8 V. Our ADC has 6-bit precision. The process of converting the signal is illustrated in Figure 17.3.

The first trial voltage is 5.0 V. Comparing the input signal to this value yields a "1" since the input exceeds the trail voltage. Accordingly, the difference of 6.8 V − 5.0 V = 1.8 V is compared with the next trail voltage of 2.5 V (one-half the initital trail voltage). In this comparison, a "0" results because the 1.8 V is less than 2.5 V. This successive approximation procedure is repeated for all 6 bits. The resulting value is 6.718 V. The error between this value and the actual input signal of 6.8 V illustrates the limitation imposed by a 6-bit ADC.

The integrating ADC method is based on converting the applied analog signal into measurable time periods. In the simplest form, the single-slope converter, the input signal is compared to a linearly increasing voltage (a ramp function), and the time required to reach this voltage is measured. This time is proportional to the level of the applied signal. The resolution of this method depends on the frequency

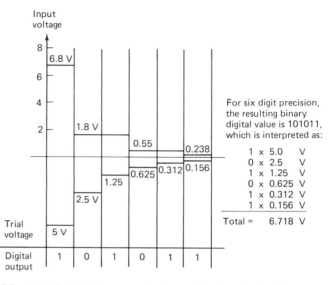

FIGURE 17.3   Successive-approximation-method analog-to-digital conversion.

of the time-counting pulse train and the slope of the ramp voltage. Highest precision is made possible by high-frequency pulse trains and low ramp slopes. Integrating ADC is slower than successive approximation, with a typical 6-bit resolution conversion taking about 14 μs. However, it tends to be more accurate in noisy environments because it is less susceptible to random electrical noise. Random noise tends to be both positive and negative, and therefore its integration tends toward zero.

Digital-to-analog converters

The digital-to-analog converter (DAC) performs the reverse function of the ADC. It receives data from the computer and generates analog voltages which can be used to drive analog devices such as data recorders and plotters. In a sense, the DAC is the computer's communication link back to the process it is measuring.

Digital-to-analog conversion takes place in two steps. First, the digital data are converted into their analog equivalent. This is called decoding. It is accomplished by a binary register in the DAC which controls the output of a reference voltage source depending on the number of bits contained in the register. Each bit controls one-half the voltage of the preceding bit. The voltage produced is determined by

$$V_o = V_{ref}[0.5B_1 + 0.25B_2 + 0.125B_3 + \cdots + (2^n)^{-1}B_n] \qquad (17.2)$$

where
$$V_o = \text{output of the decoding operation}$$
$$V_{\text{ref}} = \text{reference voltage source which equals full scale output.}$$
$$B_1, \ B_2, \ \ldots, \ B_n = \text{status (zero or 1) of successive bits in the register}$$

In the second step of the conversion, a data-holding device converts the sampled data from the binary register to an approximation of the continuous analog signal. It attempts to approximate the ideal envelope formed by the sampled data. The order of the extrapolation carried out to approximate the envelope characterizes the DAC device. A zero-order data hold approximates the continuous function with a series of step voltages derived from the binary register. In first-order approximation, the voltage changes with constant slope between sampling instants, sometimes producing a more accurate envelope. Operation of the zero-order and first-order holds is illustrated in Figure 17.4.

With relation to the output of the decoding operation V_o as defined by Eq. (17.2), we can mathematically express the result of the data-holding operation for the zero-order approximation as follows:

$$V(t) = V_o \qquad (17.3)$$

where $V(t)$ would be maintained at a constant voltage level during the sampling period.

For the first-order hold, the voltage output during the sampling interval can be expressed as:

$$V(t) = V_o + at \qquad (17.4)$$

$$a = \frac{V_o - V_o(-\tau)}{\tau} \qquad (17.5)$$

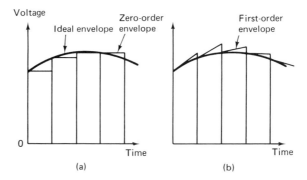

FIGURE 17.4 Data-hold function: (a) zero-order; (b) first-order hold.

where

$$a = \text{rate of change of } V(t)$$
$$\tau = \text{time interval between sampling instants}$$
$$V_o(-\tau) = \text{value of } V_o \text{ at the preceding sampling instant}$$

Operation of a DAC device will be illustrated by means of the following two examples.

EXAMPLE 17.3

A digital-to-analog converter uses a reference voltage source of 10 V and uses a binary register with six-digit precision. In two successive sampling instants 0.5 s apart, the data contained in the binary register are as follows:

Sampling	Binary data					
instant	B_1	B_2	B_3	B_4	B_5	B_6
$t-1$	1	0	1	0	1	0
t	1	0	1	1	0	1

The DAC uses a zero-order hold to maintain the voltage output between sampling instants. Determine the voltage output during the sampling interval following instant t.

The result of the decoding operation can be determined from Eq. (17.2) for the values of $B_1, B_2, \ldots, B_n$. For interval t,

$$V_o = 10[0.5(1) + 0.25(0) + 0.125(1) + 0.0625(1) + 0.03125(0) + 0.015625(1)]$$
$$= 7.03125 \text{ V}$$

The operation of the data-hold device, as defined by Eq. (17.3) would be to maintain this voltage level for the duration of the sampling interval.

EXAMPLE 17.4

Using the same data as given in Example 17.3, determine the voltage level during the sampling interval for a DAC with a first-order hold.

To solve this problem, the voltage output of the decoding operation must be determined for instant $t-1$ as well as instant t. At instant $t-1$,

$$V_o = 10[0.5(1) + 0.25(0) + 0.125(1) + 0.0625(0) + 0.03125(1) + 0.015625(0)]$$
$$= 6.5625 \text{ V}$$

At instant t, the value of V_o was calculated in Example 17.3 to be 7.03125 V.

To determine the output of the first-order data-hold device, Eqs. (17.4) and (17.5) are used. The rate of change must first be computed according to Eq. (17.5):

$$a = \frac{7.03125 - 6.5625}{0.5} = 0.9375$$

Next, the voltage output during the time interval following instant t is given by Eq. (17.4):

$$V(t) = 7.03125 + 0.9375t$$

At the beginning of the interval the voltage would be 7.03125 V. The voltage would steadily increase until its value at the end of the sampling interval would be $7.03125 + 0.9375(0.5)$ = 7.5 V.

Multiplexers

The multiplexer is a switching device connected in series with each input channel from the process. It is used to time-share the analog-to-digital converter (and associated amplifiers) among the incoming signals. The alternative to the use of the multiplexer would be to have a separate ADC for each transducer. This would be expensive for a large installation with many inputs to the computer. Since the process variables need only be sampled periodically anyway, the multiplexer provides a very cost effective method of satisfying system design requirements.

17.5   DIGITAL INPUT/OUTPUT PROCESSING

The computer program brings together the many aspects that make up computer monitoring and control of industrial processes. The control software makes use of the computer's I/O channels. The ADC devices read data from the transducers and the DACs control the servomechanisms that effect change in the processing system. Programming for computer control differs from the conventional programming in that it must deal with:

Timer-initiated events such as the regular sampling of data from process monitoring devices

Process-initiated interrupts which cause special routines to be executed to alleviate an abnormal condition

Passing commands to real-time processes

System- and program-initiated events related to the computer system, such as the transfer of data between computers

Operator-initiated events, such as status or data printout requests, startup or halt instructions, or changes in existing programs

The following sections describe how the process control computer deals with these events. An important feature in control programming is the interrupt system.

Interrupt system

Computers which are programmed to perform control functions are equipped with an interrupt logic system. The purpose of this system is to permit program control to be switched between different programs and subroutines in response to requests by devices in the operating environment. Interrupts can, for example, signal that a work station is empty, and cause a new part to be moved into place.

External interrupts are initiated by events external to the computer, such as the operator or process devices. Internal interrupts are generated in response to an internal timer or other system program events.

Interrupts are arranged in priority structures that allow critical functions to be carried out before routine tasks. The higher the priority of a given task, the greater the number of lower-importance tasks that it can overrule. Typical priority-control assignments are made as indicated in the following table.

Priority level	Assignment
1 (lowest)	Operator-initiated interrupts
2	System-generated interrupts
3	Timer interrupts
4	Process command passing
5 (highest)	Process-generated interrupts

Control programming

Most control programming is done in either assembly language or procedure-oriented languages such as FORTRAN. Assembly language instructions are more suited to the elementary operations required in process control. The programs they make up are more efficient and faster-executing, a feature that becomes critical when high repetitive calculations are to be performed. The advantage in using FORTRAN is reduced development time. Also, many engineers and scientists are already familiar with it, in one form or another, and would probably be resistant to learning a new language for such a specific application. The following example illustrates some of the features of a control program used to monitor process variables.

EXAMPLE 17.5

The program is written in FORTRAN and run on a DEC (Digital Equipment Corp.) PDP 11/34. The purpose of the program is to collect data on cutting temperature during a turning operation. A tool-chip thermocouple is used as the sensor. It uses the tool and the work material as the two dissimilar metals of the thermocouple. The millivolt output of the thermocouple is fed through an analog-to-digital converter into the computer. The thermocouple output is sampled four times per second for 5 seconds. Then the 20 mV and temperature readings are printed out on a teletype, together with the averages of the 20 readings.

    The program is shown in Figure 17.5 and a typical printout of results is displayed in Figure 17.6. The reader who is familiar with FORTRAN will recognize the structure and syntax of the statements. After the initializing statements (lines 1 through 8), the CALL ASICLN (LUN) in line 9 identifies the input unit assigned to the ADC. Literally, ASICLN(LUN) means "assign industrial controller to (logical unit number)." In line 10,

```
0001            PROGRAM L4
0002            DIMENSION ISTAT(2),IHOLD(1),ICONT(1)
0003            DIMENSION V(300)
0004            LOGICAL *1 DUMMY
0005            DATA ICONT/"140003/, LUN/1/
0006            DATA GAIN/200./
0007            J=20
0008            SUM = 0.
     C** ASSIGN A LOGICAL DEVICE NUMBER TO THE ICS UNIT
0009            CALL ASICLN(LUN)
     C** CALCULATE THE VOLTAGE CONVERSION FACTOR TO USE
0010            FACTOR = 10.24 / (32768.*GAIN)
0011            TYPE 1000
0012    1000    FORMAT (' PUT THE TOOL AGAINST THE WORK, WHEN DONE HIT <CR>',$)
0013            ACCEPT 1001,DUMMY
0014    1001    FORMAT (A1)
0015            CALL AIRD (1,ICONT,IHOLD,ISTAT,LUN)
0016    1       CONTINUE
0017            IF (ISTAT(1).EQ.0) GO TO 1
0019            OFFSET = FLOAT (IHOLD(1)) * FACTOR
0020            TYPE 1010,OFFSET
0021    1010    FORMAT (' THE OFFSET VOLTAGE IS ',F10.5)
0022            TYPE 1025
0023            ACCEPT 1001,DUMMY
0024    1025    FORMAT (' WHEN YOU WANT TO START READINGS .HIT <CR>',$)
0025            DO 10 I=1,J
     C** INITIATE A VOLTAGE READING
0026            CALL AIRD (1,ICONT,IHOLD,ISTAT,LUN)
     C** WAIT UNTIL YOU HAVE  COMPLETED THE VOLTAGE READING
0027    49      IF (ISTAT(1).EQ.0) GO TO 49
     C** CALCULATE THE VOLTAGE
0029            V(I) = FLOAT (IHOLD(1)) * FACTOR-OFFSET
0030            CALL WAIT(15,1)
0031            SUM =SUM + V(I)
     C** IF J READINGS HAVE BEEN TAKEN, DROP THROUGH, ELSE, GO TO THE
     C**    BEGINNING OF THE LOOP
0032    10      CONTINUE
0033            DO 20 I=1,J
0034            TYPE 1021,I,V(I)* 1000,TEMP(V(I))
0035    1021    FORMAT (' READING ',I3,' IS ',F10.2, ' MV',2X,F6.1,2X,'DEG. F')
0036    20      CONTINUE
0037            AVG = SUM / 20.
0038            TYPE 1022,AVG*1000.,TEMP(AVG)
0039    1022    FORMAT(/,1X'THE AVG READING = ',F10.5 ,' MV',5X,F6.1,2X,'DEGREES')
0040            END


0001            FUNCTION TEMP(A)
0002            TEMP=-4 + 67.097*1000.*A + .9612 * ((A*1000.)**2.)
0003            RETURN
0004            END
```

FIGURE 17.5 Process monitoring program of Example 17.5.

```
>RUN L4
PUT THE TOOL AGAINST THE WORK, WHEN DONE HIT <CR>
THE OFFSET VOLTAGE IS    0.00000
WHEN YOU WANT TO START READINGS HIT <CR>

       READING    1 IS     12.88 MV    1019.2  DEG. F
       READING    2 IS     12.93 MV    1023.8  DEG. F
       READING    3 IS     12.95 MV    1026.1  DEG. F
       READING    4 IS     12.97 MV    1028.4  DEG. F
       READING    5 IS     12.95 MV    1026.1  DEG. F
       READING    6 IS     12.95 MV    1026.1  DEG. F
       READING    7 IS     13.00 MV    1030.7  DEG. F
       READING    8 IS     12.95 MV    1026.1  DEG. F
       READING    9 IS     13.05 MV    1035.3  DEG. F
       READING   10 IS     13.07 MV    1037.6  DEG. F
       READING   11 IS     13.02 MV    1033.0  DEG. F
       READING   12 IS     13.02 MV    1033.0  DEG. F
       READING   13 IS     13.05 MV    1035.3  DEG. F
       READING   14 IS     13.07 MV    1037.6  DEG. F
       READING   15 IS     13.10 MV    1039.9  DEG. F
       READING   16 IS     12.88 MV    1019.2  DEG. F
       READING   17 IS     13.17 MV    1046.8  DEG. F
       READING   18 IS     13.13 MV    1042.2  DEG. F
       READING   19 IS     13.13 MV    1042.2  DEG. F
       READING   20 IS     13.15 MV    1044.5  DEG. F

       THE AVG READING =    13.02125 MV      1032.7  DEGREES
       TT2  --  STOP
       >
```

FIGURE 17.6 Output of process monitoring program of Example 17.5.

the program calculates the scale factor to use for the analog-to-digital converter (refer back to the discussion on ADCs in Section 17.4).

In lines 11 and 12, the program instructs the machine tool operator to "put the tool against the work, when done hit carriage return." The statement in line 15,

$$CALL \ AIRD \ (1,ICONT,IHOLD,ISTAT,LUN)$$

commands the computer to accept the analog input from the ADC. This provides a zero set point for subsequent measurements taken during the cut. In lines 20 and 21, this value of the "offset voltage" is programmed to be printed.

The program then instructs the operator to start taking measurements with the statement: "When you want to start readings, hit carriage return." Lines 25 through 32 comprise a DO loop which calls for the analog input (CALL AIRD) to be read and stored in a storage location called IHOLD. This is done for 20 cycles ($J = 20$ in line 7). The readings are taken at ¼-s intervals. This is defined by the statement in line 30,

$$CALL \ WAIT \ (15,0)$$

This command makes the computer wait for 15 "ticks" before accepting the input. A "tick" in this computer system is $\frac{1}{60}$ s. Hence, 15 ticks equals ¼ s.

The remainder of the program is for printout of the data. Lines 33 to 36 print the 20 values of millivolt and corresponding cutting temperature values. The symbol V(I) is used for the 20 values of voltage. V(I) multiplied by 1000 converts the voltage measurements to millivoltage for printout. The term TEMP(V(I)) calls a function for converting the voltage output of the tool-chip thermocouple into the corresponding temperature. The conversion is based on an equation that was derived from the calibration curve for the particular tool work material combination. In this case, the calibration equation is

$$\text{TEMP} = -4 + 67.097 \, \text{MV} + .9612(\text{MV})^2$$

where TEMP is the calculated cutting temperature and MV is thermocouple output. The function that performs this calculation is shown in Figure 17.5.

Lines 38 and 39 print the average values of millivolt and cutting temperature.

Figure 17.6 illustrates the output from the program during one cut of the turning operation. The teletype is located at the lathe, while the PDP 11/34 is remotely located in a separate room.

17.6 HIERARCHICAL COMPUTER STRUCTURES AND NETWORKING

Within some corporations, partially due to developments in computer-aided manufacturing, computer systems have grown to form pyramidlike structures in which computers at the industrial process level are linked to larger, more centralized systems, all the way up to the corporate computer.

In these heirarchical structures, information flows in two directions. Process monitoring data, production status information, and other operational data are passed upward through each level of computer system, and at each step, bulk information is filtered and integrated for more efficient use. In the reverse direction, commands and schedules are passed down from the corporate computer throughout the rest of the processing structure.

In this section we discuss the various functions performed at the different levels of the computer hierarchy. We also discuss the communication networks that connect these different levels.

Levels of computer hierarchy

Each level in the computer hierarchy exerts a certain amount of control over the activities performed in a manufacturing firm. The general structure of the system is depicted in Figure 17.7. The functions of these levels of control are described below.

In the first level of computer control, computers are connected directly to the manufacturing process. They are characteristically small, usually minicomputers or microcomputers. These computers are dedicated to process control and communicate only with systems in the second level of control.

Minicomputers make up the majority of systems in the second level of computer control. Usually, several minicomputers report to a larger plant computer.

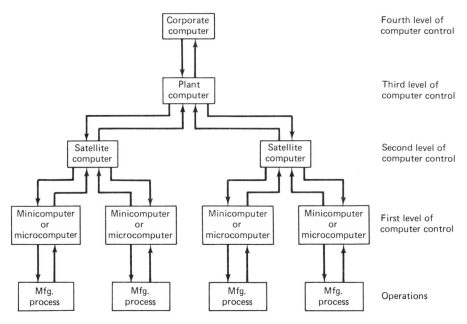

FIGURE 17.7   Hierarchy of computers in manufacturing.

The minicomputers are often referred to as satellites of the larger computer. The minis serve to coordinate the activities of the smaller computers under them. Performance data from machine tools, for example, are collected from the first-level computers, and instructions are relayed back down the line to each process. This mode of operation is known as supervisory computer control, and is the basis for direct numerical control machining.

At the third level of control are the central plant computers, which collect data from the minicomputers beneath them in the structure, and prepare daily, weekly, and monthly reports using various information. Operating instructions are relayed to the second control level for implementation. Computers at this level are larger machines, and they are usually shared by many departments.

The corporate computer resides at the fourth level of control. Different plants report information to these computers, which then summarize the data and analyze plant operations and performance over a long-term period. The corporate computer is shared by departments such as marketing, engineering, and accounting.

Several important advantages can be claimed for the hierarchical computer structure in process control:

Control can be established gradually, with each application of computers justified on its own merits. An all-or-nothing commitment to an extensive computer system is not required, nor is the large expense of implementing such a system.

Redundancy in the hierarchical structure allows other computers to assume the functions of those which occasionally fail.

Software development in these structures is made more manageable in that programming projects can be undertaken independent of one another.

Computer network structures

The term "computer network" refers to the actual physical connections between computers in the structure. This structure consists not only of the computers, but also the terminals, transmission lines, switching centers, and other devices required to accomplish the desired functions of the system. The number of possible configurations is limitless. However, the various network arrangements divide themselves into several basic categories, sometimes called topologies. Four of the more common topologies are:

1. Point-to-point
2. Star network
3. Multidrop network
4. Loop configuration

These four network topologies are illustrated in Figure 17.8 and described in the paragraphs below.

In the point-to-point configuration, a communication line is established between two devices (typically, a computer and a peripheral device such as a terminal). It is a relatively simple configuration if the number of devices is limited. However, when there are a large number of processors in the structure, the number of point-to-point communication links can become significantly large, resulting in a relatively expensive system. Other network structures share the communication links in various ways to avoid this expense. An inherent advantage of the point-to-point configuration is that whenever one of the direct links is broken, the other connections remain intact.

The star network consists of a central computer and several peripheral devices connected to it. The central computer is referred to as the master and the peripherals are called slaves. The function of the master is to delegate tasks to each of the smaller units or to act as a switch to connect various units. One of the important deficiencies of the star configuration is that the entire network system is put out of operation if the central computer fails.

In the multidrop network structure, computers and other devices are connected by a single communication line. This line is either a bus or a serial channel. The limitation of this arrangement is that only two processors are allowed to transmit data to each other at the same time. This results from the shared nature of the communication lines.

In a loop configuration, each computer is connected to a communication line

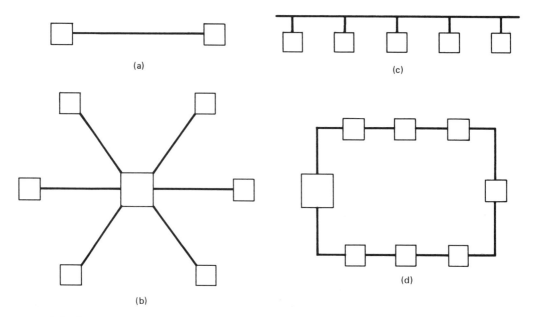

FIGURE 17.8   Types of computer network structures: (a) point-to-point; (b) star; (c) multidrop; (d) loop.

which begins and terminates at a loop controller, typically a computer that maintains control over intercomputer communications. Messages to a specific computer are usually transmitted to all computers in the network, with the addressee decoding its address and accepting the information.

REFERENCES

[1]   CADZOW, J. A., and MARTENS, H. R., *Discrete-Time and Computer Control Systems*, Prentice-Hall, Inc., Englewood Cliffs, N.J., 1970.

[2]   ENG, E., "Peripheral Equipment for the Computer," *Electronic Engineering*, November, 1978, pp. 107−116.

[3]   GROOVER, M. P.,   *Automation, Production System, and Computer-Aided Manufacturing*, Prentice-Hall, Inc., Englewood Cliffs, N.J., 1980 Chapter 11.

[4]   HARRISON, T. J. (Ed.), *Minicomputers in Industrial Control*, Instrument Society of America Pittsburgh, Pa., 1978.

[5]   OFFORD, G.,   Personal communications and special reports, February/March, 1982.

[6]   REMBOLD, U., SETH, M. K., and WEINSTEIN, J. S., *Computers in Manufacturing*, Marcel Dekker, Inc., New York, 1977.

[7]   SCHAFFER, G., "Computers in Manufacturing," Special Report 703, *American Machinist*, April, 1978, pp. 115−130.

PROBLEMS

17.1. An analog-to-digital converter is to be used to convert a continuous voltage signal into digital form. The full scale range is 110 V. An 8-bit ADC will be used. Determine the number of quantization levels, the resolution, and the spacing between adjacent quantization levels.

17.2. The selection of an analog-to-digital converter includes determining the number of bits required to achieve a certain level of precision in the converted signal. The incoming voltage signal has a full-scale range of 200 V, and it is desired to read that voltage to a spacing between adjacent quantization levels of 0.2 V. Determine the number of bits needed to provide this precision.

17.3. A successive approximation ADC is used to convert a 35.6-V signal into its equivalent digital value. A starting trial voltage of 80 V is used and the ADC possesses 8 bits of precision to accomplish the successive approximations. Construct a chart similar to that shown in Figure 17.3 and determine the final digital value as provided by the ADC.

17.4. A digital-to-analog converter uses a reference voltage of 100 V and its binary register contains eight-digit accuracy. At a particular sampling instant, the bit values in the register are as follows:

B_1	B_2	B_3	B_4	B_5	B_6	B_7	B_8
0	1	1	0	1	1	0	1

If a zero-order data-hold device is used, determine the value of the voltage during the time interval following this sampling instant.

17.5. A certain DAC uses a reference voltage of 100 V and an 8-bit binary register. In two successive sampling instants, the bit values in the register are as follows:

Sampling Instant	Binary register bit values							
	B_1	B_2	B_3	B_4	B_5	B_6	B_7	B_8
$t-1$	0	1	1	0	1	1	0	1
t	0	1	1	0	1	0	1	0

If a first-order hold is used in this DAC, determine the equation of the form of Eq. (17.4) which controls the voltage level during the interval following sampling instant t.

17.6. You have been asked to supervise a special engineering group whose function is to design and install a computer-aided manufacturing system for your company. The company consists of a corporate headquarters and two separate manufacturing plants located 25 miles apart. One plant manufactures in batches the basic components for the company's products. It has 26 numerical control machine tools. The other plant is an assembly plant, which is also a warehouse and distribution center. In the style of a memorandum addressed to the company president, describe the hierarchical computer structure you would recommend. Allow for differences between the two plants. Define the type of information that would be communicated among the various levels of the computer heirarchy. What factors affect your recommendation? Sketch a diagram in the format of Figure 17.7 showing your recommended computer system.

chapter 18

Computer Process Control

18.1 INTRODUCTION

The field of computer process control is closely associated with the process industries, such as chemicals production, oil refinery operations, and steelmaking. Many of the developments in this area have resulted from the collaborations of chemical engineers and electronics specialists. In this chapter we borrow from their accomplishments to present a survey of process control, with emphasis on how the digital computer is used in this area. In the discussion that follows we build on the presentation of Chapter 17 by assuming that the computer can be successfully interfaced to the manufacturing process.

In recent years, the use of the digital computer in process monitoring and control applications has expanded to include many production areas outside the process industries. One of the important growth areas has been in discrete-parts manufacturing: metal machining, pressworking, electronic components manufacturing, assembly, and so on. Many of these applications are limited to computer monitoring duties (see Chapter 16), but in other cases the potential has been demonstrated for using the computer for process control [1,7].

The characteristics of the product processes are different in discrete-parts manufacturing than in the process industries. In discrete-item production, the output is measured in number of parts rather than in gallons or tons. The operations are typically less complex with fewer variables. The processing time is usually of short duration compared to the process industries. Along with the differences there are similarities between the processes in discrete-parts manufacturing and in continuous-process industries. In both cases, there are economic objectives to be achieved. Many manufacturing operations are complex enough that manual optimization methods are inadequate. For example, the metal-cutting operation is characterized by a fairly large number of process variables which determine the overall performance of the operation. Although the value added per operation in discrete-item production is less than in continuous processing, the total volume of production in many manufacturing plants is significant enough and the productivity low enough to justify an investment in computer control.

Two factors contribute to the appeal of the computer in discrete-parts manufacturing. First, the physical size and cost of the digital computer have decreased dramatically in recent years. Today, it is not as difficult to justify the investment in computer hardware for process control as it was 10 to 15 years ago. The development of microprocessors has drastically altered the economics and increased the opportunities in computer process control. Second, experience in developing the software for process control has grown, thus reducing the programming costs for new process control projects. In companies with large numbers of machine tools, the same process control software can be adapted to more than one machine, so that the cost of software development per machine becomes affordable. The application of computer numerical control (CNC) is a good example of the software development economies that are possible in the metalworking industry.

18.2   STRUCTURAL MODEL OF A MANUFACTURING PROCESS

Let us begin our study of computer process control by examining the basic structure of a manufacturing process. Most production operations are characterized by a multiplicity of dynamically interacting process variables. These variables can be cataloged into two basic types, input and output variables. However, there are different kinds of input variables and different kinds of output variables. Let us first consider how the input variables might be classified. There are three categories as follows:

1. *Controllable input variables.* These are sometimes called *manipulative variables*, because they can be changed or controlled during the process. In a machining operation, it is technologically possible to make adjustments in speed and feed during the operation. In a chemical process, the

controllable input variables may include flow rates, temperature settings, and other analog variables.

2. *Uncontrollable input variables.* Variables that change during the operation but which cannot be manipulated are defined as uncontrollable input variables. In chemical processing, variations in the starting raw chemicals may be an uncontrollable input variable for which compensation must be made during the process. In machining, examples would be tool sharpness, work-material hardness, and workpiece geometry.

3. *Fixed variables.* A third category of input to the process is the fixed variable. These are conditions of the setup, such as tool geometry and workholding device, which can be changed between operations but not during the operation. Fixed inputs for a continuous chemical process would be tank size, number of trays in a distillation column, and other factors that are established by the equipment configuration.

The other major type of variable in a manufacturing process is the output variable. It is convenient to divide output variables into two types:

1. *Measurable output variables.* The defining characteristic of this first type is that it can be measured on-line during the process. Examples of variables that can be measured during process operation include flow rate, temperature, vibration, voltage, and power.

2. *Performance evaluation variables.* These are the measures of overall process performance and are usually linked to either the economics of the process or the quality of the product manufactured. Examples of performance evaluation variables in production include unit cost, production rate, yield of good product, and quality level.

The structural relationships between these different input and output variables are illustrated in Figure 18.1. The measurable output variables are determined by the input variables. The performance of the process, as indicated by the performance evaluation variable, is determined by the measurable output variables. To assess process performance, the performance evaluation variable must be calculated from measurements taken on the output variables.

The problem in process control is to control the measurable output variables so as to achieve some desired result in the performance evaluation variable. This is accomplished by manipulating the controllable inputs to the process.

There are various ways to implement computer process control of manufacturing operations, both in terms of hardware configurations and in terms of software programs. Consideration of hardware configurations includes the number and types of computers and how they are interconnected. Software programming is concerned with selecting among the available control strategies to regulate or optimize process performance. Let us first discuss the various process control strategies.

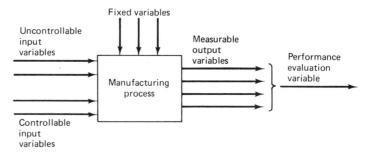

FIGURE 18.1 Structure of a manufacturing process.

18.3 PROCESS CONTROL STRATEGIES

There are a variety of control strategies that can be employed in process control. The choice of strategy depends on the process and the performance objectives to be achieved. In this section we discuss the following control strategies:

1. Feedback control
2. Regulatory control
3. Feedforward control
4. Preplanned control
5. Steady-state optimal control
6. Adaptive control

Feedback control

In the manufacturing process model shown in Figure 18.1, it is often feasible from a controls viewpoint to relate the behavior of a particular output variable to one corresponding input variable. This matching of one input variable to one output variable forms a single open-loop system. By measuring the output variable and comparing it to the input variable, it is possible to close the control loop, thereby forming an automatic feedback control system. This arrangement is illustrated schematically in Figure 18.2. In the conventional concept of a feedback control system, the value of the controlled variable is subtracted from the value of the input variable and any difference between them is used to drive the controlled variable toward its desired value. In Figure 18.2, y is the controlled variable and x is the input variable. In process control applications, the input variable is often referred to as the set point. The difference between the set point and the measured y value is called the error and becomes the input to the process controller. The manner in which the process controller is designed to operate depends on the physical nature of the process and the feedback measurement device. Consideration of

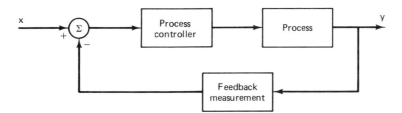

FIGURE 18.2 Feedback control system.

process control theory goes beyond the scope of the current chapter and the interested reader may want to explore this subject in Ref. [3], [9], and [10]. The feedback control system is also called a closed-loop system because the block diagram takes on the general appearance of a closed loop. By contrast, an open-loop system is one without feedback.

Regulatory control

Regulatory control is analogous to feedback control except that the objective in regulatory control is to maintain the overall performance evaluation variable at a certain set-point level. In feedback control, the objective is to control the individual output variables at their respective set-point values.

   In many industrial processes it is sufficient to maintain the performance evaluation variable at a certain level or within a given tolerance band of that level. This would be appropriate in situations where performance was measured in terms of product quality and it was desired to maintain the product quality at a particular level. In a chemical process, this quality level might be the concentration of the final chemical product. The purpose of process control is to maintain that quality at the desired constant value during the process. To accomplish this purpose, set points would be determined for individual feedback loops in the process and other control actions would be taken to compensate for disturbances to the process.

Feedforward control

The trouble with regulatory control (the same problem is present with feedback control) is that compensating action is taken only after a disturbance has affected the process output. An error must be present in order for any control action to be initiated, but this means that the output of the process is different from the desired value.

   In feedforward control the disturbances are measured before they have upset the process, and anticipatory corrective action is taken. In the ideal case, the corrective action compensates completely for the disturbance, thus preventing any deviation from the desired output value. If this ideal can be reached, feedforward control represents an improvement over feedback control.

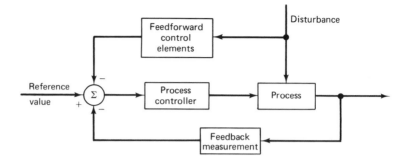

FIGURE 18.3   Feedforward control system (combined with feedback control.)

The essential features of a feedforward control system are illustrated in Figure 18.3. The feedforward control concept can be applied to the individual measurable output variables in the process or to the performance evaluation variable for the entire process. The disturbance is measured and serves as the input to the feedforward control elements. These elements compute the necessary corrective action to anticipate the effect of the disturbance on the process. To make this computation, the feedforward controller contains a mathematical or logical model of the process which includes the effect of the disturbance. Feedforward control by itself does not include any mechanism for checking that the output is maintained at the desired level. For this reason, feedforward control is usually combined with feedback control, as illustrated in Figure 18.3. The feedforward loop is especially helpful when the process is characterized by long "response times" or "dead times" between inputs and outputs. Feedback control alone would be unable to make timely corrections to the process.

Preplanned control

The term *preplanned control* refers to the use of the computer for directing the process or equipment to carry out a predetermined series of operation steps. The control sequence must be developed in advance to cover the variety of processing conditions that might be encountered. This control strategy usually requires the use of feedback control loops to make certain that each step in the operation sequence is completed before proceeding to the next step. However, feedback information may not be necessary in every control command provided by the computer.

The name "preplanned control" is not universally applied throughout all areas of industry. Other terms are used to describe control strategies which are either identical or similar to preplanned control. What follows is a listing of some of the terms most frequently used.

COMPUTER NUMERICAL CONTROL: CNC was described in Chapter 9. Essentially, it involves the use of the computer to direct a machine tool through a program of processing steps. As such, it is a form of preplanned control. Direct numerical control (DNC), although not the same as CNC, involves a similar control sequence.

PROGRAM CONTROL: This term is used in the process industries. It involves the application of the computer to start up or shut down a large complex process, or to guide the process through a changeover from one product grade to another. It also refers to the computer's use in batch processing to direct the process through the cycle of processing steps. With program control the object is to direct the process from one operating condition to a new operating condition and to accomplish this in minimum time. There are often constraints on this minimum time objective, so the strategy of program control is to determine the best trajectory of set-point values that is compatible with the constraints. In batch processing, there may be a sequence of operating conditions or states through which the process must be commanded.

The paper industry provides an example of program control. In the manufacture of various grades of paper, a slightly different operating cycle is required for each grade. The process control computer is programmed to govern the process through each phase of the operating cycle for any grade of paper produced.

SEQUENCING CONTROL: This class of preplanned control consists of guiding the process through a sequence of on/off-type steps. The variables under command of the computer can take on either of two states, typically "on" or "off." In sequencing control, the process must be monitored to make sure that each step has been carried out before proceeding to the next step.

An example of the application of sequencing control is in automated production flow lines. The sequence of workstation power feed motions, parts transfer, quality inspections which may be incorporated in the line, and so on, are all included under computer control. In addition, the computer may be programmed to perform diagnostic subroutines in the event of a line failure, to help identify the cause of the downtime occurrence. Tool change schedules may also be included as one of the computer functions. The operators are directed by the computer when to change cutters.

Steady-state optimal control

The term "optimal control" refers to a large class of control problems. We shall limit its meaning in this discussion to open-loop systems. That is, there is no feedback of information concerning the output. Instead, two features of the system must be known in advance:

1. *Performance evaluation variable.* This measure of system performance is also called the objective function, index of performance, or figure of merit. Basically, it represents the overall indicator of process performance that we desire to optimize by solving the optimal control problem. Among the performance objectives typically used in optimal control are cost minimization, profit maximi-

zation, production-rate maximization, quality optimization, least-squares-error minimization, and process-yield maximization. These objectives are general and must be specified to suit the particular application.

2. *Mathematical model of the process.* The relationships between the input variables and the measure of process performance must be mathematically defined. The model is assumed to be valid throughout the operation of the process. That is, there are no disturbances that might affect the final result of the optimization procedure. This is why we refer to the problem as steady-state optimal control. The mathematical model of the process may include constraints on some or all of the variables. These constraints limit the allowable region within which the objective function can be optimized.

With these two attributes of the process defined, the solution of the optimal control problem consists of determining the values of the input variables that optimize the objective function. To accomplish this task, a great variety of optimization techniques are available to solve the steady-state optimal control problem. These techniques include differential calculus, linear programming, dynamic programming, and the calculus of variations (13). All of these mathematical approaches have been applied to the class of problems in this category of steady-state optimal control.

Adaptive control

Adaptive control possesses attributes of both feedback control and optimal control. Like a feedback system, measurements are taken on certain process variables. Like an optimal system, an overall measure of performance is used. In adaptive control, this measure is called the index of performance (IP). The feature that distinguishes adaptive control from the other two types is that an adaptive system is designed to operate in a time-varying environment. It is not unusual for a system to exist in an environment that changes over the course of time. If the internal parameters or mechanisms of the system are fixed, as is the case in a feedback control system, the system might operate quite differently in one environment than it would in another. An adaptive control system is designed to compensate for the changing environment by monitoring its performance and altering, accordingly, some aspect of its control mechanism to achieve optimal or near-optimal performance. The term "environment" is used in a most general way and may refer to the normal operation of the process. For example, in a manufacturing process, the changing environment may simply mean the day-to-day variations that occur in tooling, raw materials, air temperature, and humidity (if these have any influence on the process operation). An adaptive system differs from a feedback system or an optimal system in that it is provided with the capability to cope with this time-varying environment. The feedback and optimal systems operate in a known or deterministic environment. If the environment changes significantly, these systems might not respond in the manner intended by the designer.

On the other hand, the adaptive system evaluates the environment. More accurately, it evaluates its performance within the environment and makes the necessary changes in its control characteristics to improve or, if possible, to optimize its performance. The manner of doing this involves three functions which characterize adaptive control and distinguish it from other modes of control. It may be difficult, in any given adaptive control system, to separate out the components of the system that perform these three functions; nevertheless, all three must be present for adaptation to occur. The three functions of adaptive control are:

1. *Identification function.* This involves determining the current performance of the process or system. Normally, the performance quality of the system is defined by some relevant index of performance. The identification function is concerned with determining the current value of this performance measure by making use of the feedback data from the process. Since the environment will change over time, the performance of the system will also change. Accordingly, the identification function is one that must proceed over time more or less continuously. Identification of the system may involve a number of possible measurement activities. It may involve estimation of a suitable mathematical model of the process or computation of the performance index from measurements of process variables. It could include a comparison of the current quality with some desired optimal performance.

2. *Decision function.* Once the system performance is determined, the next function is to decide how the control mechanism should be adjusted to improve process performance. This decision procedure is carried out by means of a preprogrammed logic provided by the system designer. Depending on the logic, the decision may be to change one or more of the controllable inputs to the process; it may be to alter some of the internal parameters of the controller, or some other decision.

3. *Modification function.* The third adaptive control function is to implement the decision. While the decision function is a logic function, modification is concerned with a physical or mechanical change in the system. It is a hardware function rather than a software function. The modification involves changing the system parameters or variables so as to drive the process toward a more optimal state.

Figure 18.4 illustrates the sequence of the three functions in an adaptive controller applied to a hypothetical process. The process is assumed to be influenced by some time-varying environment. The adaptive system first identifies the current process performance by taking measurements of inputs and outputs. Depending on current performance, a decision procedure is carried out to determine what changes are needed to improve system performance. Actual changes to the system are made in the modification function.

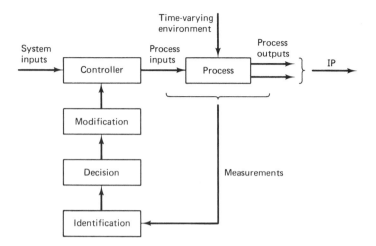

FIGURE 18.4 General configuration of an adaptive control system.

18.4 DISTRIBUTED CONTROL VERSUS CENTRAL CONTROL

There are many different equipment configurations which can be utilized to implement the control strategies described in Section 18.3. We have already discussed computer hierarchies and network topologies in Section 17.6. Equipment arrangements for computer process control are patterned closely after these more general system configurations. It should also be noted that the origins of many current-day computer control systems date back several decades, before digital computers were developed. In those times, the devices used for process control were analog rather than digital. Pneumatic devices were very common because of their inherent safety and reliability compared to mechanical—electrical devices of that period. (In many process industry installations, it is essential to avoid sparks, which might be introduced by electrical apparatus used for control.) Another characteristic of continuous process control, at least before World War II, was that the control devices were distributed. Distributed control means that the individual controllers are physically located at or near the particular operation being controlled. The opposite of distributed control is centralized control, which places the control units at some centrally located station (the control room) so that the operator can monitor many control loops at once. The sensors and actuators are, of course, still located out in the plant where the process takes place, but connections are established to communicate signals between the central controllers and their respective sensors and actuators.

The introduction of integrated circuits, computers, microprocessors, addressable computer buses, and other electronics-based technology has improved the capabilities of process control, but many of the overall concepts relating to equipment arrangement remain essentially unchanged. In the following sections we describe three common configurations used in process control in the continuous process industries. The classification is based on the suggested scheme of Kompass [6]. The three configurations are:

Type 1—Centralized control
Type 2—Optionally distributed control
Type 3—Fully distributed control

In modern process control installations, these configurations would be implemented with computer systems. However, the classification can also be applied to those existing process control systems which are not computerized.

Type 1—centralized control

This is the process control configuration in most common use today. It is organized as a "star"–type network structure (see Section 17.6) and consists of a central control station with sensors and actuators located out in the plant. Signal connections are made to input data from the sensors to the controller and to transmit commands back to the actuators. The data and command signals may be analog or digital. In earlier control systems, analog devices were more common. In modern computer control, many of the data are communicated in digital form. As we have seen in Chapter 17, it is possible to convert analog signals into their digital counterparts, and vice versa, thus providing great flexibility in selecting the most appropriate hardware for the various control loops.

In the pure form of centralized control, all control strategy functions are centralized. All of the controllers, switches, dials, recorders, and displays are located in the central control room, where the operator can monitor the process and take the appropriate action to maintain smooth operation of the plant. Traditionally, the control room equipment has been analog and required a relatively high level of attention by the operator. In today's computer-oriented control station, much of the routine checking and set-point adjustments are made automatically by the central computer, thus relieving the operator from these tasks. This transfer of responsibilities to the computer has tended to improve the overall efficiency, productivity, and product quality of plant operations in the continuous-process industries.

The general form of the centralized control configuration is illustrated in Figure 18.5. A variation of the pure centralized control which applies to computer process control involves the use of a digital bus [6]. The bus is extended into the plant and interfaced to a multiplexer, which is in turn connected to a large number of sensors and actuators. The advantage of this form of centralized control is that it reduces the number of the signal paths between the remote process devices and the central computer. The result is a lower installation cost.

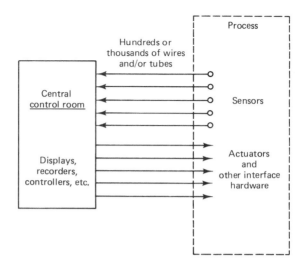

FIGURE 18.5 Type 1 configuration—centralized control.

Type 2—optionally distributed control

Types 2 and 3 are two different forms of distributed control systems in which the controllers are located closer to their associated sensors and actuators. Type 2, called optionally distributed control, is illustrated in Figure 18.6. It is basically a star network configuration but differs from centralized control by locating groups of controllers in satellite control stations closer to their associated control loops. A

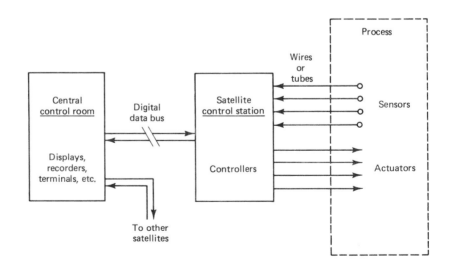

FIGURE 18.6 Type 2 configuration—optionally distributed control.

digital data bus, capable of transmission in both directions, is used to connect the central control room with the satellite stations. Signals (analog and/or digital) are transmitted between the satellite control station and the sensors and actuators in that section of the plant.

The advantage of the optionally distributed control arrangement is that it combines some of the benefits of centralized control with those of distributed control. The plant operator's console and display units are still located centrally to achieve overall process coordination and integration. At the same time, the actual control loops become physically much shorter, for improved reliability, less signal noise, and better protection against damage. Even if the connection between a satellite control station and the central computer is temporarily broken, the satellite continues to function as a local controller. Thus the affected section of the plant is maintained in operation, perhaps at a lower level of efficiency.

Type 3—fully distributed control

The fully distributed control configuration is reminiscent of the original distributed control systems before centralized analog control became commonplace. The type 3 configuration takes advantage of modern computer technology unavailable in the earlier version. The structure of the fully distributed configuration is shown in Figure 18.7. It consists of the following features. As with the optionally distributed system, there are a number of satellite stations connected to one central control room by means of a digital data bus. However, unlike the type 2 configuration, regulation of the individual process control loops is achieved by individual controllers mounted in the plant, adjacent to their respective sensors and actuators.

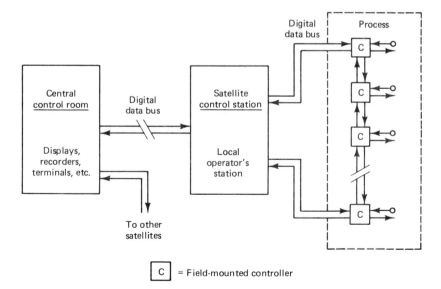

C = Field-mounted controller

FIGURE 18.7 Type 3 configuration—fully distributed control.

Communication between these controllers and the local satellite station is accomplished by means of a digital data bus in a loop configuration. The loop configuration provides increased reliability for the fully distributed system. If the data bus becomes damaged at any point, communication betweeen the satellite and its controllers can nevertheless continue through the remaining portion of the loop.

The fully distributed system offers the same basic advantage as the optionally distributed system: centralized control over the whole process but localized controllers in the plant. However, the fully distributed configuration goes one step further than type 2 control by making the control loops physically as short as possible. The controller units are located very near the operations they regulate. This improves reliability by reducing electrical noise in the shorter control loop; or if the loop is pneumatically operated, the response time and dependability are enhanced with the shorter air lines.

With the changes in control system economics resulting from developments in microprocessor technology, a trend has been started to incorporate microprocessors into the individual controller units. This allows a significant amount of intelligence and control sophistication to be brought to each control loop. The presence of microprocessors built into the sensor/actuator units and located throughout the plant represents a good illustration of hierarchical computer control in the continuous-process industries.

18.5   DIRECT DIGITAL CONTROL

Section 18.3 described the various strategies that might be used to control a production process. Section 18.4 discussed three configurations that might be used to structure the process control system. This and the following section present an alternative taxonomy of control systems which is overlaid on the three configuration types described above. The alternative organization consists of two basic categories:

1.   Direct digital control (DDC)
2.   Supervisory computer control

The present section discusses direct digital control and the following section discusses supervisory computer control.

The origins and a definition of DDC

Before computers were used to control industrial processes, analog control devices were used. These analog devices were either electrical or pneumatic, and the process control configuration was either distributed or centralized. When digital computers started to be used for control purposes, centralized control represented the state of the art because it offered the opportunity to exercise overall control and, perhaps, optimization of the entire plant. At this time in the evolution of computers (the late 1950s and early 1960s), digital computers were large and expensive. To the people responsible for designing control systems for large processing plants, it seemed a logical conclusion to use the digital computer for process control. However, the only way for this to be economically feasible, given the state of computer technology at that time, was to use one large computer to control the entire plant. These were the times and conditions in which the concept of direct digital control was born.

Direct digital control involves the replacement of the conventional analog control devices with the digital computer. The regulation of the process is accomplished by the digital computer on a time-shared, sampled-data basis rather than by many individual analog elements, each working in a continuous dedicated fashion. With DDC, the computer calculates the desired values of the input variables, and then these calculated values are applied *directly* to the process. This direct link between the computer and the process is the reason for the name "direct digital control."

DDC was originally perceived as a more efficient means of carrying out the same types of control actions as the analog elements that it replaced. However, the analog devices were somewhat limited in terms of the mathematical operations that could be performed. The digital computer is considerably more versatile with regard to the variety of control calculations that it can be programmed to execute. Hence direct digital control offers not only the opportunity for greater efficiency in doing the same job as analog control; it also opens up the possibility for increased flexibility in the type of control action, as well as the option to reprogram the control action should that become desirable.

Components of a DDC system

A complex industrial process may have a thousand variables to be monitored and regulated. An oil refinery would be an example of such a complex system. Each pair of input/output variables represents a control loop. Before digital computer control, analog controllers were used to regulate the individual loops of an industrial process. The components of each analog feedback loop included:

1. *Transducers and sensors.* These are located in the plant.
2. *Actuators.* These are the servomotors, valves, relays, and other process interface devices which operate at the command of the controller unit.
3. *Analog controller.* This controls the electronic or pneumatic devices that operate on the error signal in Figure 18.2 to drive the output variable into agreement with the set point.
4. *Recording and display devices.* These instruments provide a visual reading of the sensor measurements for the control room operator.

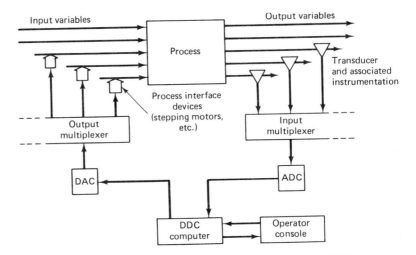

FIGURE 18.8 Components of a direct digital control (DDC) system.

5. *Set-point dial and comparator.* These would typically be part of the analog controller to allow the operator to set the desired operating level for the particular process variables. The comparator allows the feedback measurement to be compared with the set point.

In a direct digital control system, some of these components would be replaced, while others would remain the same. The transducers, sensors, and actuators are connected to the physical process and many of these devices would remain the same for DDC as for a central control plant under analog control. Other components, including the analog controllers, many of the analog-based recording and display devices, set-point dials, and comparators would no longer be needed in a DDC plant. In their place would be the central computer and its peripheral equipment.

Additional hardware elements that would be needed in a direct digital control system would include analog-to-digital converters, digital-to-analog converters, and multiplexers. Also, digital-type transducers and stepping motors, which respond to a series of computer-generated pulses, might be used as substitutes for some of the more conventional devices typically associated with analog control systems. Figure 18.8 presents the basic components and their arrangement for a direct digital control system.

18.6 SUPERVISORY COMPUTER CONTROL

The practice of simply using the digital computer to imitate the characteristics of analog controllers seems to represent a transitional phase in computer process control. Direct digital control alone is difficult to justify in terms of reduced costs.

However, the use of the computer in process control applications can be supported by improvements in the overall performance of the manufacturing operation. It is with this problem of overall process improvement and optimization that supervisory computer control is concerned. Basically, the problem is to determine the appropriate values for the set points of each control loop. In a centralized analog control system and even in the elementary concept of DDC, the decisions about set-point values are left to the operator.

Supervisory computer control denotes a computer process control application in which the computer determines the appropriate set-point values for each control loop in order to optimize some performance objective of the entire process. The performance objective of the process might be maximum production rate, minimum cost per unit of product, yield, or some other objective that pertains to the process. Based on the mathematical model of the process which is programmed into the computer, the computer calculates the set-point values that achieve the desired objectives for the process. The various control strategies used in supervisory computer control include regulatory control, feedforward control, preplanned control, optimal control, and adaptive control.

In a supervisory computer control system, adjustments in the set points for the individual control loops are accomplished in either of two ways:

1. *Analog control.* If the individual feedback loops are controlled by analog devices, the control computer is connected to these devices. The set-point adjustments are made through the appropriate interface hardware between the computer and the analog elements.

2. *Direct digital control.* If the feedback loops operate under direct digital control, the supervisory control program provides the set-point values to the DDC program. Both the supervisory control program and the direct digital control program can be contained in the same computer, or they can be in separate computers in a hierarchical configuration.

A block diagram showing the basic form of a supervisory computer control system is presented in Figure 18.9. As illustrated in the diagram, supervisory control is concerned with overall process performance, whereas DDC is concerned with the individual control loops.

In addition to set-point adjustments in the control loops, the supervisory computer may also be required to control certain discrete variables in the process. Examples of this function include starting or stopping motors, opening valves, setting switches, solenoids, and so on. When regulation of the industrial operation consists of performing a sequence of these on/off steps in a predetermined order, this type of control is called *sequencing control* (refer back to Section 18.3 on control strategies). Most industrial operations contain a mixture of analog and discrete variables. Accordingly, the supervisory control computer is called on to perform a combination of sequencing control and set-point control.

Table 18.1 summarizes some of the features and applications of direct digital control and supervisory control. The table represents the various control strategies

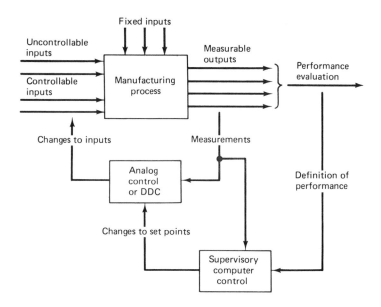

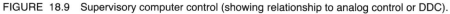

FIGURE 18.9 Supervisory computer control (showing relationship to analog control or DDC).

TABLE 18.1 Summary of Features Usually Associated with Direct Digital Control and Supervisory Computer Control

Direct Digital Control	
Control strategies:	Feedback control
	Regulatory control
	Feedforward control
Configuration types:	Type 1—centralized
Applications:	Continuous process industries
Supervisory Computer Control	
Control strategies:	Regulatory control
	Feedforward control
	Preplanned control
	Optimal control
	Adaptive control
Configuration types:	Type 2—optionally distributed
	Type 3—fully distributed
Applications:	Continuous-process industries
	Direct numerical control (DNC)
	Flexible manufacturing systems*

*Flexible manufacturing systems are the subject of Chapter 20.

416

usually identified with the two categories, the process control configurations (centralized, optionally distributed, and fully distributed), and some applications. Virtually all DDC applications are associated with the continuous process industries. Supervisory computer control is being applied more and more in discrete-parts manufacturing.

REFERENCES

[1] CASSELL, D. A., *Introduction to Computer-Aided Manufacturing in Electronics*, John Wiley & Sons, Inc. (Wiley-Interscience), New York, 1972.

[2] GROOVER, M. P., "Adaptive Control and Adaptive Control Machining," *Educational Module*, Manufacturing Productivity Educational Committee, Purdue Research Foundation, West Lafayette, Ind., 1977.

[3] GROOVER, M. P., *Automation, Production Systems, and Computer-Aided Manufacturing*, Prentice-Hall, Inc., Englewood Cliffs, N.J., 1980, Chapters 13–15.

[4] HARRISON, T. J. (Ed.), *Minicomputers in Industrial Control*, Instrument Society of America, Pittsburgh, Pa., 1978.

[5] INSTITUTE OF ELECTRICAL ENGINEERS, *Proceedings, 3rd International Conference on Trends in On-Line Computer Control Systems*, Sheffield, England, March, 1979.

[6] KOMPASS, E. J., "A Long Perspective on Integrated Process Control Systems," *Control Engineering*, August, 1981, pp. 4–9.

[7] MATHIAS, R. A., "Adaptive Control for the Eighties," *Paper MS80-242*, Society of Manufacturing Engineers, Dearborn, Mich., 1980.

[8] OFFORD, G. E.,   Personal communications and special reports, 1981, 1982.

[9] RAVEN, F. H., *Automatic Control Engineering*, McGraw-Hill Book Company, New York, 1978.

[10] RAY, W. H., *Advanced Process Control*, McGraw-Hill Book Company, New York, 1981.

[11] SAVAS, E. S., *Computer Control of Industrial Processes*, McGraw-Hill Book Company, New York, 1965, Chapter 11: "Direct Digital Control," by J. F. Hornor.

[12] SMITH, C. L., *Digital Computer Process Control*, International Textbook Co., Scranton, Pa., 1972, Chapter 1.

[13] WILDE, D. J., AND BEIGHTLER, C. S., *Foundations of Optimization*, Prentice-Hall, Inc., Englewood Cliffs, N. J., 1967.

chapter 19

Computer-Aided Quality Control

19.1   INTRODUCTION

The quality control (QC) function has traditionally been performed using manual inspection methods and statistical sampling procedures. Manual inspection is generally a time-consuming procedure which involves precise, yet monotonous work. If often requires that parts be removed from the vicinity of the production machines to a separate inspection area. This causes delays and often constitutes a bottleneck in the manufacturing schedule.   ·

Inherent in the use of statistical sampling procedures is acknowledgment of the risk that some defective parts will slip through. Indeed, statistical quality control attempts to guarantee that a certain expected or average fraction defect rate will be generated during the production/inspection process. The nature of traditional statistical QC procedures is that something less than 100% good quality must be tolerated.

There is another aspect of the traditional QC inspection process which detracts from its usefulness. It is often performed after the fact. The measurements are taken and the quality is determined after the parts are already made. If the parts are defective, they must be scrapped or reworked at a cost which is often greater than their original cost to manufacture.

There are several economic, social, and technological factors at work to modernize the quality control function. The economic factors include the high cost of the inspection process as it is currently done and the desire to eliminate inspection as a source of costly delay in production. The social factors include the ever-increasing demand by customers for near perfection in the quality of manufactured items, the growing number of expensive product-liability legal cases, and government regulations which require many firms to maintain comprehensive production and quality records. Another factor in this category is the tendency for some manual inspection tasks to involve subjective judgment on the part of the human inspector. It is considered desirable to try to remove this subjective component from inspection operations. Finally, the technological factors consist of several important advances which have been made in inspection automation. Principal among these advances have been the tremendous growth in the application of microprocessors and improvements in noncontact sensor techniques such as vision systems.

All of these various factors are driving the quality control function toward what we are calling computer-aided quality control (CAQC). Other terms that have been applied to describe this movement are ''computer-aided inspection'' (CAI) and ''computer-aided testing'' (CAT). As we describe in Section 19.3, CAI and CAT are subsets of CAQC.

The objectives of computer-aided quality control are ambitious, yet straightforward. They are:

1. To improve product quality
2. To increase productivity in the inspection process
3. To increase productivity and reduce lead times in manufacturing

The strategy for achieving these objectives is basically to automate the inspection process through the application of computers combined with advanced sensor technology. Wherever technically possible and economically feasible, inspection will be done on a 100% basis rather than sampling. It will be done on-line almost as part of the production operations. On-line 100% inspection will introduce opportunities to use the inspection measurements as feedback data to make compensating adjustments in the manufacturing process.

In this chapter we explore the topic of computer-aided quality control. The discussion includes many of the latest developments in both contact and noncontact inspection techniques. We also consider the way the quality control function should make use of the CAD/CAM data base. To begin with, let us consider some of the basic terminology in quality control.

19.2   TERMINOLOGY IN QUALITY CONTROL

In many respects, computer-aided quality control represents a significant departure from the traditional QC methods. Nevertheless, the terminology is similar, and it is appropriate to review the various terms and concepts used in this field before examining the role played by the computer.

✳ Quality in a manufacturing context can be defined as the degree to which a product or its components conform to certain standards that have been specified by the designer. The design standards generally relate to the materials, dimensions and tolerances, appearance, performance, reliability, and any other measurable characteristic of the product.

To ensure that its products adhere to the specified standards, a firm will generally organize its activities along two approaches: quality assurance and quality control. These two approaches represent the before and after in the firm's efforts to manage quality in manufacturing.

Quality assurance (QA) is concerned with those activities which will maximize the probability that the product and its components will be manufactured within the design specifications. These activities should start in the product design area, where the designer can make decisions among alternatives that might have quality consequences. For example, the decision might be between two or more materials to specify for a particular component. The designer must select the material that will achieve the best performance (in terms of properties, durability, reliability, processability, etc.) relative to its cost. QA activities continue in manufacturing planning, where decisions relative to production equipment, tooling, methods, and motivation of employees will all have an influence on quality.

Quality control is concerned with those activities related to inspection of product and component quality, detection of poor quality, and corrective action necessary to eliminate poor quality. These activities also involve the planning of inspection procedures and the specification of the gages and measuring instruments needed to perform the inspections. Included within the scope of planning would be the design of statistical sampling plans, a field of study which is usually called statistical quality control.

Statistical QC is generally divided into two categories: acceptance sampling and control charts. Acceptance sampling is a procedure in which a sample is drawn from a batch of parts in order to assess the quality level of the batch and to determine whether the batch should be accepted or rejected. A company can apply the procedure to items received from a supplier or to items of its own manufacture. Acceptance sampling is based on the statistical notion that the quality of a random sample drawn from a larger population will be representative of the quality of that population.

The application of control charts derives from the same statistical concept. Control charts are used to keep a record over time of certain measured data collected from a process. A company would use control charts to monitor its own production processes. A general form of the control chart is illustrated in Figure 19.1.

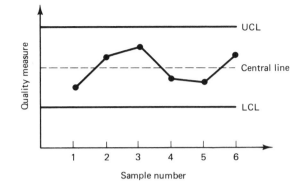

FIGURE 19.1 General form of a control chart used in a statistical quality control.

The central line indicates the expected quality level of the process. The upper and lower control limits (UCL and LCL) are statistical measures of the variation in the process which would be tolerated without concluding that the process has erred. When these limits are exceeded, it usually means that something has changed the process, and an investigation should be initiated to determine the cause. Whereas acceptance sampling is applied to a batch of product after it is completed, control charts are applied during production. This means that it is possible with control charts to make adjustments in the manufacturing process if the recorded data indicate that corrections are needed.

Both acceptance sampling and control charts can be applied to two situations in quality control: fraction defects and measured variables. The combination of possibilities is shown in Figure 19.2 together with the names commonly given to the sampling methods. In the fraction-defect case, the objective is to determine what proportion of the sample (and the population from which it came) are defective. This is often accomplished by a go/no go gage, which can quickly determine whether a part is within specification or not. In the measured-variable case, the object is to determine the value of the quality characteristic of interest (e.g., dimension, resistance, hardness, etc.). This requires the use of a measuring instrument of some kind (e.g., micrometer, ohmeter, hardness tester, etc.) and is normally a more time-consuming manual process than the go/no go case.

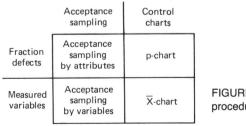

	Acceptance sampling	Control charts
Fraction defects	Acceptance sampling by attributes	p-chart
Measured variables	Acceptance sampling by variables	$\bar{X}$-chart

FIGURE 19.2 Types of statistical QC procedures.

A final distinction that should be made is the difference between inspection and testing. Although the common usage of these terms often overlaps, we will distinguish them as follows. Inspection is normally used to examine a component of a product in relation to the design standards specified for it. For a mechanical component, this would probably be concerned with the dimensions of the part. These might be checked with several go/no go gages or they might be measured with a micrometer and other instruments. Inspection should be done whenever and wherever the expected costs of not inspecting (e.g., scrap, rework, loss of customer good will) exceed the cost of inspecting. The common situations that warrant inspection are:

Incoming raw materials

At various stages during manufacturing (e.g., when the parts are moved from one production department to another)

At the completion of processing on the parts

Before shipping the final assembled product to the customer

Testing, on the other hand, is normally associated with the functional aspects of the item, and it is often directed at the final product rather than its components. In this usage, testing consists of the observation of the final product during operation under actual or simulated conditions. If the product passes the test, it is deemed suitable for sale. Harrington (5) lists several categories of tests used for final product evaluation:

Simple functional tests under normal or simulated normal operating conditions

Functional tests in which the product is tested under extreme (usually adverse) conditions

Fatigue or wear tests to determine how long the product will function until failure

Overload tests to determine the level of safety factor built into the product

Environmental testing to determine how well the product will perform under different environments (e.g., humidity, temperature)

Another type of testing that is often mentioned is destructive testing. This is a procedure that results in the destruction of the item in order to measure the property of interest. A common example is the tensile test on a specimen of metal to determine the metal's strength and ductility properties. Destructive testing is often employed as an inspection procedure, the way we have defined it, because it is often applied to raw materials, partially processed materials, and component parts. It can also be applied to the final product, as might be suggested by Harrington's list of tests. However, destructive testing for final products is expensive and would be done on a very limited sampling basis.

19.3   THE COMPUTER IN QC

Computer-aided inspection (CAI) and computer-aided testing (CAT) are merely extensions of their counterparts described above. Whereas these activities have traditionally been performed manually (with the help of gages, measuring devices, and testing apparatus), CAI and CAT are performed automatically using the latest computer and sensor technology. Computer-assisted inspection and testing methods form only part, certainly a major part, of computer-aided quality control. In our treatment of the subject we shall include the integration of the quality control function with CAD/CAM as a critical ingredient in the success of CAQC. CAI and CAT are examples of what have been called "islands of automation." They are stand-alone systems. Without their integration into larger computerized systems, CAQC will not achieve its full potential.

The implications of the use of computer-aided quality control are important. The automated methods of CAQC will result in significant changes from the traditional concepts and methods described above. Critical changes will also occur in the way the quality function is implemented within a company. We have already alluded to some of the changes. The following list will summarize the important effects likely to result from CAQC.

1.   With CAI and CAT, inspection and testing will typically be accomplished on a 100% basis rather than by the sampling procedures normally used in traditional QC.

2.   Inspection during production will be integrated into the manufacturing process rather than requiring that the parts be taken to some inspection area. This will help to reduce the elapsed time to complete the parts. Also, to incorporate online inspection into the production process will mean that inspection will have to be accomplished in much less time than with current manual techniques. How will this improvement in inspection productivity be achieved? The third point below addresses this question.

3.   The use of noncontact sensors will become much more widely used with computer-aided inspection. With contact inspection devices, the part must usually be stopped and often repositioned to allow the inspection device to be applied properly. Stopping, repositioning, and making physical contact with the part all take time. With noncontact sensor devices, the part can often be inspected "on the fly." These devices, driven by the high-speed data processing capability of the computer, can complete the inspection in a small fraction of a second. This is a rate which is certainly compatible with most production operations.

4.   The on-line noncontact sensors will be utilized as the measurement component of computerized feedback control systems. These systems will be capable of making adjustments to the process variables based on analysis of the data collected by the sensors. The data analysis would include statistical trend analysis. An

example of the need for trend analysis can be found in the gradual wear of cutting tools in a machining operation. Data would be plotted (even if only in computer memory) on a control chart similar to Figure 19.1. This would not only allow out-of-tolerance conditions to be identified, but gradual shifts in the process could also be uncovered and corrective action taken. By regulating the process in this manner, parts will be made much closer to the desired nominal dimension rather than merely within tolerance. Quality feedback control systems will help to reduce scrap losses and improve product quality.

5.   Because 100% inspection and on-line quality control systems will become prevalent, a basic assumption in statistical QC must be challenged. That is the assumption that anything less than 100% good quality is acceptable. The use of statistical quality control tolerates less than 100% perfect quality. With computer-aided inspection technology, it may no longer be necessary to settle for less than perfection.

6.   Sensor technology will not be the only manifestation of automation in CAQC. Robots will be used increasingly in future inspection applications. We discussed these applications in Section 11.8. Also, completely automated test cells will become an important component in future factories.

7.   In addition to CAI and CAT, the computer will be used in other areas of quality control. There will also be applications for the computer in quality assurance as well as QC. The CAD/CAM data base will be used to derive these various quality applications, and we discuss some of the possibilities in Section 19.8.

8.   There will be personnel implications in CAQC. To the extent that CAI and CAT take its place, manual inspection activity will be reduced. Quality control personnel will have to become more computer-wise and technologically sophisticated to operate the more complex inspection and testing equipment and to manage the information that will result from these more automated methods.

In the following three sections we consider various types of modern contact and noncontact inspection techniques. The contact methods usually involve the use of coordinate measuring machines (CMM). Most of these machines today are either controlled by NC or computers. The noncontact methods are divided into two categories for our purposes: optical and nonoptical. The optical methods usually involve some sort of vision system, although other methods, such as lasers, are also used. The nonoptical techniques are typically based on the use of electrical fields to sense the desired characteristic of the object. Ultrasonics and radiation represent other possible sensor technologies. The classification of the various types of sensors is presented in Table 19.1.

In the final two sections of the chapter we discuss computer-aided testing and the integration of CAQC with CAD/CAM, in particular, the use that might be made of the CAD/CAM data base in quality control.

TABLE 19.1   Classification of
Inspection Sensor Technologies

I.   Contact inspection methods

    A. Coordinate measuring machines

    B. Mechanical probes

II.   Noncontact inspection methods

    A. Optical techniques

       Machine vision
       Scanning laser beam devices
       Photogrammetry
       Others

    B. Nonoptical techniques

       Electrical field techniques
         Reluctance
         Capacitance
         Inductance

       Radiation techniques
       Ultrasonics

19.4   CONTACT INSPECTION METHODS

The coordinate measuring machine (CMM) is the most prominent example of the equipment used for contact inspection of workparts. A coordinate measuring machine is illustrated in Figure 19.3. It consists of a table which holds the part in a fixed, registered position and a movable head which holds a sensing probe. The probe can be moved in three directions, corresponding to the x, y, and z coordinates. During operation, the probe is brought into contact with the part surface to be measured and the three coordinate positions are indicated to a high level of accuracy. Typical accuracies of these machines are in the neighborhood of ±0.0002 in. (0.0051 mm).

Today's coordinate measuring machines are computer controlled. The operation of the machine is similar to an NC machine tool in which the movement of the measuring probe is either tape controlled or computer controlled. Programs and coordinate data can be downloaded from a central computer, much in the manner of direct numerical control. Also similar to DNC is the capability to transmit data from the CMM back up to the host computer.

Recent advances in CMM technology are based largely on greater intelligence and convenience features provided by the computer. These advances include the capability for automatic workpart alignment on the machine table, interactive programming of the CMM for inspection personnel who are inexperienced in the

FIGURE 19.3   Coordinate measuring machine. (Courtesy of Automation and Measurement Division, Bendix Corp.)

use of computers, and conversion routines between polar and cartesian coordinate systems.

Savings in inspection time by using coordinate measuring machines are significant. Typically, between 5 and 10% of the time is required on a CMM compared to traditional manual inspection methods. Other advantages include consistency in the inspection process from one part to the next which cannot be matched by manual inspection, and reductions in production delays to get approval of the first workpiece in a batch.

Although the reductions in inspection time are significant with a CMM, there is nevertheless wasted time associated with the fact that the coordinate measuring machine is physically located away from the production machine, usually in a separate area of the shop. Accordingly, the parts must be transported from the production area to the CMM. In fact, if inspection is required at several different stages of production, several moves will be involved. One possible approach to overcome this problem is to use inspection probes mounted in the spindle of the machine tool. These inspection probes are contact sensing devices that operate with the machine tool much like the coordinate measuring machine. We have already discussed inspection probes in connection with numerical control and the reader is referred back to Section 9.8 and to Figure 9.10, which shows an example of an inspection probe.

19.5 NONCONTACT INSPECTION METHODS— OPTICAL

Noncontact inspection of items is an attractive alternative to the types of methods discussed in the preceding section. Among the advantages of noncontact inspection are [11]:

It usually eliminates the need to reposition the workpart.

Noncontact inspection is usually much faster than contact inspection.

It eliminates mechanical wear encountered with the contacting inspection probe because it eliminates the probe.

It reduces potential danger to people, who must touch a hazardous material if contact inspection is used.

It removes the possibility of damage to the surface of a part which might result during contact inspection.

We divide the varieties of noncontact inspection schemes into two categories: optical and nonoptical. This section covers the different types of optical inspection devices available. The next section covers the nonoptical category.

Optical systems are the dominant type of noncontact inspection method. These systems generally rely on the use of microelectronics technology and computer processing of the sensing signals. Improvements in performance and reductions in cost in these two areas are making optical systems more and more economically feasible. There are a variety of optical sensing techniques used for inspection work. We shall discuss three types:

1. Machine vision
2. Scanning laser beam devices
3. Photogrammetry

All of these optical systems use some form of light sensor or photosensitive material. Simple light-sensing systems use photocells, photodiodes, and photographic paper.

Machine vision

The use of machine vision systems for inspection is an exciting area which holds the promise of significant improvements in both the productivity of the inspection process and the quality of the resulting product. Other names given to these systems include microprocessor-based television and computer vision. The typical machine vision system consists of a TV camera, a digital computer, and an interface between them that functions as a preprocessor. The combination of system hardware and software digitizes the picture and analyzes the image by comparing it with data stored in memory. The data are often in the form of a limited number of models of the objects which are to be inspected.

The technology of machine vision inspection is one in which advancements and refinements are continually being made. At the time of this writing, there are several limitations of machine vision which are imposed principally by current computer speed and storage technology. The first limitation is concerned with the problem of dividing the picture into picture elements. This is very similar to the problem encountered in the development of graphics terminals for computer-aided design (Section 5.3). Most machine vision systems in use today have a picture area consisting of roughly 240 by 240 pixels. (There are variations in these numbers for commercially available systems.) In terms of image resolution, this represents a very limited capability to accomplish precise measurements and analyze complex images. Future improvements in vision technology will allow the number of picture elements in machine vision systems to be substantially increased for better image resolution.

A second limitation is that the object in front of the camera must be capable of being divided into areas of contrasting lightness and darkness. Most commercial systems today divide the image into two states, black and white. Gray areas must be interpreted as being either black or white, depending on their relative level of brightness. This is done by selecting a threshold brightness level and assigning each picture element to one of the two states depending on whether its brightness is greater than or less than the threshold level. This limitation imposes requirements on the lighting that must be used to illuminate the object. The requirement is sometimes satisfied by backlighting the object to accentuate the contrast between the object and surrounding areas.

Third, there are limitations on the capability of machine vision systems to recognize the object in the viewing area. For example, the number of distinct objects that can be recognized by the system is limited by its ability to discriminate features of different objects and by its computer storage capability. The features of an object which can be determined by a typical vision system include area, perimeter, center of gravity, the dimensions of an enclosing geometric form such as a circle, and certain directional features such as the line passing through the two centers of an ellipse. The storage capability restricts the amount of data and the number of separate models that can be compared with the image. Another difficulty is in the ability of the system to deal with variations in the image. Such variations might represent defects in the object that should be identified in the inspection process, or they might be variations in part orientation and position. Part orientation and position problems can be readily solved by today's vision inspection technology. For example, the orientation problem can be solved by using the directional features of the object to adjust for differences in rotation of the part. One of the problems which still remains is when parts in the viewing area overlap one another so that the system is unable to identify the outline of each part.

These various limitations are expected to be gradually reduced as the technology develops during the next several years. Future machine vision systems will have better image resolution, greater ability to distinguish grey areas and even color, and more intelligence and memory for improved object recognition capability.

There are a wide variety of inspection problems that can be solved by current day machine vision systems. The solutions are often individualized to the particular inspection problem. Collectively, they represent the scientific and engineering efforts of many different companies and research laboratories in this fast developing industry. Machine vision inspection problems can be divided into two categories [13]:

1. Noncontact gaging of dimensions
2. Inspection based on pattern recognition of object features

Noncontact gaging in machine vision involves the inspection of part size and other features where it is not necessary to process the image of the entire part outline, only those portions that must be examined for dimensional accuracy. During setup for an inspection, a parts-training program is used to view the workpart of interest on a TV monitor. With the image in fixed position on the screen, the operator manipulates a cursor to define the edges of interest and to apply an appropriate scale factor to establish the correct units of measure. During actual inspection, the vision system identifies the relevant boundaries so that the desired dimensions of the part can be scaled for automatic gaging. The inherent limitation of the scaling accuracy of the system is determined by such factors as the density of picture elements in the viewing area (the first limitation of present day machine vision systems discussed above) and the field of view of the TV camera.

The second category of machine vision inspection is based on pattern recognition techniques. In this category, the attributes of the object to be inspected are typically more subjective and in some respects more complicated than part dimensions. The machine vision pattern recognition process can be conceptualized as involving a comparison of features (for example, area, perimeter, and so on) between the object being inspected and the model of the object stored in computer memory. One of the techniques that is sometimes applied in pattern recognition is called automatic edge detection [6]. In machine vision edge detection, the problem is to distinguish the boundaries between light and dark areas in the image. These boundaries indicate prominent part features such as edges and holes. Based on the results of the edge detection process, the system can be programmed to compute object features such as surface area, number of holes, hole area, perimeter, and center of gravity. Correction routines must be programmed into the pattern recognition software to compensate for nuisances such as statistical variations in the images, part misalignment, and variances in part orientation. Examples of the kinds of inspections in this second category include:

Inspection of labels on bottles and cartons. In the case of clear bottles, machine vision systems have been programmed to inspect the level of the contents in the bottles.

Optical character recognition problems

Inspection of the gross outline of parts

Inspecting for the presence or absence of features in an assembled product.

The electronics manufacturing industry has made significant contributions to the development of these techniques [3]. Printed circuit boards, for example, can be inspected for potential defects such as short circuits, missing holes, over-etching and under-etching. Other applications in electronics include inspection of microcircuit photomasks and semiconductor chip inspection.

Inspecting for cracks and other imperfections in work surfaces

An example of the kind of commercially available vision system that can perform inspections based on pattern recognition is shown in Figure 19.4. The TV camera in the foreground is focused on a small portion of the bottle label and the image is displayed on the television monitor.

Automatic vision systems and other types of optical sensors used for inspection are often built into the production line to operate with some form of parts rejection mechanism. As the image is processed, the system makes a determination as to whether the part is good or reject. If good, the part proceeds to the next pro-

FIGURE 19.4 Vision system showing camera (in foreground), control console, and TV monitor. (Courtesy of Object Recognition systems, Inc.)

cessing station. If reject, the automatic rejection mechanism is triggered to eject the part into a separate location.

Machine vision systems using pattern recognition techniques have been applied in areas other than manufacturing inspection. Two primary examples are the processing of satellite images and the analysis of medical (for example, x-ray) images. Another application is in robotics, where the vision system is utilized as a sophisticated feedback control sensor for actions that are to be carried out by the robot. We have previously discussed this application in Section 10.9.

Scanning laser beam devices

Not all scanning beam devices use lasers as the light source, but most of them do. The advantage of the laser is that it is a coherent light beam which can be projected great distances without significant diffusion. Lasers have found many applications in industrial measurement problems.

The scanning laser beam device relies on the measurement of time rather than light, although a light sensor is required in its operation. The schematic diagram of its operation is pictured in Figure 19.5. A laser is used to project a continuous thin beam of light. A rotating mirror deflects the beam so that it sweeps across the object to be measured. The light sensor is located at the focal point of

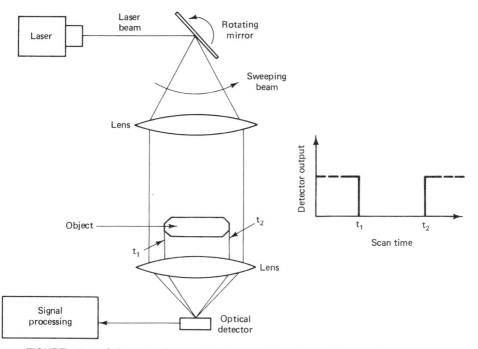

FIGURE 19.5 Schematic diagram showing operation of scanning laser beam system.

the lens system to detect the interruption of the light beam as it is blocked by the object. The time lapse corresponding to the interruption of the light beam is measured to determine the desired dimension of the part. Typically, a microprocessor is programmed to make the conversion of the time lapse into a dimensional value and to perform other functions, such as signaling an automatic parts-rejection mechanism to eject a defective part from the line.

Photogrammetry

Photogrammetry is a technique which may gain in usage in inspection work as it is perfected. The term refers to a procedure which was borrowed from aerial reconnaisance and geological mapping applications. A more recent application of the procedure is in aerospace plants to measure large airframe assembly fixtures.

Photogrammetry involves the extraction of three-dimensional data from a pair of photographs taken at different angles. The two photographs can be combined much in the way that a stereoscope uses a pair of photographs to form a three-dimensional image for the viewer.

In the measurement process used for inspection, the two photographs are read by a device called a monocomparator to establish coordinates and positions of objects. These data are then computer-analyzed to extricate the desired information.

The drawback of the conventional photogrammetry technique is the need for photographs, an inconvenient and time-consuming step in the procedure. An improvement in the technique which is being developed will delete the photographic step. Instead, the images from two cameras set up in a stereoscopic configuration will send visual data directly to a computer for mathematical analysis and real-time extraction of dimensional data. This arrangement is illustrated in Figure 19.6.

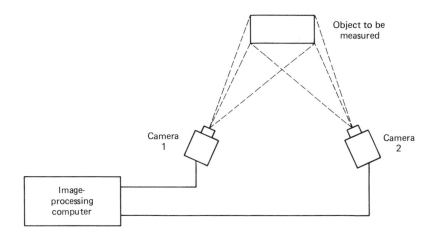

FIGURE 19.6 Measurement system based on photogrammetry principles.

There are a wide variety of approaches to optical sensing systems for non-contact inspection. We have described only three of the techniques to give an idea of the variety available. A particularly good review of the range of available systems is presented in an article by Schaffer [11]. Written in 1979, the article may be slightly dated at present in such a fast-moving technology. Comforting, however, is the fact that the physical principles remain fairly constant.

19.6 NONCONTACT INSPECTION METHODS— NONOPTICAL

In addition to noncontact inspection methods based on optical systems, nonoptical approaches can also be used. We will describe three general types which are quite representative of the current technology in this area. The three general types are:

1. Electrical field techniques
2. Radiation techniques
3. Ultrasonics

Electrical field techniques

Various types of electrical field techniques can be applied to noncontact inspection. Three types of electrical fields are employed:

1. Reluctance
2. Capacitance
3. Inductance

The reluctance transducers are proximity devices that indicate the presence and distance from the probe of a ferromagnetic substance. The obvious limitation of the device is that the object being inspected must be electromagnetic.

A capacitance-based transducer can also be used to measure the distance of an object from the face of a probe. The measurement is based on the variable capacitance from part/probe coupling. This capacitance is inversely proportional to the distance between the probe face and the part, and thus the distance can be calculated. The capacitance transducer can be used to detect a variety of materials. The material must be an electrical conductor.

Inductance systems operate by subjecting the object to an alternating magnetic field by means of an electromagnetic coil. The result is that small circulating currents (eddy currents) are generated in the object. These eddy currents, in turn, create their own magnetic field, which interacts with the primary field. This interaction affects the impedance of the coil, which can be measured and analyzed to determine certain characteristics about the object.

In all three cases, the object to be inspected is placed in the electrical field and its effect is observed and analyzed. Typically, the location of the object is measured in reference to the probe which is generating the electrical field. In some arrangements, two probes must be used. Location, part dimensions, part thickness, and other characteristics can be measured depending on the inspection setup. Eddy-current-based systems can be utilized to inspect below the surface of an object to detect cracks, voids, and other flaws in metals.

Radiation techniques

X-ray radiation techniques are employed for purposes of noncontact inspection in the metals and metalworking industry. The amount of radiation absorbed by a material can be used to measure its thickness and other quality characteristics. In a typical application in a rolling mill, an X-ray scanning unit measures the thickness of the plates or strips going through the rolls so that the proper adjustments can be made in the rollers. X-ray techniques are also used to inspect weld quality in fabricated steel and aluminum pressure vessels and pipes. In this case the radiation can be used to detect flaws and voids in the weld.

Ultrasonics

Ultrasonics in inspection work involves the use of very high frequency (above 20,000 Hz) sound waves to indicate quality. A principal application is in nondestructive testing of materials. Ultrasonic techniques can also be applied to the problem of determining dimensional features of workparts. One approach, called acoustical phase monitoring, involves the analysis of sound waves reflected from the surface of an object. The sound waves are produced by an emitter and directed against the object. Assuming that all else remains constant, the reflected sound pattern from the object should always be the same. During inspection, the sound pattern from the part is analyzed by a computer program and compared to the pattern of a standard part, one that is known to be of acceptable quality. If the pattern of the test part differs significantly from that of the standard, it is rejected.

19.7 COMPUTER-AIDED TESTING

As described in Section 19.2, testing is generally applied to assess the functional performance of a final product. It may also be applied for major subassemblies of the final product, such as the engines and transmissions of automobiles. Testing may also be performed on individual components in which some functional aspect of the component must be examined and cannot be implicitly determined by means of a mechanical inspection. An example of this might be the case of a brake lining in which the dimensions are correct, but the functional performance must be determined through a testing procedure.

Computer-aided testing is simply the application of the computer in the testing procedure. There are different levels of automation which can be found in CAT. At the lowest level, the computer would be used simply to monitor the test and analyze the results, but the testing procedure itself is manually set up, initiated, and controlled by a human operator. In this case the computer receives the data from a data logger or a data acquisition system (refer to Section 16.6) and prepares a report of the test results.

At a much higher level of automation are computer-integrated test cells, which consist of a series of testing stations (a dozen or more stations is not uncommon) interconnected by a materials handling system. An automated test cell has most of the earmarks of a computer-integrated manufacturing system of the type to be discussed in Chapter 20. These cells are often interfaced directly to the assembly line so that the products flow automatically from final assembly to final testing. All facets of the operation of the test cell are under computer control. The individual stations typically operate independently of each other. During operation, a product is transferred by the handling system to an available test station. The test station automatically registers the product in the proper location and orientation, and attaches the required connecting apparatus to conduct the test. The testing then begins with the computer monitoring the data and analyzing the results. If the product passes the test, it is automatically moved to the next assembly operation or final packing. In the event the product fails to pass the test, there is often the provision to transfer the product to a manual station for examination by a human operator. The computer can often be helpful in this regard by indicating the reason the test failed, or even diagnosing the problem and recommending the most promising repair alternative. Another feature of some test cells is the capability to make adjustments in the product during the test cycle to fine-tune its functional operation.

Computer-aided test cells of the type described above are applied in situations where the product is complicated and produced in significant quantities. Examples include automobile engines, aircraft engines, and electronic integrated circuits. Advantages of these cells include higher throughput rates, greater consistency in the test procedure, and less floor space occupied by the automated cell as compared to a manual facility of similar capacity.

EXAMPLE 19.1

An example of a large test cell is the automatic hot test system for automobile engines. The system was designed and built by Scans Associates, Inc., of Livonia, Michigan, for the Ford Motor Plant in Cleveland, Ohio. A layout plan of the facility is shown in Figure 19.7. The system consists of 40 individual automatic test stations arranged along two conveyor loops. A feed loop is used to receive the engines from the engine assembly department and route them to one of the two testing loops. The engines are moved on pallets within the system. Upon completion of testing, the engines are routed back to the feed loop and transferred out of the system for final dress-up.

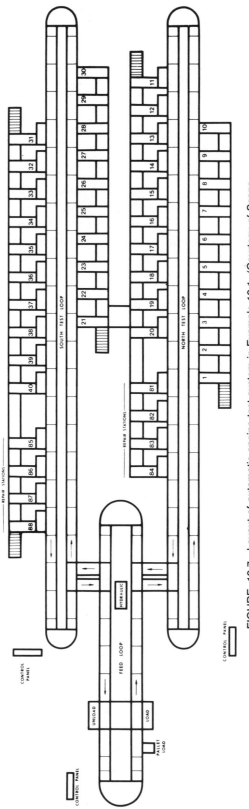

FIGURE 19.7 Layout of automatic engine test system in Example 19.1. (Courtesy of Scans Associates, Inc.)

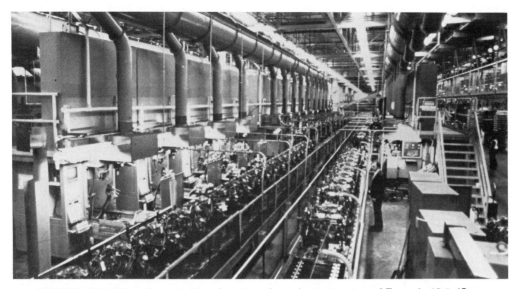

FIGURE 19.8   View of one test loop in automatic engine test system of Example 19.1. (Courtesy of Scans Associates, Inc.)

Each testing loop has 20 test stations plus four minor repair stands, all of which are loaded and unloaded automatically. Figure 19.8 presents a view of one of the test loops, looking along the conveyor line with repair stations shown on the immediate left.

One of the testing stations is illustrated in Figure 19.9. Each station is equipped to supply the required connections for fuel, coolant, electrical ignition, and exhaust. The engines are automatically started and gages indicate the operating characteristics, such as coolant temperature, pressure, and manifold vacuum. The test stations are each interfaced to a programmable controller (refer to Chapter 3) which is connected to a CRT and keyboard console. Figure 19.10 shows the results of an engine testing sequence as they would be displayed on the CRT. The programmable controller cycles the engine through its test sequence by making throttle adjustments to control engine speed versus time.

The system is equipped with several sensors which check engine test results and display them on the CRT. In the event that the measured test parameters do not fall within the limits programmed into the PC for that model, the engine is rejected and is routed to one of four minor repair stations on the loop. These stations are manually operated and are used to make minor repairs and adjustments to the engine. The engine is than retested in the repair station automatically by identical equipment and with the same test sequence.

A foreman's console is interfaced with all 48 test and repair stations, and display lights which indicate what is happening at each station. The system is designed to be interfaced to a central computer which maintains records of each engine and prepares statistical reports covering the tests. The system operates at a rate of 320 engines per hour, with each test loop operating at 160 units per hour. This amounts to an average testing and handling time of 7.5 min per engine on each test station.

FIGURE 19.9   One test station in system of Example 19.1. (Courtesy of Scans Associates, Inc.)

```
                    TEST RESULTS
    ENGINE NO                          : : : :2=549
    CODE TAG REF NO                    : :3
    TRANSACTION CODE                   : :2
    INSPECTOR CODE                     357
    TEST RUN TIME                      :71
    ENGINE STATUS                      REJECT

    DEFECT                             :21
    ADJ OIL PRESS            NN.N      223
    OVER HI LIM              NN.N      200
    PART THROT MODE

    DEFECT                             :32
    ADJ OIL P DIFF           NN.N      :15
    UNDER LO LIM             NN.N      :20

    DEFECT                             :45
    H20 TEMP DIFF                      : :6
    OVER HI LIM                        : :5
    IDLE MODE

                   CONTINUED
```

FIGURE 19.10   CRT display of engine test results from Example 19.1. (Courtesy of Scans Associates, Inc.)

19.8   INTEGRATION OF CAQC WITH CAD/CAM

Although many important benefits result from the use of computer-aided quality control, additional benefits can be obtained by integrating CAQC with CAD/CAM. Throughout the book we have emphasized the merits of an integrated CAD/CAM data base because of the need for both design and manufacturing to use the same basic information about the product. The design department creates the product definition and the manufacturing department makes use of and supplements this definition to develop the manufacturing plan. It is important to add the QC connection to the CAD/CAM framework. The quality control department must use the same CAD/CAM data base to perform its function. Indeed, quality was defined earlier in this chapter as the degree to which a product or its components conform to the standards specified by the designer. These standards are all contained in the CAD/CAM data base, available for QC to use.

One way in which the data base can be used is to develop the NC programs to operate the tape-controlled or computer-controlled coordinate measuring machines. These programs can be generated automatically or interactively by the same methods described in Section 8.8. These programs would then be downloaded to the CMM through a DNC link from the central computer to the controller unit for the CMM. The same sort of downloading process is possible for some of the noncontact inspection methods discussed earlier.

Another way in which a common data base is helpful to QC is when engineering changes are made to the product. Of course, engineering changes are liable to have an influence on inspection and testing. It is helpful for any changes to be recorded in a common data file for all departments, including QC, to use.

Finally, another area where CAD/CAM benefits the QC function is in computer production monitoring. The types of production records that are generated during computer monitoring are sometimes useful to the quality control department in tracing the cause of poor quality in a particular production lot.

REFERENCES

[1] BALLARD, D. H., AND BROWN, C. M., *Computer Vision*, Prentice-Hall, Inc., Englewood Cliffs, N.J., 1982.

[2] CHASE, R. B., AND AQUILANO, N. J., *Production and Operations Management*, 3rd, Richard D. Irwin, Inc., Homewood, Ill., 1981, Chapter 10.

[3] CHIN, R. T., "Automated Visual Inspection Techniques and Application: A Bibliography," *Pattern Recognition*, Vol. 15, No. 4, 1982, pp. 343–357.

[4] "Giving Machines Vision," *Quality Progress*, May, 1982, pp. 40–41.

[5] HARRINGTON, J., JR., *Computer-Integrated Manufacturing*, Industrial Press, Inc., New York, 1973, Chapter 9.

[6] HILDRETH, E. C., "Edge Detection for Computer Vision Systems," *Mechanical Engineering*, August, 1982, pp. 48–53.

[7] HUGHES, T., "Winning with Better Quality: Bull's Eye for Digital Inspectors," *Production Engineering*, July, 1982, pp. 38–44.

[8] "Laser-Based Automatic Inspection Systems," *Quality Progress*, December, 1981, pp. 26–27.

[9] Monks, J. G., *Operations Management*, McGraw-Hill Book Company, New York, 1977, Chapter 10.

[10]   PUMA, M., "Quality Technology in a Changing Manufacturing Environment," *Quality Progress*, August, 1980, pp. 16–19.

[11]   SCHAFFER, G., "A New Look at Inspection," Special Report 714, *American Machinist*, August, 1979, pp. 103–126.

[12]   STAUFFER, R. N., "Inspection Makes Its Mark," *Manufacturing Engineering*, August, 1980, pp. 168–172.

[13]   ZUECH, N., Personal correspondence dated November 1, 1982.

chapter 20

Computer-Integrated
Manufacturing Systems

20.1   INTRODUCTION

In the discrete product manufacturing industries, the most automated form of production is the computer-integrated manufacturing system. A variety of other names have been given to these systems, including flexible manufacturing system (FMS), variable mission manufacturing (VMM), and computerized manufacturing system. The different titles all refer to a production system which consists of a group of NC machines connected together by an automated materials handling system and operating under computer control.

Computer-integrated manufacturing systems (CIMS) incorporate many of the individual CAD/CAM technologies and concepts which we have discussed throughout this book. These include:

Computer numerical control (CNC)
Direct numerical control (DNC)
Computer process control
Computer-integrated production management
Automated inspection methods
Industrial robotics

Each CIMS is designed to meet the particular production requirements of the user company. Accordingly, there are differences among the designs of these systems. For example, many of the current CIMS installations do not include robots. The use of industrial robots as a component in these manufacturing systems is a somewhat recent approach.

Computer-integrated manufacturing systems are designed to fill the gap between high-production transfer lines and low-production NC machines. The relative position of the CIMS concept is illustrated in Figure 20.1. Transfer lines are very efficient when producing parts in large volumes at high output rates. The limitation on this mode of production is that the parts must be identical. These highly mechanized lines are inflexible and cannot tolerate variations in part design. A changeover in part design requires the line to be shut down and retooled. If the design changes are extensive, the line may be rendered obsolete. On the other hand, stand-alone NC machines are ideally suited for variations in workpart configuration. Numerically controlled machine tools are appropriate for job shop and small batch manufacturing because they can be conveniently reprogrammed to deal with product changeovers and part design changes. In terms of manufacturing efficiency and productivity, a gap exists between the high-production-rate transfer machines and the highly flexible NC machines. This gap includes parts produced in midrange volumes. These parts are of fairly complex geometry, and the production equipment must be flexible enough to handle a variety of part designs. Transfer lines are not suited to this application because they are inflexible. NC machines are not suited to this application because their production rates are too slow. The solution to this midvolume production problem is the computer-integrated manufacturing system.

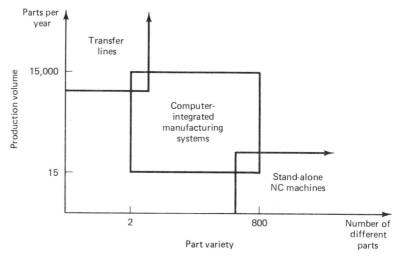

FIGURE 20.1   General application guidelines for the computer-integrated manufacturing system.

In this chapter we describe the types and components of computer-integrated manufacturing systems, as well as their application benefits.

20.2   TYPES OF MANUFACTURING SYSTEMS

The middle range in Figure 20.1, covering the medium part variety and medium production volume, can be further divided into finer categories. These categories represent different levels of compromise between the objective of flexibility versus production capacity. Kearney & Trecker Corporation defines three types of manufacturing systems to satisfy the variety of processing needs within this middle range [9,12]. They are:

1. Special manufacturing system
2. Manufacturing cell
3. Flexible manufacturing system (FMS)

Figure 20.2 illustrates the general application guidelines of each of the three types.

The special manufacturing system is the least flexible computer-integrated manufacturing system. It is designed to produce a very limited number of different parts (perhaps two to eight) in the same manufacturing family. The annual production rate per part would typically lie between 1500 and 15,000 pieces. The configuration of the special system would be similar to the high-production transfer line. The variety of processes would be limited, and specialized machine tools would not be uncommon.

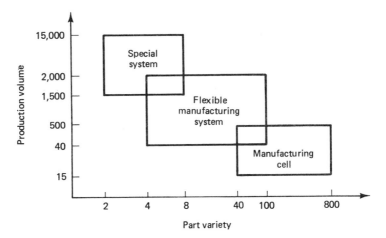

FIGURE 20.2   Application guidelines for the three types of computer-integrated manufacturing systems.

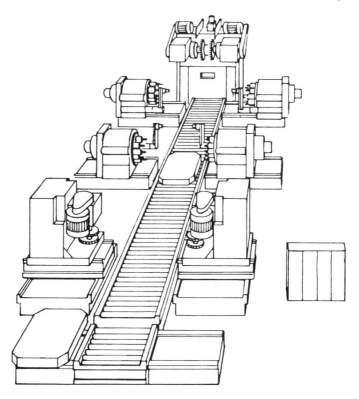

FIGURE 20.3   Special manufacturing system. (Reprinted from Ref. [9].)

At the opposite end of the midvolume range is the manufacturing cell. It is the most flexible but generally has the lowest production rate of the three types. The number of different parts manufactured in the cell might be between 40 and 800 and annual production levels for these parts would be between 15 and 500.

Figures 20.3 and 20.4 illustrates the special system and the manufacturing cell, respectively. The highly integrated and in-line flow is evident in the workpart handling system of Figure 20.3. As pictured in Figure 20.4, the manufacturing cell might consist of several separate NC machines without an interconnecting materials handling system.

The flexible manufacturing system covers a wide middle territory within the midvolume, midvariety production range. A typical FMS will be used to process several part families, with 4 to 100 different part numbers being the usual case. Production rates per part would vary between 40 and 2000 per year. Figure 20.5 illustrates a representative layout for a flexible manufacturing system.

Workparts are loaded and unloaded at a central location in the FMS. Pallets are used to transfer workparts between machines. Once a part is loaded onto the handling system it is automatically routed to the particular workstations required in its processing. For each different workpart type, the routing may be different, and

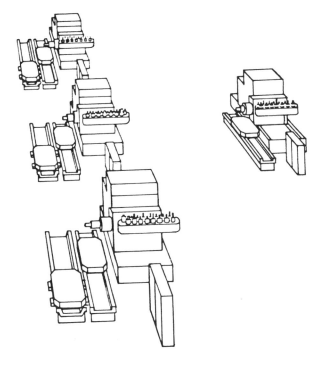

FIGURE 20.4   Manufacturing cell. (Reprinted from Ref. [9].)

the operations and tooling required at each workstation will also differ. The coordination and control of the parts handling and processing activities is accomplished under command of the computer. One or more computers can be used to control a single FMS. The computer system is used to control the machine tools and materials handling system, to monitor the performance of the system, and to schedule production. We cover these computer functions in more detail in Section 20.5.

Human labor is required to operate the CIMS. Among the functions performed are loading and unloading of workparts, changing tools, tool setting, and programming the computer system. Section 20.6 dicusses these human activities related to the operation of these manufacturing sytems.

A computer-integrated manufacturing system consists of the following basic components:

1.   Machine tools and related equipment
2.   Materials handling system
3.   Computer system
4.   Human labor

We discuss each of these components in the following four sections. The comments will be most applicable to the FMS, which should be considered as the gen-

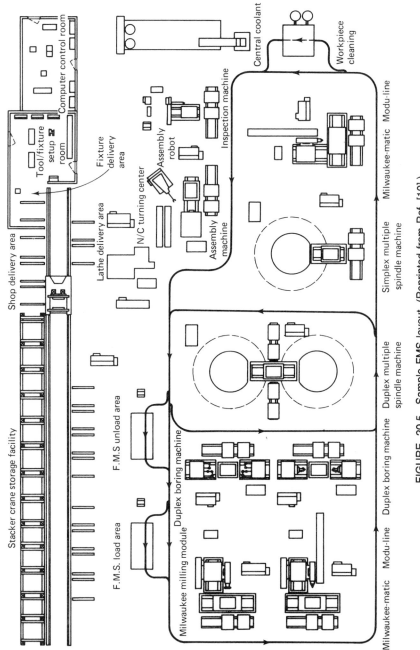

FIGURE 20.5 Sample FMS layout. (Reprinted from Ref. [10].)

eric computer-integrated manufacturing system. However, the components and operation of the special system and the manufacturing cell are generally similar to those of the FMS.

20.3 MACHINE TOOLS AND RELATED EQUIPMENT

The machine tools and other equipment that comprise a computer-integrated manufacturing system include the following:

Standard CNC machine tools
Special-purpose machine tools
Tooling for these machines
Inspection stations or special inspection probes used with the machine tools

Some of the standard NC machines used as FMS components are shown in Figures 20.6 and 20.7.

The selection of the particular machines that make up a CIMS depend on the processing requirements to be accomplished by the system. These processing needs also influence the design of the parts handling system. Some of the factors that define the processing requirements are the following:

FIGURE 20.6 Duplex multiple spindle head indexer used as module on CIMS. (Courtesy of Kearney & Trecker Corp.)

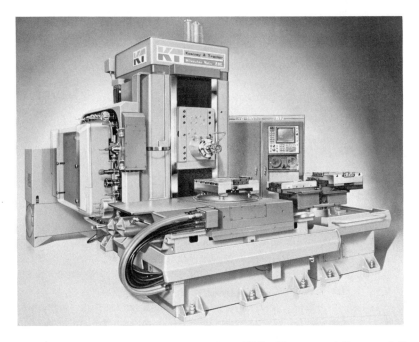

FIGURE 20.7 Machining center module used on CIMS. (Courtesy of Kearney & Trecker Corp.)

1. *Part size.* The size of the workparts to be processed on the CIMS will influence the size and construction of the machines. Larger parts require larger machines.

2. *Part shape.* Machined workparts usually divide themselves naturally into two types according to shape: round and prismatic. Round parts, such as gears, disks, shafts, requiring turning and boring operations. Prismatic workparts, which are cube-shaped and nonrotational, require milling and drilling operations.

3. *Part variety.* If the part variety is limited, the machine tools would be more specialized for higher production. The CIMS would be designed as a special system. If a wide variety of parts are to be processed, standard machine tools which are more versatile would be selected.

4. *Product life cycle.* The influence of product life cycle is similar to that of part variety. If the product life is relatively long, the CIMS can include more specialized and less flexible machine tools.

5. *Definition of future parts.* Another factor that affects the versatility of the CIMS is the level of knowledge about parts which are to be processed. We can distinguish two cases. The first case is where the manufacturing system is designed to process a family of parts that are completely known in advance. An example of this case is the dedicated manufacturing system for machining components of the XM-1 tank. The range of components is precisley known at the time of CIMS

design and the system can be configured to meet these specific needs. The other case is where the future parts are not known in advance. New part designs must be accommodated by the system; therefore, its machine tools must possess a significant degree of flexibility.

6. *Operations other than machining.* Most computer-integrated manufacturing systems are designed for machining exclusively. In some cases the processing requirements include other operations, such as assembly or inspection.

20.4 MATERIAL HANDLING SYSTEM

The material handling system in a CIMS must be designed to serve two functions. The first function is to move workparts between machines. The second function is to orient and locate the workparts for processing at the machines. These two functions are often accomplished by means of two different but connected materials handling systems. We shall refer to them as the primary handling system and the secondary handling system.

The primary work handling system is used to move parts between machine tools in the CIMS. The requirements usually placed on the primary material handling system are:

> It must be compatible with computer control.
>
> It must provide random, independent movement of palletized workparts between machine tools in the system.
>
> It must permit temporary storage or banking of workparts.
>
> It should allow access to the machine tools for maintenance, tool changing and so on.
>
> It must interface with the secondary work handling system.

The term "random, independent movement of parts" means that the parts must be able to flow from any one station (machine tool) to any other station. This requirement is not always necessary in the case of the special manufacturing system, which may involve an in-line flow of parts based on a fixed processing sequence.

The secondary parts handling system must present parts to the individual machine tools in the CIMS. The secondary system generally consists of one transport mechanism for each machine. The specifications placed on the secondary materials handling system are:

> It must interface with the primary handling system. Parts must be transferred automatically between the primary system and the secondary system.
>
> It must be compatible with computer control.
>
> It must permit temporary storage of workparts.

It must provide for parts orientation and location at each workstation for processing.

It should allow access to the machine tool for maintenance, tool changing, and so on.

An illustration of the primary and secondary work handling systems is found in Kearney & Trecker's FMS concept shown in Figure 20.8. The primary work handling system is an under-the-floor towline system which pulls a series of carts between the different workstations. The carts have four wheels which roll on the floor surface. Slots in the floor define the permissible pathways for the carts. These paths may branch and merge in a manner similar to that depicted in Figure 20.5. The layout of the pathways depend on the design of the FMS. Guide pins at the front and back of the carts engage the slot in the floor to follow the correct path. The front guide pin engages a moving chain in the slot, which propels the cart.

The secondary work handling system is the shuttle system at each machine. The shuttles are designed to transfer palletized parts to and from the towline carts, and to move them onto and off the machine tool table for accurate registration with the cutting tools. If the processing cycle at the workstation is relatively short, the

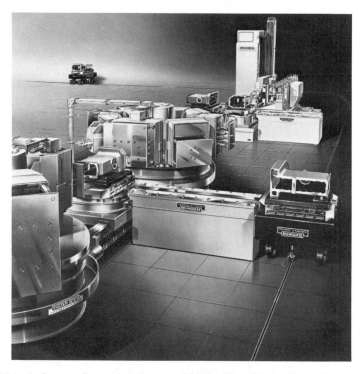

FIGURE 20.8 In-floor towline cart system combined with pallet shuttle system on a flexible manufacturing system. (Courtesy of Kearney & Trecker Corp.)

shuttle system may also be required to maintain an inventory of parts ready for machining. Both the towline cart and shuttle system are shown in Figure 20.8.

The use of both a primary and a secondary work handling system is not always required to satisfy the material handling objectives of a CIMS. In some cases, the objectives can be satisfied with one work handling system. In the special system illustrated in Figure 20.3, the fact that all parts are processed in the same sequence means that the same roller conveyor system can fully accomplish the in-line flow of parts needed for this manufacturing system.

A relatively recent innovation in the design of flexible manufacturing systems involves the use of industrial robots to perform a portion of the parts handling chore. Figure 20.9 shows the layout for a robotic manufacturing cell. The geometry of the workparts to be processed on the CIMS has a big influence on the type of work handling system. Round workparts have been found to be ideal candidates for handling by industrial robots. The robot is physically located in the center of a group of machine tools and transfers parts from one machine to the next in an independent,

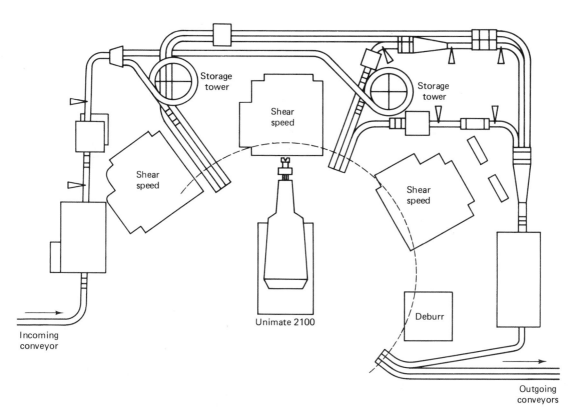

FIGURE 20.9 Layout of manufacturing cell with robot used for parts transfer between machines. (Reprinted from Ref. [14].)

random fashion depending on processing requirements for each part. It has been estimated that 75% of all rotational parts can be accommodated in a machining cell which uses a robot as the handling mechanism [8]. Most round parts are within the lift capacities of the common commercially available robots. The design problem is to develop a gripper device that is suited to the variety of parts handled in the cell.

Prismatic parts which are processed by machining are usually heavier and larger. Robots cannot generally be used to transport these parts within the manufacturing system. The handling problem is typically solved by mounting the part on a pallet fixture and designing a materials handling system (primary system and secondary system) which will transport the standard-size pallet as required. Typical solutions include the following types of materials handling systems:

Towline cart combined with shuttle system as illustrated in Figure 20.8, in a layout similar to that of Figure 20.5.

Roller conveyor system. Figure 20.3 illustrates this form of handling system. Guided vehicle combined with shuttle system. A wire-guided cart is illustrated in Figure 20.10.

A means of coordinating the activities of the materials handling system with those of the machine tools in the CIMS must be achieved. This is done by the computer control system.

FIGURE 20.10 Wire-guided cart used in variable mission manufacturing system. CNC machining center in background. (Courtesy of Cincinnati Milacron.)

20.5 COMPUTER CONTROL SYSTEM

This section describes how the digital computer system is used to manage the operation of a complex manufacturing system. We discuss the various functions performed by the computer, the data files needed to carry out these functions, and the various types of reports that the computer can be programmed to prepare.

Functions of the computer in a CIMS

The functions accomplished by the computer control system can be divided into eight categories. The following descriptions apply best to the case of the flexible manufacturing system. To a slightly lesser extent, they also apply to the special system and the manufacturing cell.

1. *Machine control.* This is usually accomplished by computer numerical control (CNC). The advantage of CNC is that it can be conveniently interfaced with the other elements of the computer control system.

2. *Direct numerical control (DNC).* Most computer-integrated manufacturing systems operate under DNC. In some of the special systems which are dedicated to a limited part variety, CNC may be a sufficient control method for the system. The purpose of direct numerical control is to perform the usual DNC functions, including NC part program storage, distribution of programs to the individual machines in the system, postprocessing, and so on.

3. *Production control.* This function includes decisions on part mix and rate of input of the various parts onto the system. These decisions are based on data entered into the computer, such as desired production rate per day for the various parts, numbers of raw workparts available, and number of available pallets. The computer performs its production control function by routing a pallet to the load/unload area and providing instructions to the operator to load the desired raw part. A *data entry unit* (DEU) is located in the load/unload area for communication between the operators and the computer.

4. *Traffic control.* This term refers to the regulation of the primary workpiece transport system which moves parts between workstations. This control can be accomplished by dividing the transport system into zones. A zone is a section of the primary transport system (towline chain, conveyer, etc.) which is individually controlled by the computer. By allowing only one cart or pallet to be in a zone, the movement of each individual workpart is controlled. The traffic controller operates the switches at branches and merging points, stops workparts at machine tool loading points, and moves parts to operator load/unload stations.

5. *Shuttle control.* This is concerned with the regulation of the secondary part handling systems at each machine tool. Each shuttle system must be coordinated with the primary handling system, and it must also be synchronized with the operations of the machine tool it serves.

In cases where there is only one parts handling system (rather than a primary and secondary handling system) the functions of traffic control and shuttle control may be combined. This would be the case in some of the special systems and certain robotic work cells.

6. *Work handling system monitoring.* The computer must monitor the status of each cart and/or pallet in the primary and secondary handling systems as well as the status of each of the various workpart types in the system.

7. *Tool control.* Monitoring and control of cutting tool status is an important feature of the computer system. There are two aspects to tool control: accounting for the location of each tool in the CIMs and tool-life monitoring.

The first aspect of tool control involves keeping track of the tools at each station in the system. If one or more tools required in the processing of a particular workpart are not present at the workstation specified in the part's routing, the computer control system will not deliver the part to that station. Instead, it will determine an alternative machine to which the part can be routed, or it will temporarily "float" the part in the handling system. In the second case, the operator is notified via the data entry unit what tools are required in which workstation. The operator then manually loads the tools and notifies the computer accordingly. Any type of tool transaction (e.g., removal, replacement, addition) must be entered into the computer to maintain effective tool control.

The second aspect of tool control is tool-life monitoring. A tool life is specified to the computer for each cutting tool in the CIMS. Then a file is kept on the machining time usage of each tool. When the cumulative machining time reaches the life for a given tool, the operator is notified that a replacement is required.

8. *System performance monitoring and reporting.* The system computer can be programmed to generate various reports desired by management on system performance. The types of reports are discussed later in this section.

These computer functions can be accomplished by any of several different computer configurations. One computer can be used for all components of the CIMS, or several different computers can be used. Up to three levels are practical in a given manufacturing system. CNC would be used for control of each individual machine tool. DNC would be appropriate for distribution of part programs from a central control room to the machines. A third level would concern itself with production control, the operation of the work handling system, tool control, and generation of management reports.

CIMS data files

To control the operation of the manufacturing system, the computer relies on data contained in files. The principal data files required are of the following six types:

1. *Part program file.* The part program for each workpart processed on the system is maintained in this file. For any given workpart, a separate program is required for each station that performs operations on the part.

2. *Routing file*. This file contains the list of workstations through which each workpart must be processed. It also contains alternate routings for the parts. If a machine in the primary routing is down for repairs or there is a large backlog of work waiting for the machine, the computer will select an alternate routing for the part to follow.

In the special system, where the routings would be identical or similar and the parts proceed through the system in an in-line flow, the importance of the routing file is substantially reduced.

3. *Part production file*. A file of production parameters is maintained for each workpart. It contains data relative to production rates for the various machines in the routing, allowances for in-process inventory, inspections required, and so on. These data are used for production control purposes.

4. *Pallet reference file*. A given pallet may be fixtured only for certain parts. The pallet reference file is used to maintain a record of the parts that each pallet can accept. Each pallet in the CIMS is uniquely identified and referenced in this file.

5. *Station tool file*. A file is kept for each workstation, identifying the codes of the cutting tools stored at that station. This file is used for tool control purposes.

6. *Tool-life file*. This data file keeps the tool-life value for each cutting tool in the system. The cumulative machining time of each tool is compared with its life value so that a replacement can be made before complete failure occurs.

System reports

The data collected during monitoring can be summarized for preparation of performance reports. These reports are tailored to the particular needs and desires of management. The following categories are typical:

1. *Utilization reports*. These are reports that summarize the utilization of individual workstations as well as overall average utilization for the system.

2. *Production reports*. Management is interested in the daily and weekly quantities of parts produced from the CIMS. This information is provided in the form of production reports which list the required schedule together with actual production completions. One possible format for the production report is illustrated in Figure 20.11.

3 *Status reports*. Line supervision can call for a report on the current status of the system at any time. A status report can be considered an instantaneous "snapshot" of the present condition of the CIMS. Of interest to supervision would be status data on workparts, machine utilization, pallets, and other system operating parameters.

4. *Tool reports*. These reports relate to various aspects of tool control. Reported data might include a listing of missing tools at each workstation. Also, a tool-life status report can be prepared at the start of each shift, similar to the one

PART SUMMARY

SHIFT 1   PAGE 01 02/13   17:35:44

02/13 09:28 TO 02/13 17:35   8.1 HRS.

	P/S	SCHED.	COMPL.	DIFF.	PCT.
268923	1	7	6	−1	−14.3
268923	2	7	9	2	28.6
268315	1	7	8	1	14.3
268315	2	7	8	1	14.3
268315	3	7	7	0	0.0
267171	1	7	5	−2	−28.6
267171	2	7	9	2	28.6

FIGURE 20.11 Production shift report, indicating CIMS performance. (Reprinted from Ref. [10].)

TOOL LIFE STATUS FOR STATION 3   PAGE 01   07/16   11:40'22

TOOL NUMBER	EST. LIFE	ACT. USE	PCT. OF EST.
00003	500.0	600.6	120.1
00165	50.0	3.0	6.0
00166	200.0	3.0	1.5
00173	91.5	22.0	24.0
10011	800.0	0.0	0.0
10014	135.0	3.9	2.9
10017	225.0	236.0	104.9
10020	300.0	59.0	19.7
10021	115.0	59.0	51.3
10023	210.0	0.0	0.0
10025	142.0	118.0	83.1
10027	100.0	0.0	0.0
10032	300.0	0.0	0.0
10034	999.0	0.0	0.0
10035	320.0	112.0	35.0
10036	380.0	1188.7	312.8
10037	35.0	85.0	242.9
10056	600.0	6.0	1.0
10057	400.0	118.0	29.5
10061	400.0	472.0	118.0
10062	350.0	472.0	134.9
10064	65.0	91.0	140.0

FIGURE 20.12 Tool status report for one machine tool in a CIMS. (Reprinted from Ref. [10].)

illustrated in Figure 20.12. This listing shows that several of the tools have been used well beyond their anticipated lives and are in need of replacement.

20.6 HUMAN LABOR IN THE MANUFACTURING SYSTEM

The computer-integrated manufacturing system is a highly automated production facility. However, human resources are required to operate the system. In the majority of CIMS installations, the individual machines are operated under CNC or DNC control (or a combination of these). The machines are not manually operated except in certain special operations, such as assembly. Personnel are required principally to manage, maintain, and service the CIMS. Paprocki [12] lists the following personnel requirements for a computer-integrated manufacturing system:

1. *System manager*. This person has overall responsibility for the operation of the CIMS. The functions include production planning, responding to deviations and exceptions to normal operations, and supervision of the other human resources which support the system.

2. *Electrical technician*. This person is often a member of the plant's electrical maintenance crew. Duties performed include maintenance and repair services on the electrical components of the machine tools and materials handling system.

3. *Mechanical/hydraulic technician*. Again, this person is likely to be a regular member of the plant maintenance department. Technical services consist of maintenance and repair of the mechanical and hydraulic components of the CIMS.

4. *Tool setter*. The tool setter is responsible for the tooling inventory and making the tools ready for production.

5. *Fixture setup and lead man*. This person is responsible for setting up the fixtures, pallets, and tools for the system.

6. *Load/unload man*. This person is responsible for loading raw workparts and unloading finished parts. This is typically done according to instructions and schedules generated by the computer. The load/unload area is at a convenient central location in the manufacturing system.

7. *Rover operator*. The duties of the rover operator include reacting to unscheduled machine stops, identifying broken tools or tools in need of immediate replacement, tool adjustments, and so forth. This person may also be responsible for certain manual production tasks or inspection operations.

Although each of these functions must be accomplished, this does not necessarily mean that seven people are required full time to operate the CIMS. The electrical technician and the mechanical/hydraulic technician are required on an on-call basis. In most companies, these workers would report to the plant maintenance department. In the startup of a new manufacturing system they might be required full time to solve the many problems which are typically associated with a new installation of complex equipment. Under normal operation, these technicians would devote only part of their time to the CIMS.

The remaining five functions might be combined to some extent to reduce the actual number of people required. The duties of the tool setter, lead man, and rover operator might be shared by two persons rather than three. The amount of human resources needed to operate the manufacturing system will depend to a large degree on its size, number of processing machines, and level of sophistication and automation.

In addition to these seven operational functions, NC part programmers, computer programmers, and related support staff are required to prepare the programs by which the computer system will control production and monitor performance.

20.7 CIMS BENEFITS

Computer-integrated manufacturing systems are intended for midvolume, midvariety production situations, as illustrated in Figure 20.1. Competing methods of production include group technology manufacturing cells[1] and conventional process layout shops which utilize stand-alone NC machines or special non-NC machines. When applied in the proper circumstances, a CIMS offers the following benefits:

1. *Increased machine utilization.* Most NC machines may operate at about 50% utilization or less. Because of minimum setup times, efficient workpart handling, simultaneous workpart processing, and other features, the utilization of a flexible manufacturing system may run as high as 85%. Downtime on a computer-integrated manufacturing system typically results from problems similar to those that plague a stand-alone NC machine. These include tooling problems, electrical and computer problems, and mechanical and hydraulic problems. Other reasons for downtime on a CIMS include scheduled maintenance and tool changeovers.

2. *Reduced direct and indirect labor.* Given the higher utilization and greater productivity of a CIMS, the cost of direct and indirect labor per unit of production is lower for this form of production than for the alternatives. In the operation of a CIMS, the labor allocation may be three or four workers for 6 to 10 machine tools. In the typical operation of many stand-alone NC machines, one operator is used for each machine. Accordingly, the ratio of direct labor cost to machines is reduced for the CIMS. Indirect labor for the CIMS is reduced compared to job shop operation through automated material handling, as opposed to manual parts handling in conventional batch production.

3. *Reduced manufacturing lead time.* Most workparts require processing in batches through several different work centers. There is setup time and waiting time at each of the work centers. With computerized manufacturing systems, the nonoperation time is drastically reduced between successive workstations on the line. Also, setup time is minimized in the CIMS operation. The setup for a traditional production machine consists of two main elements: tooling setup and workpart setup. Tooling setup means collecting the required tools from the tool crib and setting them in the machine. Workpart setup involves adjusting the workholding fixture, getting the raw materials ready, and so on. The tooling for a CIMS is preset off-line. The tool setup for a particular workstation consists of loading the preset tools required for the job into the tool drum at that station. Each tool drum may be capable of holding up to 60 or more cutting tools. The workpart setup is

---

[1]In many respects, a CIMS is a group technology manufacturing cell.

performed external to the CIMS. Since pallets are used to transport parts from station to station, the setup consists of adapting the pallet to the particular part for holding it. This setup is accomplished in the load/unload area before the part and pallet are launched onto the system. Fixtures are designed that can adapt to various part configurations within a part family. Several different adaptable fixtures may be required to handle the various part families. The parts are clamped in place onto the fixture, and the fixture is attached to the pallet. The CIMS has a large number of these pallets (the pallets may outnumber the workstations by 5 to 1), so some pallets can be off the system being loaded or unloaded, while other pallets are being used on the system. These features of the manufacturing system allow the processing lead time to be significantly reduced.

4. *Lower in-process inventory.* The float of workparts in process on a CIMS is significantly below the corresponding level for conventional batch manufacturing. In-process inventory is related closely to production lead time. Because the parts spend a relatively short time in the system, the number of parts being processed is low.

5. *Scheduling flexibility.* Random launching of workparts onto the system means that any workpiece handled by the CIMS can be introduced without downtime for setup. When the part is launched onto the system, the computer system routes it to the proper machines. The only limitation is that the workstations must be equipped in advance with the tooling required to process the particular workpart. This method of operation gives the system considerable flexibility to deal with changes in the production schedule.

It should be noted that the three types of CIMS vary in their scheduling flexibility. The special system is the least flexible and the manufacturing cell is the most flexible. In the case of the flexible manufacturing system, it is desirable to maintain a mixture of parts on the system. The FMS is designed to process various part configurations simultaneously by using its workstations concurrently. If only parts of the same type are launched onto the system, the workstations required for the particular part type will tend to be fully utilized, while other stations will be underutilized. As a result of this consideration, it is desirable to schedule a mix of parts on the FMS.

REFERENCES

[1] CINCINNATI MILACRON, *Concept for Variable Mission Manufacturing Systems* (marketing brochure), Cincinnati, Ohio, 1979.

[2] CINCINNATI MILACRON, *Variable Mission Modular Manufacturing Systems* (marketing brochure), Cincinnati, Ohio, 1980.

[3] CURTIN, F. T., "Manufacturing Systems Moving Stage Center in Dramatic Search for Better Productivity," *N/C Commline*, January/February, 1979, pp. 34–35.

[4] GROOVER, M. P., *Automation, Production Systems, and Computer-Aided Manufacturing*, Prentice-Hall, Inc., Englewood Cliffs, N.J., 1980, Chapter 19.

[5] HATSCHEK, R. L., "Guided Carts Link Machines into Systems," *American Machinist*, August, 1980, pp. 97−100.

[6] HUBER, R. F., "90% Uptime from Unmanned Machining Systems," *Production*, June, 1979, pp. 68−73.

[7] HUBER, R. F., "Planning for a Flexible Machining System," *Production*, August, 1981, pp. S-67 − S-85.

[8] JABLONSKI, J., "Aiming for Flexibility in Manufacturing Systems," Special Report 720, *American Machinist*, March, 1980, pp. 167−182.

[9] KEARNEY & TRECKER CORP., *K & T's World of Manufacturing Systems*, Milwaukee, Wis., 1980.

[10] KEARNEY & TRECKER CORP., *Understanding Manufacturing Systems* (a series of technical papers). Vol. I, Milwaukee, Wis.

[11] MATTOX, J. E., "Industrial Robots and Integrated Manufacturing Systems," *N/C Commline*, January/February, 1979, pp. 41−45.

[12] PAPROCKI, J. T., "Flexible Manufacturing Systems Automating the Factory," technical paper presented to the 2nd International Manufacturing Management and Technology Conference, Munich, Germany, June, 1979.

[13] SPROW, E. E., "FMS: On the Launch Pad," *Tooling and Production*, January, 1982, pp. 74−80.

[14] UNIMATION, INC., *Unimate Industrial Robot System Planbook*, Danbury, Conn.

PART VIII

CAD/CAM Implementation

chapter 21

Implementing a CAD/CAM System

21.1 INTRODUCTION

In previous chapters we have explored the many uses of computer systems in design and manufacturing. We have discussed computer-aided design—the hardware, software, and how CAD systems are applied to the design of a product. And we have discussed computer-aided manufacturing—numerical control, robots, computerized production management systems, computer-controlled manufacturing cells, and a host of other topics related to CAM. Throughout the discussion, the importance of integrating the design and manufacturing functions with CAD/CAM has been emphasized.

In this chapter we attempt to develop the guidelines and criteria for implementing CAD/CAM. By this we mean the implementation of a CAD/CAM system. The system would include the following features:

An interactive graphics system and associated software for design.

Software packages for manufacturing. Typically, these would have to be customized to the needs of the user, but would be likely to include programs such as NC part programming, automated process planning, fixture design, and various other aids for production.

A common CAD/CAM data base organized to serve both design and manufacturing.

A CAD/CAM system with these general features could be acquired by a user company in the following ways:

1. Purchase the system from one of the CAD/CAM vendors.
2. Extend an existing corporate computer system by adding graphics devices and software for CAD/CAM.
3. Develop a CAD/CAM capability from scratch within the user company. This would require the user company to acquire computer and graphics hardware and to develop the associated software.

Whereas the third alternative may have been a feasible approach a decade or more ago, the CAD/CAM technology has since progressed to the point where this is probably no longer feasible except under special and unusual circumstances. Our discussion on implementation of CAD/CAM will emphasize the first alternative: purchasing a system from one of the CAD/CAM system vendors. Such a commercially available system is referred to as a turnkey system. However, the criteria we will develop are applicable to all three approaches.

It is quite appropriate at the outset for the prospective user company to question whether it needs a CAD/CAM system. The question should be answered by analyzing current and projected work loads in the company. The analysis might be based on the following general approach.

1. Determine the size of the work load in the various applicable functions of the company (e.g., design, manufacturing planning, etc.)
2. Estimate the amount of this work load that is applicable to a CAD/CAM system.
3. Establish priorities among the different areas in item 2.
4. Use the prioritized list of work load areas developed in item 3 to develop the features of an appropriate CAD/CAM system for the company. These features should be divided into two major categories:
   "Must" features
   "Desirable" features

This general approach may lead to the conclusion that a CAD/CAM system is not appropriate for the company. Either the work load applicable to CAD/CAM is not sufficient to warrant the expense of the system or the list of features ("must" features, in particular) includes a significant number of items which are not available on existing CAD/CAM systems. In the second case, the company may be faced with circumstances under which it would want to develop its own CAD/CAM capability or to extend the capability of an existing computer system.

Assuming that the company's conclusion is that a CAD/CAM system is needed, the following general procedure and guidelines are appropriate for implementing the system. Our procedure is based on the selection of a turnkey system rather than in-house development.

1. Develop the criteria for selecting a turnkey CAD/CAM system. The criteria should be defined with the specific needs of the the user company in mind.
2. Study and visit other companies using CAD/CAM systems with similar needs. Assess their effectiveness in similar applications.
3. Study the CAD/CAM system vendors. Invite them in to make presentations on their companies and product lines.
4. Reduce the list of vendors down to the three or four most attractive candidates.
5. Determine a benefit/cost ratio (a system value per cost for the specific needs of the user company) for each system under consideration.
6. Invite the vendor with the highest ratio to run a benchmark. A benchmark consists of one or more user-specific problems which are representative of the typical applications expected of the CAD/CAM system by the user company.
7. If the benchmark test is successful, the vendor is officially selected. If not successful, the second choice, based on benefit/cost ratio, is selected for benchmark testing and potential contract award.

In the remainder of this chapter we discuss turnkey CAD/CAM systems and examine the foregoing procedure in more detail.

21.2 TURNKEY CAD/CAM SYSTEMS

The term "turnkey system" indicates that a vendor will deliver the product, install it, check it out, and provide it in a form ready to use. A turnkey CAD/CAM system would be installed with certain basic software to run on the system. Additional software programs would typically be available at extra cost. This allows the user company to purchase only those programs that are appropriate for its applications. The cost of a turnkey CAD/CAM system can run from several hundred thousand dollars to several million. Nevertheless, turnkey systems are often preferable to in-house development when the many costs and disadvantages of internal development are considered.

Disadvantages of in-house CAD/CAM development

In order for a user company to develop a complete graphics system internally, the company would require a staff with expertise in computer graphics. This staff would include graphics programmers. A good graphics programmer must be competent in computer programming, must understand graphics, and must be able to relate the two skills to solving applications problems. This combination of skills is not widely available, and most of the best people are in great demand by companies that are developing and implementing CAD/CAM systems.

Another problem for the potential in-house developer is the time factor. Even if a sophisticated programming staff is available, it takes time to develop a project as sophisticated and difficult as a CAD/CAM system. There is a learning process involved, and moving up the learning curve is a long, expensive process. To develop an in-house capability could take years of effort. By contrast, the CAD/CAM vendors have already invested this time and effort and have already experienced the learning process. If a capability in computer-aided design is desired by the user company in the minimum possible time, the vendors are ready to satisfy that desire with an available product line.

A CAD/CAM vendor's business is CAD/CAM systems. The vendor can and must exert considerable effort toward continued development in the technology in order to remain competitive. Hundreds of person years are devoted annually by these vendors to develop the software and hardware for CAD/CAM. The scope and intensity of this activity cannot be matched by the normal user company. To the user, CAD/CAM is a means to an end. To the vendor, CAD/CAM is the means and the end.

Extending a turnkey CAD/CAM system

Many user companies desire features in a CAD system, particular in the form of application programs, which are not provided by the system vendor. Accordingly, it is common for the user to purchase a commercial CAD/CAM system and customize it with extensions developed in-house. The resulting system becomes ideally suited to the user's specific requirements.

User programs for a CAD/CAM system are usually written either in a graphics programming language (GPL) that is provided by the vendor or in another high-level language (such as FORTRAN). The first alternative is best utilized for applications which are basically graphics routines. FORTRAN is suited for analysis modules or for interfacing the new system with previously written programs.

In attempting to extend the turnkey system, modification of the vendor's source code should be avoided. This modification is not possible in some cases because the vendors are reluctant to provide the source code for the system. This reluctance is based on proprietary considerations of the vendor and the difficulty in maintaining the software when modifications are made to it. What is compatible with today's version of source code may not be compatible with future software releases. To avoid the maintenance problem, some users have expended substantial sums to develop software extensions only to turn them over to the original CAD/CAM vendor with the understanding that the vendor would maintain the software. Both user and vendor benefit from this arrangement. The vendor receives a marketable software package at no cost, and the user is relieved of the maintenance problem.

Software development is expensive. A figure typically cited when referring to software development cost is $8 to $15 per line of FORTRAN code. This range includes all aspects of design, coding, debugging, testing, and documenting the program. Although this figure is inexact and depends on program complexity and length per line of code, it serves to indicate that large software development projects are expensive and time consuming. Large-scale application programs can cost hundreds of thousands of dollars.

When it becomes necessary to develop application software, efforts should be made to make the code as independent of the particular CAD/CAM system as possible. This can be done by writing the program in standard code (e.g., an ANSI standard version of FORTRAN) to ensure its portability. Table 21.1 presents a recommended standard outline for documenting the software. The value of a clearly written user's manual cannot be overstated.

Turnkey system prices

Prices from different vendors are similar for turnkey CAD/CAM systems of comparable features and capabilities. The CAD/CAM market is quite competitive, and this prevents significant variations in system pricing. The price of a typical three-

TABLE 21.1   Recommended Standard Outline for Software Documentation

Abstract

Theory and principles of the method used in the program

Scope and limitations of program

Hardware and software requirements

Output produced by the program

Numerical techniques employed

Overall program structure, including flowcharts and other similar documentation

Subprogram logic

Dictionary of variables used in the program

Sample runs showing major program features; these should include input/output examples

dimensional mechanical design system would be in the vicinity of $350,000. This price would be for a basic system with two or three workstations and a plotter.

The price of the CPU, disk and tape storage, and system console is typically in the range $130,000 to $250,000. A design workstation is priced at $40,000 to $50,000, and plotters of various types would range in price from $20,000 to $150,000. These various hardware units are discussed in Chapter 5. The reader can appreciate how the price of a CAD/CAM system would vary depending on the components used to comprise the system. To illustrate, let us specify a minimum and a maximum three-dimensional mechanical design system with the types of units indicated above. The minimum system would include the following:

CPU, storage, and console	$130,000
Plotter	20,000
Design workstations (two at $40,000)	80,000
Applications software	20,000
Total	$250,000

A larger, more powerful system with four workstations might include the following:

CPU, storage, and console	$250,000
Plotter	100,000
Design workstations (four at $50,000)	200,000
Applications software	50,000
Total	$600,000

CAD applications software is included in both cases. It should be noted that price bargaining between vendor and potential customer often revolves around the software.

The prices indicated above are approximate and are applicable at the time of this writing. These prices are, of course, subject to change due to inflation, the impact of new technologies, economies of scale in future manufacturing, and other factors. For example, a reduction in the price of CPUs and certain other hardware items can be expected if the future follows the trends of the past.

21.3 SELECTION CRITERIA

It is a difficult decision to select among all the competing hardware and software alternatives available in CAD/CAM systems. Assuming that the decision has been made to purchase a turnkey system, the first step is to develop the selection criteria. Rather than having a single person attempt to develop these criteria, a CAD/CAM evaluation committee should be set up for this purpose.

The purpose of the CAD/CAM evaluation committee would be not only to develop the selection criteria, but also to evaluate the alternative systems, and make the recommendation of which vendor to select. This committee would follow the selection procedure from beginning to final decision. The evaluation committee should have representatives from interested departments throughout the user company. It would typically include members from design engineering, manufacturing engineering, industrial engineering, management information systems, and purchasing.

Development of the selection criteria involves determining the precise needs of the user company for a CAD/CAM system. However, this should not be a process performed exclusively for current functions within the company. The committee should consider future needs as well as current needs. The use of a CAD/CAM system will present opportunities for improving the procedures of the user company, and these opportunities should not be overlooked. The committee should include among its sources of ideas the pertinent manuals and technical information from each vendor interested in bidding on the project.

It is important to conduct technical discussions with experts from the vendors and to visit several installations with configurations similar to the system under consideration. During the visits, the opinions and experiences of the personnel who manage and operate the system should be noted. In addition to visits to user companies, a trip to the corporate headquarters of each vendor should be arranged. This will provide information regarding the plans and philosophies of the various CAD/CAM companies.

From all of these various sources it should be possible to develop a comprehensive list of applicable criteria for the prospective user company. A representative checklist of considerations and criteria for selecting a CAD/CAM system is presented in Table 21.2. This table was compiled principally from References [4] and [5]. The criteria and considerations that should be used for each company are different, and any general checklist such as Table 21.2 should be adapted to the specific needs of a particular company. Nevertheless, this table should be useful to a prospective user company in developing its own criteria for evaluating alternative CAD/CAM systems.

Some of the more difficult issues to deal with in evaluating a CAD/CAM system are [4]: first, how will the user company evaluate the ease or complexity of interfacing with the current information systems that are concerned with product data or product logistics data? These data include purchasing, MRP, vendor tracking, quality, and scrap-and-rework accounting systems. Second, how will the user company accommodate system growth to encompass more users, more workstations, networking, new products, and new production technologies, without abandoning the short term need to achieve early benefits? These issues are different for each business, and no general recommendation can be given. A prospective user

TABLE 21.2 Checklist of Considerations and Criteria for Selecting a CAD/CAM System

A. *General Considerations*

1. Cost

   Hardware
   > CPU
   > Added Stations
   > Peripherals

   Software
   > Turnkey basic package
   > Added specialty packages

   Hardware maintenance
   Software maintenance
   Specials
   Spares
   Documentation
   Training
   Transportation
   Facilities
   Supplies
   Field support
   Support personnel

2. Service

   Contract
   Parts location
   Turnaround
   Warrantee and discontinuance clause
   Software bug service
   Software service

3. Quality

   User group existence and support
   Corporate quality
   > Responsiveness
   > Finanical stability
   > Number of installations
   > Growth

   Reliability
   Simultaneous operation
   Crash recovery
   Power loss recovery
   Environmental sensitivity
   Human factors considerations
   > Hardware
   > Software
   > Response time

   Output device speed
   Interfaces
   > Communication
   > Plotters

   Product documentation
   Training

4. Delivery and logistics support

   Staging/in-plant benchmark
   Packing
   Installation aid
   Pre-delivery inspection
   Installation guide
   Receipt acceptance
   Supplies
   Revisions
   Billing
   Proprietary agreement
   On-site debugging

5. System management

   High speed peripherals
   User diagnostics
   File management software
   Expense reports and logging

6. Programming

   High level vendor language
   Standard language
   Assembly language
   Source availability
   Clear documentation
   User protection
   Diagnostic aids
   Data base access
   Clear interfaces

7. Miscellaneous other considerations

   Compatibility with existing equipment
   Graphic terminal control
   Operator input methods
   Prompting
   Plotting control

B. *Applications—Electrical*

1. Electrical design applications

   Logic diagrams
   Circuit diagrams
   Schematics
   Electrical wiring
   > Cables
   > Harnesses
   > From/To lists

   Integrated circuits
   Printed circuit boards
   Hybrid circuits
   Rules checking
   Bill of materials

TABLE 21.2 (*continued*)

2. Electrical interfaces
   Circuit analysis
   Logic and timing
   Field analysis
   Router
   Test vector generator
3. Electrical numerical control packages
   Drill
   Board router
   Autoplacement/Insertion
   Cable weaving
   Wire wrap
4. Documentation
   Handbooks
   Perspective
   Organization charts
   Flow diagrams
   PERT charts and scheduling
5. Interface
   Photoplotter
   Line plotters/scribe
   Point plotters

C. *Applications—Mechanical*

1. Mechanical design
   Logic diagrams
   Process diagrams
   Schematics
   Architectural drawing
     Plant layout
     Structural steel design
     Piping
     From/To lists
     Mapping
   Sheet metal design

Two-dimensional mechanical design and drafting
   Point functions
   Line functions
   Arc functions
   Conic functions
   Dimensioning
   Line type
   Text, arrows
   Witness line suppression
   Character height
   Flag note
Three-dimensional mechanical design and drafting
   Wire-frame or solid modeling
   Hidden line removal capabilities
   Surfaces and planes
   Curves and curved sections
Tool design
Jig and fixture design
Nameplates
Dimensionless drawings

2. Mechanical interfaces
   Stress analysis
   Finite element modeling (FEM)
   Modal analysis
   Flow analysis
   Mechanism analysis
   Mass properties determination
   Geometric properties
   Spline analysis
3. Manufacturing planning packages
   Computer-aided process planning capabilities
   Metal forming NC programming
   Nesting of sheet metal parts for efficient cutting
   Machining NC programming
   Automatic NC functions
   Other NC functions
   NC default values
   NC interfaces

Source: Adapted from References [4],[5].

company must incorporate these kinds of questions as subjective factors in their decision process.

21.4   EVALUATION OF ALTERNATIVE SYSTEMS

At this point in the procedure, there should be three or four most attractive vendors and their respective systems remaining. The rest have been dropped due to lack of interest on their part, or a sense that their technical emphasis was less compatible with the user company business, or for various other subjective reasons. For the remaining candidates, it is appropriate to devise some sort of quantitative measure which relates the value of each system to its cost. We might construct this measure in such a way as to refer to it as a benefit/cost ratio.

The benefit/cost ratio represents an attempt to deal with the selection process in an orderly quantitative manner. It is not an objective measure because its components are determined by the collective judgments of the individuals who comprise the CAD/CAM evaluation committee. Despite this unavoidable imperfection, the ratio offers the virtue of being an organized, systematic, and quantitative procedure for dealing with the selection problem.

The CAD/CAM benefit/cost ratio is determined by the following method. The list of criteria and considerations from Table 21.2 will be used as a starting point in the following discussion. For a given user company this list may vary depending on the company's specific needs. For example, a company which manufactures mechanical components will want to emphasize mechanical applications and minimize the importance of electrical applications in their list of CAD/CAM criteria. For each of the three major headings in the list, a weight is assigned that reflects the importance of that general factor to the user company. For example, the weights might be assigned as follows:

A.	General Considerations	50
B.	Applications — Electrical	15
C.	Applications — Mechanical	35
		100%

Within each of the categories a total of 1000 possible points are allocated among the various criteria within the category. For example, under General Considerations, the point allocation might be as follows:

1.	Cost	200
2.	Service	150
3.	Quality	250
4.	Delivery and logistics support	100
5.	System management	50
6.	Programming	150
7.	Miscellaneous other considerations	100
	Total possible points	1000

Then, within each of these headings, points would be assigned to the various subdivisions in a similar manner. The process of allocating points to the various criteria is based on the judgment of the committee as to the relative importance of each particular criterion.

After this allocation process is completed, it is the task of the committee to rate the candidates relative to each individual criterion. Out of the possible number of points for a given criterion, the committee must decide how many points each candidate should receive based on its merits in that category.

Upon completion of the scoring procedure, the scores for each candidate are summed within each of the three major headings. The sums are then multiplied by their weighting factors for that heading and added together to get the final total score for each candidate. For example, the scores for a hypothetical vendor X might be as shown in the following table:

	Criteria	Score		Weight		Weighted Score
1.	General Considerations	650	by	50%	=	325
2.	Applications — Electrical	600	by	15%	=	90
3.	Applications — Mechanical	800	by	35%	=	280
	Total CAD/CAM system score					695

Each of the three or four candidates would have a total score which presumes to reflect the relative value of the system.

It might be tempting at this point to select the CAD/CAM system on the basis of this score. However, we have not yet taken the price of the system into account. This is accomplished by forming the benefit/cost ratio. The ratio is determined by dividing the total CAD/CAM system score by the associated price of that system. It is most convenient to express the price to the nearest $1000 and to truncate the three trailing zeros. To illustrate, suppose that there are three candidates with scores and prices as shown in the following table.

		Vendor		
		X	Y	Z
1.	Total system score	695	737	495
2.	Price of the system	$375,400	$430,200	$312,000
3.	Truncated price	375	430	312
4.	Benefit/cost ratio (1/3)	1.85	1.71	1.59

In this illustration, candidate X has the most favorable benefit/cost ratio. Although candidate Y has a total system score which is greater, its higher cost leads to a lower relative value of benefit/cost ratio which is being used here as the final decision criterion.

After selecting the system with the highest ratio, the next step in the evaluation procedure is to invite the vendor to run a benchmark to analyze the performance of its system. A benchmark is a group of specific problems representative of user company applications. The problems in the benchmark should include an appropriate mix of design and manufacturing problems. The problems selected should represent the intended applications of the CAD/CAM system by the user company. The total number of problems in the benchmark depends on their complexity, with perhaps three or four problems being typical. The design problems should require the CAD/CAM system to accomplish the necessary engineering analysis and to prepare the appropriate design documentation (e.g., engineering drawings, bills of material, etc.). The manufacturing problems should also force the system to accomplish the desired analysis or planning function and to produce the required documentation in hard-copy or soft-copy form (e.g., NC punched tape or program stored in computer memory).

Benchmarks can generally be divided into two categories, synthetic and live. Synthetic benchmarks are problems that have solutions known by the user company and are intended to exercise certain features and resources of the CAD/CAM system. Live benchmarks are problems that reflect the actual user company workload. Live benchmarks are usually preferred over the synthetic type, but it may be difficult for the user company to develop these without prior experience. Also, there is often a question of how representative a live benchmark really is.

The purpose of the benchmark is to validate the vendor's claims for its system. The reason for benchmarking only the system with the highest benefit/cost ratio is that this procedure is very costly for both vendor and prospective user. In addition, lengthy delays can result from trying to evaluate more than one benchmark. The vendor should know that if it has been requested to do a benchmark, it is very close to a purchase order. If the benchmark is successful (as it should be), the system is selected. If not, the second-best system, in terms of benefit/cost ratio, is benchmarked.

REFERENCES

[1] ALLAN, J. J., III,   Editor, *A Survey of Commercial Turnkey CAD CAM Systems*. 2nd Edition, Productivity International, Inc., Dallas, Tex. 1980.

[2] CHASEN, S. H., AND DOW, J. W., *The Guide for the Evaluation and Implementation of CAD/CAM Systems*, CAD/CAM Decisions, Atlanta, Ga., 1979.

[3] MCMILLAN, B. E., "How to Select a CAD/CAM System," in *The CAD/CAM Handbook*, (C. Machover and R. Blauth, Eds.), Computervision Corp., Bedford, Mass., 1980.

[4] QUANTZ, P., personal correspondence, February, 1983.

[5] QUANTZ, P., *CAD/CAM—Promise and Pitfalls in Planning, Systems Selection and Implementation*, Quantz Associates, Inc., Stratford, Conn., 1983.

[6] QUANTZ, P., "CAD/CAM — The Computer Invasion of Engineering and Manufacturing", *Tech. Paper ER81-02*, Socieity of Manufacturing Engineers, Dearborn, Mich., 1981.

[7] ZIMMERS, E. W., JR, *Computer-Aided Design Module*, General Electric Company CAD/CAM Seminar, Lehigh University, 1982.

chapter   22

The Future of CAD/CAM

Over the past decade, CAD/CAM has provided hope and excitement about the prospects for the manufacturing industries which have been in sharp contrast with recent reports of slow growth in U.S. productivity. CAD/CAM technology has responded to industry needs for sophisticated interactive graphics, computer-controlled machine tools, intelligent robots, improved inspection techniques, and a host of other innovations to do manufacturing better. It is contingent upon management to make the most of this new technology so that its full promise can be realized in the future.

Our purpose in this concluding chapter is to explore some of the future developments in computer-aided design and manufacturing. Most of the comments are based on trends in the technology which are already clearly recognizable.

Future prospects for CAD/CAM are greatly enhanced by developments in communications, microprocessors, and associated software. Improved communication techniques will result in greater exchange of information among people, machines, and computers. One of the manifestations of better communication will be systems that permit engineers and operating personnel to access powerful computing techniques from a terminal which can be far removed from a large computer. The terminal might be as small as a conventional pocket calculator but will have the capability to communicate with a large computer. Even sophisticated computer applications, such as interactive graphics, are possible. At least one manufacturer has built a pocket-size television prototype that could be used for this purpose. It includes a built-in zoom feature that can be used to enlarge any of the four screen quadrants for close-up viewing.

Another clear trend that will have an impact on CAD/CAM is the greater use of microcomputers and microprocessors to construct a new generation of machines (e.g., machine tools, inspection devices, robots, and computer terminals) with built-in intelligence. The motivation behind this is improved utilization of equipment. For example, in computer-aided design, a greater amount of local intelligence built into the design workstations translates into a larger number of these terminals that can be shared by one minicomputer. The same result occurs in the case of plotters and other peripheral devices. If the plotter contains sufficient local intelligence, it is capable of drawing complicated shapes based on relatively simple concise instructions from the minicomputer. The trends in this direction foretell that, within a few years, all the intelligence and computer power now resident in today's CAD/CAM systems will be available at every terminal in the system. The use of these intelligent terminals in distributed systems will consistitute the new family of CAD/CAM systems. With the trend toward lower computational costs, future CAD systems based on local intelligence will be cost competitive with current systems. At the same time, the capability of the CPU, enhanced by distributed processing, can be expected to increase considerably. There will also be redundancy in future systems to shift the work load to another part of the system when a component malfunctions or breaks down.

The use of localized intelligence through the use of microprocessor-based systems will also influence manufacturing. The use of intelligent robots, machine tools, and inspection devices, connected to a host computer, will provide an important boost to automation. It will provide greater flexibility in production systems to deal with a variety of different products [1]. Manufacturing and inspection instructions (e.g., part programs) which have been prepared automatically on the host computer can be downloaded to the appropriate machine on the shop floor for execution.

The cost of computer storage continues to drop and this will have implications in CAD/CAM. It will become feasible to store tens of thousands of drawings on-line instead of the limited number characteristic of present systems. At some point in the future, the computer itself may become the principal storage component in file systems rather than relying so heavily on secondary storage. Secondary storage will be relegated principally to a backup fail-safe role.

Graphics display technology is improving and this will affect other areas of operation within a company in addition to CAD. Refresh displays are becoming cost competitive with storage tube displays. The price/performance of raster systems will become more and more favorable. High performance (i.e., high resolution and fast response) is costly in these systems currently, but it is expected that hardware costs will continue to decline in the future. According to Machover [4], raster systems with resolutions of 2000 lines should become common during the middle to late 1980s. By the end of the decade the technology will provide raster tubes with 4000-line resolution. With this level of performance, the raster-type CRT will be the dominant graphics display device during the present decade and perhaps beyond. Competing display technologies include flat panels based on plasma or liquid crystal displays. The advantages offered by flat panel displays over the CRT are [3]:

Much less depth and volume; greater ratio of viewing area to depth

Better linearity and accuracy

Lower voltage required to operate

Potentially greater resolution and contrast

Because of the tremendous market potential in home television, research in flat-panel technology will probably yield commercial products which are eventually competitive with raster-type CRTs.

The use of color and solid modeling in computer graphics will become significant in design and other applications (industrial art, movie making, technical, and other publications). New plotters and hard-copy units with enhanced color capabilities will emerge to support the growth in color and solids.

Another future trend involves the combination of data base management systems (DBMS) with computer-aided design systems. Although fairly common in business applications, the use of DBMS for CAD systems is very limited at the present time. Advances in storage technology will influence this trend.

Voice recognition and vision systems technology will be refined and improved over the next decade. Computer terminals will be equipped to recognize and accept speech input as a means of speeding the input process. We have already discussed speech input for NC part programming in Chapter 9. Future speech input systems will be included in the CAD/CAM environment. As suggested in Chapter 19, vision systems will be used increasingly in computer-aided inspection systems. Vision is also an important emerging technology in robotics. Many future intelligent robots will be furnished with vision capability to perform their various industrial tasks.

Robotics itself is a fast-moving technology. Other sensors besides vision will be included within the capabilities of future robots. Speech recognition, tactile and proximity sensing, variable-pressure grippers, plus the intelligence to use these senses will open up large opportunities for robots, not only in production work, but also in the service industries.

Accompanying the technological innovations and improvements described above, there must also be a change in the way business is done in the manufacturing industries. With new communication techniques, there will be opportunities to have computers from different companies place purchase orders and communicate engineering data and specifications. With improvements in computers, there will be opportunities for nontechnical persons to use the computers. English-like commands will make the machines more user friendly.

Among the many changes in the operations of a manufacturing firm which are forced by the introduction of computer-aided design and computer-aided manufacturing, there will be a gradual dissolution of the traditional separation between design and production. Indeed, at some time in the distant future, we may look back at the impact of CAD/CAM on industrial progress and conclude that it was the integration of the design and manufacturing functions that was the most significant achievement of this technology.

REFERENCES

[1] GROOVER, M. P., "The Role of Robotics and Flexible Manufacturing in Production Automation," *Fall Industrial Engineering Conference*, Institute of Industrial Engineers, November, 1982.

[2] GROOVER M. P., and ZIMMERS, E. W., JR., "Automated Factories in the Year 2000," *Industrial Engineering*, November, 1980, pp. 34–43.

[3] HOBBS, L. C., "Computer Graphics Display Hardware," *IEEE Comptuer Graphics and Applications*, January, 1981, pp. 25–39.

[4] MACHOVER, C., "What's New? What's to Come?," in *The CAD/CAM Handbook* (C. Machover and R. Blauth, Ed.), Computervision Corp., Bedford, Mass., 1980, pp. 257–260.

Index